The Arabic Hermes

basis several of these questions could be answered—if they could ever be answered—would also require extensive prolegomena. Many preliminary problems required attention before a direct study of the Arabic Hermetica would be feasible. Some of these problems were already present in early Arabic literature, but others were the result of unintentionally misleading modern scholarship. My attempts to obtain a clear basis of understanding have resulted in this book.

A second book, well into preparation, is to follow this one, including an inventory of the actual texts attributed to Hermes in Arabic, most of which are still in manuscript, an outline of their chronology, and descriptions of the contents of the majority of them. In short, the work to follow will be a study of the Arabic *Hermetica*, whereas the present book is a study of the Arabic *Hermes*, the legendary author and his myth, and the conduits through which Hermes became known in Arabic literary tradition.

Acknowledgments

It is my sincere pleasure to thank the many individuals and institutions that have helped me in this project. Thanks first to Ralph Mathisen for accepting the book in his series, to Stefan Vranka of Oxford University Press for facilitating the entire process, and to the copy editors at the press for their careful work. For various assistance by supplying bibliographical or primary source references or discussing with me matters dealt with herein, I am grateful to Mohsen Ashtiany, Persis Berlekamp, Vahid Brown, John Darnell, Racha El-Omari, Garth Fowden, Beatrice Gruendler, Jaakko Hämeen-Anttila, Stephan Heilen, Tariq Jaffer, Richard Jasnow, Aaron Johnson, Todd Krulak, Martin Mulsow, Naʿama Pat-El, David Porreca, Annette Reed, David Reisman, Gabriel Reynolds, Ramzi Rouighi, Shaul Shaked, Isabel Toral-Niehoff, Paul E. Walker, Steven Wasserstrom, Edward Watts, Christian Wildberg, Witold Witakowski, and all who attended the Yale Arabic Colloquium on December 11, 2002. I would like to single out for special thanks Emilie Savage-Smith for providing detailed information about several Hermetic manuscripts that had escaped my attention; Shawkat Toorawa for his successful search on my behalf, in the middle of his own research and restrictions of time, in the Asafiyah library in Hyderabad; and Hossein Kamaly for his invaluable delivery of reproductions of Hermetic manuscripts in Iran, also while pressed by his own research and time limits. The anonymous readers for the press deserve thanks of their own for helpful suggestions on drafts of the book. Some of these colleagues will see the full fruit of their assistance not in this book, but in the one to follow.

The late Franz Rosenthal from the outset gave his encouragement and allowed me to photocopy the relevant section of his handwritten and unpublished edition of al-Mubaššir ibn Fātik's *Muḫtār al-ḥikam* (on this abortive edition see Rosenthal 1960–1961), which helped my work on the wise sayings of Hermes. Paolo Lucentini introduced me to current European scholarship on Hermetic traditions by generously facilitating my visit to the Convegno Internazionale sulla Tradizione Ermetica dal Mondo Tardo-Antico all'Umanesimo in November of 2001. I thank him again for his hospitality then. Everett Rowson, Charles Burnett, and Alexander Treiger deserve special thanks for reading different drafts of the manuscript very closely and offering numerous critical suggestions for improvement and crucial bibliographic assistance. Most of all, I am grateful to Dimitri Gutas, whose historical understanding and linguistic expertise made him an ideal mentor for me as I wrote the PhD dissertation from which this book has grown (*Hermes Arabicus*, Yale University 2004). His demand for precision and his eye for detail saved me from countless mistakes.

Despite the assistance of so many friends and teachers, this book remains my own, along with all the errors lingering in it.

My gratitude is also due to the Beinecke Rare Book Library for a summer fellowship and to the Yale Graduate School for support with a John F. Enders Fellowship and a Robert M. Leylan Fellowship in the early stages of the work. A National Endowment for the Humanities Summer Stipend awarded in 2007 helped me to collect quite a few more reproductions of Arabic Hermetic texts in manuscript. Any views, findings, conclusions, or recommendations expressed in this publication do not necessarily reflect those of the National Endowment for the Humanities. A semester as a member in residence at the Institute for Advanced Study in Princeton, in the fall of 2007, and my term as a Visiting Research Scholar at New York University's Institute for the Study of the Ancient World beginning in January of 2008, made for excellent conditions for thoroughly revising and increasing substantially what I had written.

Shilarna Stokes, my wife, has helped to make my life a better one than it was when I began this project. Without her encouragement and patience this would have taken even longer to publish. Her suggestions to improve the clarity of the text have touched on many pages. And to my parents, who never dissuaded me from a career in such arcane matters, I dedicate the work.

Contents

Note on Documentation and Transliteration

Scholars of classical languages usually cite works according to standard editions, a set of standard abbreviations, and numbers keyed to a universally accepted division of the sections of a text. Using these, anyone who wishes to check a citation can easily find it in any research library. For those working on the Greek and Latin authors studied most during the nineteenth and twentieth centuries, these standards work well. However, a large number of the texts cited in the following pages—even some of the Greek ones—are available only in publications hard to find even in a well-stocked research library. When it comes to Arabic works, there are few if any standard editions and fewer still critical editions. Middle Eastern presses regularly reprint old editions with new typesetting and pagination so that standardized enumerations of a text's sections are practically impossible. Many editions of Arabic works represent no more than the readings of one or a small number of manuscripts, without any editorial method other than to arrive at a legible text. Typographical errors abound.

Much of the research presented here naturally depends on the precise wording of texts in Arabic and other less well known languages of late antiquity. In order to make my arguments convincing, I have deemed it important to save the scholarly reader the trouble of searching for dozens of rare books by providing as much documentation in the original words as possible without, I hope, overwhelming the main body of the text. For the reader who is not concerned with this documentation, it may simply be passed over. In many instances I have added the proper chapter and section numbers of Arabic works, in addition to the page and line numbers of the text that I had at hand, because these are a surer guide to the location of a passage than the pagination of any given printing.

Words and names transliterated from Arabic and other Semitic languages follow most of the conventions of European scholarship on Arabic, adhering as much as possible to the principle of one sign for every phoneme. Those who do not know Arabic may note the following basic pronunciations of transliterated signs.

ṯ	as in "think"
ḏ	as in "this"
ğ	as in "gymnasium"
ġ	a voiced velar fricative similar to French "r"
ḫ	a voiceless velar fricative, as in Scottish loch
š	as in "shake"
ʾ	a glottal stop as in Cockney "bottle" (bo'el)
ʿ	the *ʿayn*, a voiced pharyngeal approximant, for which a glottal stop may be substituted as a foreigner's convenience

Most other diacritics on letters are for the Arabist and the student of Arabic; they may be ignored by the rest.

All dates are given according to the Common Era. To rely on the Muslim *hiğrī* era, counting lunar years beginning in 622 CE, makes no sense in a work dealing as much with the period before as after the *hiğra*. There are many books of tables, computer programs, and Internet sites that can convert any Common Era date into the *hiğrī* equivalent for the interested reader.

PART I

Background

1

Introduction

EARLY ARABIC ACCOUNTS TELL OF AN ANCIENT EGYPTIAN SAGE named Hermes Triplicate-in-Wisdom, the founder of the pagan religion before the great Flood, a lawgiver who ascended to the heavenly spheres of the planets and returned to teach astrology, a true prophet like Jesus and Muḥammad. The teachings of this wise Hermes circulated in Arabic translation, having been handed down, as the accounts said, over long ages by the great philosophers of the past, such as Aristotle and Apollonius of Tyana, or after being rediscovered inscribed on tablets in subterranean chambers of Egyptian ruins. The Arabic books of Hermes claim to explain the hidden workings of the universe, the secrets of creation, and the mysteries of nature. They offer means to predict the future from astral omens, instructions for the manufacture of potent talismans, and encrypted directions for manufacturing the elixir of the alchemists. To this day many of these works survive intact, waiting for modern readers in dozens of Arabic manuscripts. Most of them have never been printed.

Both the legends about Hermes and the works attributed to him in Arabic present complicated puzzles for the historian. They are part of the rich product of late antiquity out of which early Arabic culture, in all its aspects, grew, and one piece of the enormous body of knowledge translated into Arabic from the books of the ancients (as they called them), on which much of Arabic literature was based. To unravel this tradition of many strands is a manifold task. It is not enough to understand the Arabic context, even if that alone is difficult and the subject of ongoing research. One must also look back to the ancient background from which the Arabic Hermetica came into being, and forward to their celebrated reception in fifteenth- and sixteenth-century Europe, which has endowed them with a special importance for modern scholarship in European languages. At the same time, interest in the Hermetica reached far beyond Europe, where there are totally unexplored avenues of research and untold histories. Works of Hermes never known in Europe were copied and studied from Morocco to India, surviving today in mosque collections and national and private libraries of North Africa and Asia. For example, as late as the nineteenth century, an Iranian intellectual could still find it worth his while to translate an alchemical revelation of Hermes claiming ancient Egyptian origins from Arabic into his mother tongue, Persian.[1] When the British government of India purchased the remains of the Royal Library of the fallen Muġal Emperors in Delhi in 1859, there were Arabic works of Hermes among those thousands of volumes.[2] What these Hermetic texts, and the figure of Hermes Trismegistus himself, meant to their numerous audiences in far-flung lands remains to be discovered. It is certain, however, that the European Hermetic movement that began in the late fifteenth century is only one part of something much greater, at the center of which, geographically and chronologically, is the Arabic tradition.

1.1 The Ancient Greek Hermetica of Roman Egypt and Their European Reception

Sometime in the first centuries of the Common Era, certain anonymous sages in Roman Egypt, apparently over several generations, wrote treatises on science and philosophy in Greek while professing to be part of a distinctly Egyptian tradition.[3] A number of these works portray dialogues between various Egyptian gods, some given Greek names and others retaining their Egyptian names. The principal character in these texts is Hermes, the Greek syncretic equivalent of the Egyptian god of knowledge and writing, Thoth. Hermes usually appears as the teacher of the other characters in the dialogues, including Tat, Ammon, and Asclepius. From the main teacher we take the name of the texts as a group: *Hermetica*, writings having to do with Hermes. The Hermes here is often dubbed Τρισμέγιστος, Trismegistus, the "thrice-greatest," an epithet in loan translation from an old Egyptian title of Thoth, found first in Greek in the usage of Egyptian priests of the Ptolemaic period as μέγιστος καὶ μέγιστος θεὸς μέγας, "greatest and greatest great god."[4] In this way the Egyptian Hermes is distinguished from the mischievous Olympian messenger god Hermes of Greek mythology.

The Greek Hermetica of Roman Egypt deal with a wide range of topics, including astrology, cosmology, and medicine. They describe a philosophical education, revealed by Hermes and others on the basis of divine insights, through which his disciples could know and understand the soul, the universe, and God. Garth Fowden has neatly summarized the cosmos taught in the Hermetica:

> God is one, and the creator of all things, which continue to depend on God as elements in a hierarchy of beings. Second in this hierarchy after God

1. Translation by Mīrzā Āqā Muḥammad ʿAlī Garakānī, manuscript in the Kitābḫāna-yi Millī-yi Malik 805/6, 1875 CE. The Arabic text is edited with a German translation (Vereno 1992: 160–181).
2. On the Delhi collection see Ursula Sims-Williams 1982.
3. On the search for the authors of the Hermetica see Fowden 1993a: 155–195.
4. Festugière 1.73–74, Copenhaver 1992: xiv–xv, Fowden 1993a: 216–217.

> himself comes the intelligible world, and then the sensible world. The creative and beneficent powers of God flow through the intelligible and sensible realms to the sun, which is the demiurge around which revolve the eight spheres of the fixed stars, the planets and the earth.[5] From these spheres depend the daemons, and from the daemons Man, who is a microcosm of creation. Thus everything is part of God, and God is in everything, his creative activity continuing unceasingly. All things are one and the pleroma of being is indestructible.[6]

Human beings are constrained in the hierarchy of cosmic sympathies and subject to the force of the daemons; that is, they are bound in the cosmos by fate. Altogether, these ideas were not unique to the authors of the Hermetica. One could say they were, in varying forms, popular ideas around the Mediterranean under Roman rule. But what makes the Hermetica special is their response to the situation of humans in the world. The Egyptian Hermes imparts two kinds of knowledge in his books. First, it is possible to gain a degree of power in and control over the terrestrial world through a knowledge of the sympathetic harmony of the celestial forces with particular terrestrial things. Thus Hermes teaches, for example, the interpretation of heavenly signs for the prognostication of the future, as well as the occult properties of substances, such as different plants, and the astral powers they possess from their planetary sympathies. Modern scholars have called such treatises *technical* because they explain *techniques* for the application of the knowledge of the principles and forces at work in the universe to daily life and practical problems. Second, and more famously, Hermes teaches his disciples in the dialogues how to transcend their bodily circumstances and to overcome fate through purification of the intellect and contemplation, leading to a true knowledge of God.[7] Much of the appeal in the discourses of Hermes lies in this promise of higher knowledge of a greater, divine self liberated from worldly conditions through a developed understanding, and this goal must partly have motivated the anonymous community that composed these texts.

The Greek Hermetica best known today are of the latter type, the philosophical dialogues. They came to the attention of scholars in the Latin tradition of Western Europe in 1460, when a monk named Leonardo brought a manuscript of Greek Hermetic texts to Cosimo de' Medici in Florence from somewhere in Macedonia.[8] The scholar Marsilio Ficino was asked to set aside his translation of the works of Plato and to give precedence to Hermes, who was thought to be much more ancient

5. For the apparently heliocentric view, which has no role in the Arabic reception of Hermetica, see CH 16.7. Scott 2.444–445 suggests that the view is rather that the sun occupies the middle of the seven geocentric spheres, but it is not clear.

6. Fowden 1993a: 77.

7. Fowden 1993a: 75–115.

8. Rigo 2002 offers evidence for scholarly interest in works attributed to Hermes in Byzantium from the fourteenth century.

than the classical Greek authors. His translation was published in 1471, and over the next century and a half this Latin version of the Greek Hermetica went through twenty-five editions.[9] The *editio princeps* of the Greek original was produced by Adrianus Turnebus in 1554. Greeted with extraordinary interest, the Greek Hermetica were thought during this period to derive from before the great Flood or at least to be contemporary with Moses, and to represent the primordial, original wellspring of philosophy. They were therefore regarded as authorities on a par with the book of Genesis, a discovery of utmost importance. Renaissance authors used them as source texts for a new movement that blended Neoplatonism, Jewish Cabbala, and the teachings of the Hermetica into a system of thought called *Hermetism* today. Historians have amply documented the deep influence of these works on scholars in Western Europe from the end of the fifteenth through the seventeenth century, and more marginally thereafter until the present.[10]

The beginning of the end of Hermetic popularity was in 1614, when Isaac Casaubon showed that the Greek Hermetica were not so ancient as had been thought, but rather were apparently derivative products of the early Christian era. Although his arguments did not become widely known immediately, the prestige of the Hermetica declined markedly by the end of the century as Casaubon's research had its effect.[11] Hermann Conring, a professor of medicine at the University of Helmstedt, argued in 1648 against the claims to a Hermetic legacy made by the alchemist followers of Paracelsus.[12] Casaubon, Conring, and others after them effectively dispelled the aura of primordial authenticity around the Hermetica themselves as well as the supposed ongoing heritage of those texts. Following these and other attacks, the eighteenth and early nineteenth centuries witnessed a low ebb of interest in the Hermetica, though Egyptomaniacs, alchemists, members of secret societies, and those involved in currents of esoteric thought continued to use and study them. Between 1630 and 1856 there were no new editions or reprints of the Greek text of the Hermetica brought to light by Ficino, a sure sign of attenuated scholarly interest.[13] The Hermetic movement created in the sixteenth century diminished greatly during the subsequent period of "scientific revolution," when scholars relied less and less on ancient writings for scientific authority.

It was only with the new methods and purposes of classical scholarship developed in the nineteenth century that the Greek Hermetica found a restored place in

9. Yates 1964: 12–13, Mahé 1982: 3–5, Faivre 1995: 183.

10. For overviews of the reception of this Greek corpus in Western Europe until Isaac Casaubon disproved its antiquity in 1614, see Mahé 1982: 5–9, Fowden 1993a: xxi, Faivre 1995: 183–188, Yates 1964, to be tempered by Copenhaver 1990, Kühlmann 1999, and Ebeling 2007: 59–141. These works provide copious bibliography.

11. Mahé 1982: 8–9, Fowden 1993a: xxii, Faivre 1995: 186, Yates 1964: 398–403.

12. Ebeling 2007: 97–100.

13. Copenhaver 1992: li.

respected research, now not as sources of inspiration and esoteric teachings but as historical artifacts to be fitted into a detailed and comprehensive record of Greek and Roman "civilizations," then a new framework of study. For some practitioners of the newly developing field of religious studies, the Hermetica became documents of an alleged perennial philosophy and sources for structuralist schemes. The occultists of the Romantic and Modern eras also found their own interest in these texts.

For much of the twentieth century the scholarly effort to explain the Hermetica as products of a historical context remained inconclusive. The debate about them effectively centered on determining to which "nation" they belonged, whether they were essentially Greek and therefore European or "Western," or Egyptian and Oriental and therefore foreign, and secondarily on whether or not a particular defined community produced them. If a group of authors could be found, then ascertaining their relationship to other religious movements, construed broadly as Judaism, Christianity, Gnosticism, and paganism, would help to give them their historical place. (Today these broad categories seem somewhat simplistic.) The monumental labor of Festugière in the 1940s and '50s seemed to show definitively that the Hermetica were, in fact, totally Hellenistic in content and form and owed nothing essential to non-Greek cultures. He wrote four volumes to prove his point.[14]

In the late 1970s and the 1980s, studies by Jean-Pierre Mahé and Garth Fowden argued that there is indeed something distinctively Egyptian in them, and that they can be really understood only when we place them in the context of Hellenistic Egypt and indigenous Egyptian cultural tradition.[15] Mahé hypothesized that the Greek texts are descendants of the Egyptian genre of aphoristic instructions.[16] Fowden showed convincingly that one cannot understand the genesis of the Hermetica without locating them within the social contexts of Roman Egypt. His well-documented arguments make the Hermetica neither essentially Greek nor Egyptian but the product of a Roman Egyptian society that had synthesized practices and ideas current in both of these linguistic, cultural, and religious traditions, a society that must be understood without an easy separation of "Greek" and "Egyptian" characteristics. It seems safe to say that our present understanding of the Greek Hermetica stands with Fowden.

14. Festugière 1944–1954.

15. A convenient history of the reception of and scholarship on the Greek Hermetica in Western Europe is found in Mahé 1982: 3–43, culminating in Mahé's own work. For some later work see Fowden 1993a: xiii–xix. Gonzáles Blanco 1984 is also useful as an overview of the history of scholarship on the subject of the Hermetica and Hermetism.

16. Mahé 1978–1982: 2.407–436. Mahé argues that what began originally as sentences or aphorisms of Hermes were developed into the Hermetic *logoi* of the dialogues. Against Mahé on this see Fowden 1993a: 69–74.

Further research in Egyptology may well be able to elucidate more tangible local roots of the Hermetica in the scrolls of the Egyptian priests of the Roman period.[17] Indeed, direct evidence of a possible Egyptian literary background to the Hermetica has recently emerged. In 2005 two Egyptologists, Richard Jasnow and Karl-Theodor Zauzich, published an edition of a Demotic Egyptian text containing a dialogue of instruction between the god Thoth and a student, called *mr-rḫ*, "lover of knowledge" or perhaps "philosopher." The manuscripts, many incomplete pieces of what was once apparently a long text kept in Egyptian temple libraries, have been dated approximately to the period from the first century BCE to the second century CE. This is just the period during which the Greek Hermetica are generally supposed to have had their genesis. Because the contents of this Demotic *Book of Thoth* (so dubbed by its editors) do not exhibit an "admixture of hellenistic philosophy," as the Greek Hermetica clearly do, the editors of the text have prudently concluded that the work is not a direct ancestor of the Greek Hermetica, though it may be rather "one of the grandfathers" of the Greek Hermetica.[18] Both are dialogues between Thoth-Hermes and a pupil on theological subjects, so it is easy to imagine a connection between them, but no convincing parallel has yet been discovered, either in wording or in doctrine, despite the editors' best efforts to do so. The assessment of the ancient author of the *De mysteriis Aegyptiorum* was that "those documents . . . which circulate under the name of Hermes contain Hermetic doctrines, even if they often employ the terminology of the philosophers; for they were translated from the Egyptian tongue by men not unversed in philosophy."[19] This view now seems more likely to win the day than ever before, but it will certainly require a careful comparative study by a scholar at home in both Egyptology, particularly in Demotic Egyptian, and ancient Greek. Perhaps undiscovered Egyptian texts will contribute further to solving the problem of the formation of the Greek Hermetica.

Even within Greek literature itself, there are still unexplored avenues for scholarly research. The title *Corpus Hermeticum* is normally used to refer only to the Greek philosophical Hermetica brought to light in fifteenth-century Italy (though often in practice it also includes the text surviving complete only in Latin translation

17. Boylan 1922 is still a standard treatment of the ancient Egyptian tradition of Thoth. Iversen's essay (1984) is a notable attempt to connect Egyptian literature with the Hermetica; although finding some important parallels in content, it admittedly does not show how the older Egyptian ideas could have been transmitted into Greek. Fowden 1993a: 57–68 offers much interesting information to illustrate the Egyptian background. The arguments of Kingsley 1993 for the etymology of the Hermetic name Poimandres (based on a proposal by Francis L. Griffith in 1925) are interesting although probably impossible to verify; one would like the etymological portions of the article to be reviewed by an Egyptologist-Coptologist knowledgeable in Egyptian phonology.

18. Jasnow and Zauzich 1998: 618. Edition by Jasnow and Zauzich 2005. Cf. Mahé 1996.

19. Usually the text is attributed to Iamblichus (Saffrey 1971). Trans. by Emma C. Clarke, John M. Dillon, and Jackson P. Hershbell 2004. τὰ μὲν γὰρ φερόμενα ὡς Ἑρμοῦ ἑρμαϊκὰς περιέχει δόξας, εἰ καὶ τῇ τῶν φιλοσόφων γλώττῃ πολλάκις χρῆται· μεταγέγραπται γὰρ ἀπὸ τῆς Αἰγυπτίας γλώττης ὑπ' ἀνδρῶν φιλοσοφίας οὐκ ἀπείρως ἐχόντων (*De Mysteriis*, 8.4.265).

known as the *Asclepius*, and other related fragments), for it is these texts that enjoyed such celebrity when they came to the attention of early modern Europeans. However, there are quite a few more works attributed to Hermes surviving in Greek than just these philosophical discourses, contrary to the implication in the authority-granting and comprehensive-sounding title *Corpus*.[20] Although specialists do take them all into account, the recent English translations of the *Corpus Hermeticum* do not include the important and substantial Greek excerpts of Hermetic dialogues found in the *Stobaean Anthology*.[21] The Greek texts attributed to Hermes dealing with omens, the use of the occult properties of substances, and related practices—as opposed to theoretical discussions, philosophy, and descriptions of meditations such as those found in the Greek *Corpus Hermeticum*—are even more neglected, practically unknown. The modern view of the Greek Hermetica is largely focused on a limited section of what was and indeed is still available.[22]

Understanding the Hermetica requires also that works attributed to Hermes in languages other than Greek be taken into account, and again specialists have not failed to notice them. The Coptic translations of Greek Hermetica, discovered in the Nagʿ Ḥammādī codices, have by now received much well-deserved attention.[23] Some of these Hermetica in Coptic do not survive in their original Greek, and they were not published until the 1970s. One of these previously unknown texts is a dialogue in which Trismegistus initiates his disciple into the ineffable mysteries of the mind, instructing him to inscribe the teaching in hieroglyphic characters and to head the text with an oath to protect its contents from misuse. The Coptic codex also includes a substantial portion of the *Perfect Discourse*, surviving also in the Latin translation known as *Asclepius*. The latter version has attracted students in the Latin tradition in ancient, medieval, and modern times.[24] The Armenian Hermetic text called *The Definitions of Hermes*, translated from Greek and with later interpolations, has been

20. A survey of the extant Greek Hermetica is found in Fowden 1993a: 1–11, most of the discussion there being devoted to the collection of the Greek philosophical Hermetica.

21. Scott vol. 1 contains English versions of these.

22. On the inseparability of the ancient Hermetica prescribing techniques from the better-known ones concerned with theory, see the remarks of Fowden 1993a: 116–120. Examples of published "technical Hermetica," include the *Iatromathematica* (Ideler 1841: 1.387–396); two short works: *On Earthquakes* (Pitra 1888: 275–278 and CCAG 7.167–171) and *To Asclepius on the Plants of the Seven Planets* (Pitra 1888: 279–282 = CCAG 8/3.153–165); the Ἱερὰ Βίβλος (Pitra 1888: 284–290; Ruelle 1908); *To Asclepius on the Plants of the Twelve Zodiacal Signs* (Pitra 1888: 291); a *Brontologion* (CCAG 7.226–230); and the Μέθοδος εἰς πᾶσαν καταρχήν (CCAG 8/1.172–177). Festugière, vol. 1, studied several of these. There is still no historical study of these texts more advanced than the notes by Pingree 1978: 2.429–432. Alchemical excerpts from Hermes are listed by Festugière 1.251 and Letrouit 1995: 81. There are probably more short Greek Hermetica of these types to be found in Greek manuscripts.

23. Mahé 1978–1982; the standard edition is *Nag Hammadi Codices V, 2–5 and VI with Papyrus Berolinensis 8502,1 and 4*, 1979.

24. On the Latin readership of the *Asclepius* in the twelfth and thirteenth centuries see most recently Porreca 2001, which contains an up-to-date bibliography for the medieval Latin reception of this text.

edited and studied by Mahé; Armenian authors from the sixth century onward cited it.[25] There is also a large number of medieval Latin Hermetica; these need research not only on their contents and doctrines, but also on their relationship to the Hermetica in other languages. Fortunately a new series of studies organized by Paolo Lucentini under the name *Hermes Latinus* has appeared, seeking to fill this gap, published in the series *Corpus Christianorum: Continuatio Mediaevalis.*[26] With the help of these new editions and studies of the European Latin Hermetica of the Middle Ages, one can begin to reach an idea of what was already available in Latin Europe before the Greek *Corpus Hermeticum* was published in Latin in the late fifteenth century.

The present book is aimed at the least understood part of Hermetic tradition. There are probably more works attributed to Hermes surviving in Arabic than in any other language, and the majority of them are still unknown and unpublished. Some of them definitely derive from ancient Greek sources through translation. Others, like many of the Latin Hermetica, are later works originally composed in Arabic. Yet even where the texts themselves are not of ancient origin, the *idea* of Hermes is. The problem then is to establish the means and continuity of tradition from the ancient Hermetica, and what people thought about Hermes, to the time of their attestation in Arabic.

The first principle for researchers in Graeco-Arabic studies, as Dimitri Gutas defined it, is a matter of common sense: nothing was translated from Greek into Arabic unless it was available in manuscripts to the translators.[27] Put differently, the Arabic reception of works in translation from Greek, as well as from other languages, is a product of the period now generally called late antiquity. It is fundamental therefore to trace the history of the Hermetica and their reception, insofar as we can discern them, down to the time of the Arabic translators in the eighth, ninth, and tenth centuries. One way to sketch the earlier history of the texts is to notice who mentions them and for what purpose. At the same time it is possible to arrive at a general idea of the interest readers had in the Hermetica down to that time.

1.2 The Reception of the Greek Hermetica in Antiquity

A complete history of the reception of the Hermetica is still required, but for now a brief chronological outline of the early readership of these texts in antiquity will

25. Mahé 1978–1982: 2.273–406; his English translation with bibliography is found in Salaman et al. 2000: 99–124.

26. Lucentini and Perrone Compagni 2001 present an outline of all the Latin Hermetica, editions, manuscripts, and basic bibliography. The *Hermes Latinus* Internet site outlines future plans and past projects.

27. Gutas 1994: 4941–4944.

have to suffice. What follows is based largely on the useful collection of testimonia by Scott, completed by A. S. Ferguson; the reader may find complete textual references there.[28] The nature of these testimonia is largely doxographic, so they tell us mostly about the reception of the philosophical Hermetica by theologians, whose works survive in large number, whereas exceedingly little is known about the history and use of the technical Hermetica. One may assume that works of Hermes on astral divination and the occult properties of substances continued to circulate among scientists and astrologers without hiatus.

We know little for certain about the original community or communities that produced the Greek Hermetica, except that they were pagans in Roman Egypt who saw their teachings as distinctly Egyptian.[29] It seems likely to me that they included members of the Egyptian priesthood.[30] Whether one imagines them as a philosophical sect or a religious movement depends on how one defines those terms.[31] The earliest external references to Hermes Trismegistus as an authoritative figure come into view in early Christian writings of the second and third centuries. Athenagoras (late second cent.) cites a passage of an unknown Hermetic work to support the well-known antipagan argument that the beings worshipped as gods by the polytheists were all really men of bygone times. Tertullian (fl. ca 200) points to the limits of human knowledge when he says that not even Hermes, "the master of all natural philosophers," could explain the origin of matter. One can see from such a remark that even from the beginnings of the attestation of the Hermetica, authors who knew them believed them to be very ancient. The pseudo-Justin who wrote the *Cohortatio ad gentiles* (third cent.) provides the earliest citation of what would become Hermes Trismegistus' most popular words for theologians: "It is difficult to understand God, but for the one who does, he is impossible to describe." Non-Christians read the Hermetica, too. Around 300, in Egypt, the early alchemist

28. Scott, vol. 4, remains an exceptional resource for studying the reception of the Hermetica not only in Greek and Latin but also in Syriac and Arabic. Yates 1964: 1–19, presents an impressively lucid discussion of the ways in which Renaissance scholars treated the authority of Latin Church Fathers on the Hermetica. Fowden 1993a: 196–212 deals with the late antique reception of the Hermetica outside of Egypt down to the fifth century.

29. Fowden 1993a: 168–176.

30. Bagnall 1993: 261–273 shows that traditional Egyptian temple life began to disintegrate in the third century, and that this is therefore apparently not caused directly by the success of Christianity. Bagnall suggests rather that the decline was due to lack of financial support from the government. For context see also Frankfurter 1998.

31. Mahé 2007: 797 says that the Egyptian Hermetic community constituted "an actual religious movement with a deep social impact, based on congregations quite similar to Gnostic communities." The claim that they had ritual congregations appears to be based on the idea that the descriptions of Hermes' initiations of his students in the Hermetic dialogues reflect the actual practices of the people who wrote and read these texts. Thus also Salaman et al. 2000: 84: "A balanced view of the Corpus will recognize its indebtedness to the vocabulary of its milieu. One can imagine men and women of diverse backgrounds joining Hermetic associations in Alexandria." I am less willing to make these assumptions, though the theory that the Hermetic dialogues describe actual practices remains possible.

Zosimus of Panopolis demonstrated his familiarity with "sacred books" of Hermes and quoted Hermetica in his works, which survive in the original Greek and in Syriac translation. According to some evidence, works of Hermes were also translated into Middle Persian already in the third century, in the court of the Iranian emperor Šāpūr I (d. 270), along with works of Ptolemy, Dorotheus, and other astrologers. Chapter 2 deals with this Persian reception of the Hermetica and the attempt to establish its date.

The Christian Arnobius, at the beginning of the fourth century, included Mercurius (Latin for Hermes) among the pagan philosophers against whom he was arguing. His student Lactantius took a different approach, citing philosophical Greek Hermetica to make the case that God is one and truly above the world. This is the earliest example of what became a common use of the Hermetica: citing the pagan philosopher Hermes in *support* of Christian theology. For centuries thereafter, late antique and medieval Christians used the Hermetica for apologetic purposes.[32] The Hermetic treatise *The Perfect Discourse*, which survives complete only in a fourth-century Latin compilation known as *Asclepius*, contains a prophecy about the lamented decline of Egyptian religion and culture, but it was used by Christian writers to prove that paganism was doomed to extinction. Meanwhile the pagan theurgist who wrote the *De mysteriis Aegyptiorum*, thought to be Iamblichus,[33] also mined the Hermetica for materials suitable to his own pagan theology. The fourth-century Syriac poet Ephraem claimed that the Manichaeans, too, considered Hermes one of their prophet's forerunners. Coptic translations of Hermetica are found in the Nagʿ Ḥammādī codices from the mid-fourth century, showing that Egyptian sages read Hermetica in translation in their own language. The fifth-century anthology of literature and wisdom of the pagan Stobaeus includes some of the most important extant Greek Hermetica. Hypatia's father, Theon of Alexandria (fl. ca 364), is supposed to have taught the astronomical works of Hermes, and her friend Synesius, bishop of Cyrene, listed Hermes as one of the four greatest sages of ancient times. Hermes appears at the end of the sixth century in more generic collections of the sayings of ancient philosophers in Greek and in Syriac. The Syriac collection is especially interesting because it is directed against the obstinately pagan inhabitants of the Syrian city Ḥarrān (Carrhae). (On this work, see section 3.4.1, page 83 in this volume.) In the early fifth century, Cyril of Alexandria had used Hermes to similar effect in his book refuting the emperor Julian and his treatise against the Christians.

Such a rapid survey cannot do justice to the use and abuse of the Hermetica in their reception. Each author had his own purposes in using Hermes for an argument, and these need to be understood individually according to their proper contexts. What should be clear, however, from the collection of these references is that

32. Ebeling 2007: 38–44 offers a brief summary of the medieval Christian uses of the Hermetica.
33. Saffrey 1971.

Hermetica, not all of which survive today, were read, excerpted, and appreciated widely during the third, fourth, and fifth centuries, and applied to a variety of tendentious theological arguments often not in harmony with their original purposes. The Hermetica had authority as seemingly authentic documents of an ancient Egyptian wisdom, and that authority is what made them valuable to authors beyond the unknown Egyptians that produced them. It is not clear, however, how many readers really saw the Hermetica directly themselves, as opposed to mere citations handed down through other authors or excerpts in anthologies. Moreover, as Fowden puts it, the "way of Hermes" created and followed by the Roman Egyptian authors of the Hermetica might have been "appreciated outside Egypt, as by Iamblichus, but was not so far as we know put into practice."[34] In other words, the Hermetica really had become documents of a philosophical and spiritual training the specifics of which belonged to a bygone era. After the fourth century, there is no more sign of a living community of devoted successors to the "original" Graeco-Egyptian authors and students of the Hermetica.[35]

References to Hermetica as authoritative sources of theological arguments continued to appear even when Byzantine power was replaced by the Arab kingdom in the Middle East during the seventh century. For example, by the year 700, Jacob of Edessa quoted with approval the views of Hermes Trismegistus on God's creation of the sun in his Syriac *Hexaemeron*.[36] In this case it is unlikely that he was familiar with Hermetica directly. Probably he found these passages in the treatise of Cyril just mentioned. Yet for him, it was worth citing Hermes to make a point. Also by this time someone had translated the philosophical definitions of Hermes into Armenian.[37] Hermes had become a mysterious ancient sage, whose wise words of higher understanding were known partly from citations and were used to bolster religious arguments. He was known in many languages at different times: Greek, Latin, Syriac, Coptic, Armenian, and Middle Persian.

The Arabic reception of Hermes, as an idea and an author, began in the eighth century, in a world very different from that in which the Hermetica had first circulated. By that time the Byzantine polity was reduced to parts of Anatolia and the Balkans, replaced everywhere else by the new Arab empire of unprecedented size. There was a new ruling religion, as well as a new common language, and a new tradition of scholarship was taking shape. To understand the Arabic Hermes, it is necessary to have some idea of the larger context of this early Arabic scholarship.

34. Fowden 1993a: 212.

35. Fowden 1993a: 175–176 describes papyrus letters of the fourth century mentioning Hermes Trismegistus with reverence.

36. Jacob of Edessa, *Hexaemeron*, 149b11–150a17. Cf. Cyril, *Contra Julianum* 588A-B; cited also in Scott 1.546, 31 and 33; again with the entire passage from Cyril in Scott 4.214.3–216.1; CH 4.138–141, 32b and 33.

37. Mahé, in Salaman et al. 2000: 101, dates the Armenian translation to the second half of the sixth century.

1.3 The Birth of Arabic Scholarship and the Books of the Ancients in Baġdād

The Arab conquests and the resulting empire have been treated in many studies, and historians have proposed many theories, not all of them mutually compatible, to explain these momentous events. Readers new to the subject will find excellent narrative overviews elsewhere.[38] What follows here is only the briefest possible background to orient readers unfamiliar with the world of letters in which the Arabic Hermetica first appear.

The 620s and 630s were marked by devastating wars between the Persian and East Roman Empires and by apocalyptic sentiments across the Eastern Mediterranean and Southwest Asia. In these decades, the diverse populations of Arabia united as a new community following the teaching of the prophet Muḥammad of Mecca, who warned through his revelation of God's imminent judgment at the end of the world and offered divine laws to his community. After Muḥammad's death, his community followed the prophet's successors, the caliphs. Sedentary and pastoralist alike, the Arabian peoples, along with allies joining them at every stage, collaborated in the conquest of one half of the former Roman Empire and the entire Sasanian Persian Empire, both of which had weakened defenses after the previous three decades of intense warfare. The Arab conquests of the lands from the Iberian peninsula to Central Asia were not nearly as rapid as some histories make them seem, but were carried out through several generations of institutionalized warfare. The conquerors had to adapt to new and different physical and social environments and to manage their ever-increasing wealth, property on a scale totally unfamiliar to their recent ancestors. This first century (the seventh to eighth centuries CE) therefore brought many changes to the Arabic-speaking populations now scattered in a far-reaching diaspora as a ruling élite. The pattern of caliphal succession, though always contested, especially by supporters of leaders from the family of the prophet, became in the first century essentially dynastic under the Umayyads, a different albeit related family, and the caliphs became effectively kings of an immense empire. The religion of the community following Muḥammad came soon to be defined as Islām—a term that had earlier referred to conversion Muḥammad's community—but its doctrines, law, and theology were still only beginning to take their canonical shapes. Arabic remained the language of the Muslims. By official order Arabic replaced other languages as the medium of administration across all the conquered territory around the year 700. The process was not accomplished overnight, but it was nevertheless the new policy to use Arabic in government and administration.

38. For newcomers I recommend the shorter introductions of Fred Donner 1999 and H. Kennedy 1999. Hoyland 1997: 11–31 offers a concise introduction to the cultural background of the first "Islamic century." Donner 1998: 1–31 provides an especially clear introduction to the recent historiographic debates about the Arabic historical sources for the earliest part of the period.

In 750 the Umayyad dynasty of caliphs, weakened by factionalism, was overthrown by the ʿAbbāsids, who claimed to represent better the family of the prophet and who were backed by strong supporters from the prosperous and relatively recently conquered distant provinces of eastern Iran and Central Asia, territories that had previously been outside even the Sasanian Empire. These ʿAbbāsid caliphs were distinguished from their predecessors in following older models of Iranian kingship. One way in which they did this was to encourage scholarship and to fund scholars, particularly where it could promote their own authority and the prosperity of their administration. For example, the caliphs required astrologers, not only to advise them on the most favorable times for new undertakings, but also to demonstrate that ʿAbbāsid power was determined by long-established patterns in the heavens. Mathematics and medicine had their obvious practical applications. The patronage of and participation in scholarship became an increasingly prestigious and normal interest for members of the caliphal court and the privileged classes under the ʿAbbāsids. In this patronage they employed many non-Arabs, including many scholars learned both in Arabic and in the languages of their communities of origin, such as Persian, Greek, and Aramaic.

The foundation of the new capital of Baġdād in 762, about twenty miles upstream from the former Persian capital, represented the ʿAbbāsid al-Manṣūr's centralizing policies and physically brought together populations from all regions and ethnicities of the far-flung empire, representing many different creeds. The pace of conversion to Islam by members of the conquered populations, who now played a greater role in the administration of government, rapidly increased in the following century. Many converts had to learn Arabic prayers;[39] regardless of their first language, when their children were schooled, they learned Arabic writing and memorized Arabic scripture and verse, often attaining fluency. Arabic became the normal intercommunal medium of learned discussion, as the languages of conquering colonists have repeatedly done in history. The class of literate bureaucrats, many of them of non-Arab background from both the former Roman and Persian empires, grew larger and created a new, urbane literature and richly developed sciences of grammar and religious tradition, all in Arabic. The introduction of paper technology, almost simultaneous with the rise of the ʿAbbāsids, facilitated the composition and compilation of written works on multivolume scales previously unknown in the days of the much more expensive parchment. In such circumstances, in which intense intercultural contact in a new arena coincided with a sharply rising regard for advanced scientific learning, patrons commissioned a huge number of translations from the literary languages of those ruled—primarily Greek, Aramaic, and Persian—into the language of the rulers. Arabic, spoken by the Muslim kings and all

39. The status of the language of prayer was under dispute, but it was widely held that Muslim prayers were efficacious only in Arabic.

their followers, eclipsed competition in the public life of Muslim urban centers. Book traditions in languages such as Middle Persian and Aramaic were preserved for later ages only among religious scholars in their particular communities.

Of course, Arabic literature, oral and written, was never truly separate from contact with other literatures and translation. The Qur'ān itself retells stories from biblical and other literature that its audience was expected to know. What makes the ʿAbbāsid literary efflorescence remarkable is its scale, including the scale of translation, the widespread support for translation, and the prestige of the translated works. Translation was a full-time profession for a select few, well paid by rich patrons. Several translators became enduringly famous for their work. The translated texts became models for all later writings on the subjects they covered, standards to emulate, react against, or build upon. In a word, they became *classical*, but exclusively in the new language. They were not studied in their original, now obsolete, languages.[40]

The Arabic Hermes appears in this context of fresh exchange, in which authors in many fields of study and inquiry from many different ethnic backgrounds and countries met together, all using Arabic, in the young capital city of the largest empire the world had ever seen. Hermes is cited as an ancient authority already in the eighth century in some of the earliest known Arabic scientific works, the writings of Iranian court astrologers of the ʿAbbāsids. In the ninth century, Hermes came to be considered in Arabic literature as not just an ancient scientific authority, but an antediluvian prophet whose revelation formed the very beginnings of science. His works were therefore thought to have been transmitted from the extremely distant past as a part of the heritage of the ancient world, recovered by Muslim scholars in this great epoch of translations. With respect to the chronological error—the attribution of very ancient dates to the works of Hermes—the reception of Hermetica in Arabic bears close resemblance to the Western European reception of the Hermetica that began in the fifteenth century. This is not fortuitous, for the European concept of Hermes Trismegistus as a very ancient sage was derived from Arabic tradition as well as from the Church Fathers.[41] The Arabic Hermetica appeared in the midst of the bloom of Arabic literature in every field of human inquiry.

40. Gutas 1998 offers the best current introduction to the widespread translation of works from other languages, especially Greek, into Arabic. Readers wanting more information on the transmission of pre-Islamic literature into Arabic can consult Rosenthal 1975a (an anthology of translated texts) and Endress 1987–1992 (a comprehensive bibliographical essay) for the secular Greek scientific and philosophical traditions; Graf 1944–1953 for Christian Arabic translations from Greek, Syriac, and Coptic; for translations from Middle Persian into Arabic there is Panaino 2001 as well as F. de Blois, "Tardjama. Translations from Middle Persian (Pahlavī)," *EI*², 10.231b–232b, and the works of Shaul Shaked listed in the bibliography. For Sanskrit into Arabic, see the works of David Pingree and van Bladel (forthcoming). For an introduction to Arabic literature as such, the volumes in the *Cambridge History of Arabic Literature* are recommended.

41. Burnett 1976 and Delp 2006: 5–9.

There are very many classical Arabic works attributed to Hermes registered today in the world's manuscript catalogs,[42] including treatises on magical talismans, astrology, and alchemy, and ranging in length from short works of a single page to books of a couple of hundred. Just six of these texts have been published (dealing with astrology, alchemy, talismans, and snake venom), and only a few of them have known origins.[43] At least two of the Arabic Hermetica are certainly translations from Middle Persian, one of which at least is in turn based on known Greek material.[44] Another is a translation from Greek, perhaps a direct translation.[45] Where did the rest of the Arabic Hermetica come from and who wrote them? Perhaps more of the Arabic Hermetica will turn out to be translations of Greek originals, even works lost in the original. Each text promises potentially to preserve in Arabic material deriving from Greek as early as the third century. The history of these Arabic Hermetica within Arabic tradition itself will require extensive study, as well: Hermes is cited or discussed in at least seventy individual Arabic works by different authors from Andalusia to India, dating from the eighth to the eighteenth century; this is based only on preliminary gleanings of the bibliographical sources.[46] The massive Arabic Hermetic tradition—if the use and reuse of Hermes as an authority can be called Hermetic tradition—still awaits its modern students.

1.4 Understanding the Terms *Hermetic* and *Hermetism* in the Arabic Context

Before going on, it is necessary to deal with a problem of definition. In modern scholarly usage, the terms *Hermetic* and *Hermetism* are both used to refer to several totally different things in various traditions in different languages and periods. One author uses these terms to refer to anything seeming esoteric, whereas another uses it to refer to anything having to do with alchemy. Another calls any work that even mentions Hermes in passing *Hermetic*, and still other uses of the word confuse the matter more. These names have been employed in so many ways that there cannot

42. Strictly speaking, there are about fifty different titles of works reported in manuscript catalogs. However, some of these titles have turned out to be different designations of the same work, so a precise enumeration of the actual works is not possible at this point. Still it is possible to say that there are more than twenty different Arabic Hermetica.

43. Badawī 1955: 55–116, Vereno 1992: 136–159 and 160–181, Ullmann 1994, Kunitzsch 2001, Toral-Niehoff 2004. One can add the short Emerald Tablet of Hermes from at the end of the *Kitāb Sirr al-ḫalīqa* of pseudo-Apollonius, ed. Weisser 1979: 524–525 (book V, section 33.2) (cf. Ullmann 1972: 170–172), and the wise sayings of Hermes (on which see section 5.3 in this volume).

44. See section 2.2.

45. This is the text known as *Kyranis*; see Ullmann 1975, edited by Toral-Niehoff 2004. Although it was associated with Hermes in its Greek version, the extant Arabic does not mention Hermes at all.

46. The standard bibliographic manuscript surveys show how widely Hermes was cited and mentioned. See the references in Ullmann 1970 and 1972, indices under Hermes; GAS 3.171, 4.42.5, 7.5.7, 7.57.28–29, 7.58.30, 7.58.31; Fahd 1966: 396.

be any consensus among scholars today about what works can be considered as Hermetica. Is a work *Hermetic* if it deals with alchemy, or merely if it claims to be a secret text revealed to an elect few? Sometimes *Hermetism* is treated as a literary genre, and other times it seems to describe a supposed doctrine. Some modern authors also refer to the *Hermetists*, meaning those who followed an alleged doctrine set out in Hermetic works or who practiced Hermetic disciplines. What they practiced is usually called *Hermetism*, as if they held a definite and specific doctrine, yet the nature of this Hermetism has not been adequately addressed in discussion. Even when European alchemists during the sixteenth and seventeenth centuries self-consciously describe their doctrines as *hermetic*, the meaning of the term for them is not entirely clear.[47] Yet the terms, as historians employ them, are crucially important, because the moment we use them, we implicitly assert historical and social realities behind them and propose distinctive categories where there may have been none.

The debate about the social reality behind the ancient Greek Hermetica in Egypt has lasted for a long time. Richard Reitzenstein, for example, wanted to see a school or community of Hermetists as the source of the Greek Hermetica.[48] This idea was rejected by the authoritative Festugière as groundless. Instead, he argued that attribution to Hermes was just a feature of a literary genre of the Roman Empire in the first centuries of the Common Era. Thus works attributed to Hermes could be grouped together with similar works falsely attributed to other legendary sages such as Zoroaster.[49] Nevertheless, he did not hesitate to use the word *hermétisme* to describe the doctrines of the Hermetica. Despite Festugière's criticism, the idea of *Hermetists* as a historical group in Roman Egypt has been championed by several later scholars, and this now seems to have been victorious. One of the main arguments of Garth Fowden's *The Egyptian Hermes* is that there were ancient Hermetists as such practicing a Hermetism as such, and that the Greek *Corpus Hermeticum* represents their coherent school of thought. Although he did not state explicitly exactly when these Hermetists are supposed to have been practicing their way, testimonia of the discourses of Hermes (the "philosophical" Hermetica) would point to the second or third century. I find Fowden's arguments about the character of the circles of spiritual aspirants who produced these texts mostly convincing, though, as he himself admits, his conclusions necessarily involve some uncertainty.[50] But even if we concede that the term *Hermetists is* suitable to describe the authors and original readers of the ancient Hermetica, what do we do with the name when we try to describe the Arabic Hermetica, half a millennium later and more, the doctrines and practices described within them, as well as their authors and readers?

47. Copenhaver 1990: 280–290 and Ebeling 2007: 1–2.
48. Reitzenstein 1904: 248.
49. Festugière 1.81–87.
50. Fowden, 1993a: 155–195.

When it comes to the Arabic Hermetica, one finds in modern scholarship a highly inconsistent use of the terms *Hermetic, Hermetist*, and *Hermetism*. Because the words have ordinarily been used to refer to a putative religious group in early Roman Egypt and its works, or to occultists of early modern Europe, one might expect some critical reflection on the meaning of these terms when applied to the very different background of early Arabic culture, and to the social realities assumed behind those terms. I have not found any such reflection on *Arabic Hermeticism*. Moreover, in the context of Arabic literature, modern scholars generally use the word *Hermetic* to refer not just to works associated with the name Hermes, but also to all manner of works with allegedly "Gnostic," "Neoplatonic," "Neopythagorean," or "esoteric" tendencies or practically any early Arabic pseudepigraph.[51] Above all, it is used to refer to the beliefs of the Ṣābians of Ḥarrān, about whom very little is known, but who have been the subject of seemingly endless speculation. (Their relationship to the Hermetica is the subject of chapter 3.)

In fact, a remarkable number of the modern studies discussing *Arabic Hermeticism* emphasize just how vaguely defined it must have been. It is as if a vague definition of the supposed phenomenon proved that the thing itself was vague. The general unwillingness to define the object means that obscure, meaningless, and sometimes simply false assertions about Hermeticism have prevailed. Some have claimed that a given Arabic author or work was Hermetic with hardly any reference to Arabic works of Hermes or any explanation of the term.[52] Others allege the great "influence of Hermetism on Islam,"[53] or oppose the "rational Aristotelianism" to the "irrational Hermeticism" in Arabic philosophy.[54] Some find Hermetists in figures that they see as marginal to an imagined Islamic orthodoxy, such as certain Šīʿites, alchemists, and unconventional individuals such as the tenth-century physician ar-Rāzī, or sometimes in any author who so much as cites Hermes.[55] Hermetism has even been imagined as a risky belief system dangerous for its adherents because of the

51. The terms *Gnostic, Neoplatonic*, and *Neopythagorean* are also all equally ill-defined for Arabic literature.

52. E.g., Joosse 1993. Walbridge 2001: 17–50 attempts this for as-Suhrawardī, although maintaining nevertheless that "The Hermetica do not have clear, distinctive, and consistent doctrines" and that their teachings "can be related to almost every significant intellectual movement in antiquity" (42). Walbridge's conclusions about the "Hermetic" aspect of as-Suhrawardī's thought are not based on reference to Arabic or Persian works of Hermes that as-Suhrawardī knew.

53. Affifi 1951, who calls the Hermetic vagueness "eclecticism" and thinks that Hermeticism might therefore help to explain the "eclectic character of Moslem thought" (841).

54. Massignon 1950: 389–390; Nasr 1967: 76: "The Hermetic school ... considered inner purification as the means of attaining the truth as opposed to the rationalistic tendencies of the Peripatetics. In the sciences of nature, this school distrusted the syllogistic method and sought to rely on a knowledge of concrete causes which could be observed and experienced ... In the Muslim world, Hermeticism must be considered as one of the most important factors which aided in the construction of the Muslim world view."

55. Massignon 1950: 395–399; Grignaschi 1975: 62–63; Peters 1990; Green 1992: 169.

supposedly oppressive "orthodox" Muslims, who, it appears, uniformly hate Hermetists, even though Hermes came to be regarded by Muslims as a prophet.[56] Another approach has been to apply the definition of the Hermetic genre created by Festugière for the Greek Hermetica to Arabic literature.[57] Festugière's definition of "the hermetic" was, however, fundamentally tied to the motif of revelation of a hidden text. Consequently, it is of little value in the context of the literature of Muslims, whose very scripture, the Qur'ān, would have to be described as Hermetic on the same grounds.[58] Characteristic of all such studies is a total or near total lack of attention to the books actually attributed to Hermes in Arabic, not to mention the chronological order of hundreds of years of sources.

Much of what has been written so far about the Arabic Hermetica and Hermetism in Arabic is therefore misleading. It is inappropriate to apply terms, the utility of which is debatable within the study of Greek literature of the Roman period, and which are loaded with connotations derived from the Hermetic movement of early modern Europe, to early Arabic literature without careful discussion. Calling something Hermetic in this context does more to obscure than to explain its meaning and history. We are left only with a vague impression of a supposedly distinctive intellectual movement and increasingly elaborate speculations on the same well-known but inconclusive textual references. The majority of Arabists having no special expertise in the problems presented by the Hermetica themselves are led by works such as these to notions of "the influence of Hermetism in Islam," and these have subsequently been repeated in more general works as accepted facts. A survey of the views on Arabic Hermetica on the Internet, including reference works as well as the Web sites of modern alchemists, astrologers, and those of "esoteric" interests, easily found with any search engine, gives an even more disappointing picture of misinformation—often based, however, at least partly on scholarly works like those described above. What is needed is not more pure theorizing about Hermetism in Arabic or in Islam, but studies devoted to the Arabic Hermetica and the Arabic works discussing them, the primary sources themselves which have so far not been subjected to sufficiently critical scrutiny.

After all this speculation about *Arabic Hermetism*, it still remains to be shown precisely what the Arabic Hermetic texts themselves say, who used them and how, from where they originated, how they were transmitted, in what sense they are Hermetic, and indeed who Hermes was thought to be and why. To assume the

56. Scott, 1.106–109; Stapleton, Lewis, and Sherwood Taylor 1949: 69–70; Peters 1990: 188: "There were few outright and confessed Hermetists in Islam. Such confessions were likely to prove dangerous in a professedly religious society such as Islam..."

57. Festugière 1.309–354; Plessner 1954a: 47; Klein-Franke 1973.

58. The Qur'ān itself says that it was kept on "a guarded tablet" and that it was revealed by God (85:21–22); Muḥammad would therefore be a "Hermetist" classed with Pseudo-Apollonius of Tyana, who discovered revelations of Hermes on a hidden tablet.

existence of a special sect of Hermetists who produced the Arabic Hermetica, as has been done with reason for the Greek Hermetica, is not only so far unsupported, but could further confuse the issue by suggesting implicitly the identity of, or at least a connection between, the ancient Hermetists—whose characteristics and even existence have long been the subject of debate—and the authors of the Arabic Hermetica, the Arab Hermetists, a modern invention. That is not to say that no such community in Arabic will ever be discovered, as unlikely as it may be. For the purposes of this study, however, I am staying close to the texts.

The term *Hermetic* will be applied only to texts *attributed by name to Hermes.*

The expression *Hermetic tradition* is used here only to mean the tradition of textual transmission of writings *attributed to Hermes.*[59]

The terms *Hermetism* and *Hermetists* will be avoided as much as possible.

Some deficiencies in this working definition of *Hermetic* are immediately clear. It excludes much literature that *might* be related to the Arabic Hermetica but is not attributed to Hermes by name, such as works associated in Arabic with the names Agathodaemon, Asclepius, and Pseudo-Apollonius of Tyana. I have indeed had to make some exceptions to this definition.[60] However, given the absence of any well-founded idea of what *Hermetic, Hermetist,* and *Hermetism* might mean in the context of Arabic literature, I am making these restrictions to give the project some coherency. One should keep in mind that the term *Hermetism* did not exist in Arabic literature; other terms and criteria appropriate to this context are needed and remain to be determined. To move forward, one must discard all of the assumptions commonly made in modern scholarship and rely as much as possible on primary sources, not to talk vaguely about *Hermetism*, but to study the Arabic Hermetica themselves and relevant works contemporary with them. Only later will it be possible to take texts attributed to other figures related to Hermes, such as Agathodaemon, into account, and to determine whether a coherent set of doctrines that might be called "Arabic Hermetism" can be found, and whether such a designation will actually help us to understand the history of the texts.

Despite the picture just painted, there are a number of notable and praiseworthy exceptions to the pattern of obfuscation in the study of the Arabic Hermetica during

59. Copenhaver 1990: 289 urges roughly the same application of the term for early modern European hermetic tradition: "In my view, the term 'hermetic' should be used primarily to name a set of texts: either the philosophical *Hermetica* published by Ficino and Lodovico Lazarelli . . . or popular *Hermetica*, such as the *Kyranides* . . ." Copenhaver also calls for research on how the term *Hermetic* was used in the early modern period; the same is required for any period in which like terms are used.

60. The only extant Greek Hermeticum to survive in Arabic translation, the *Kyranis*, actually has lost all reference to Hermes even though it survives in manuscripts including other Arabic Hermetica (Beeston 1962–1967, Ullmann 1975, Toral-Niehoff 2004).

the last half century. When scholars have dealt directly with the texts themselves, they have made real advances. Such studies are the foundation of everything that follows here, and their importance will be obvious at every stage.[61] I have benefited much from these predecessors in this work, and I wish to express my debt to them here at the outset.

61. Fuat Sezgin's bibliographical *Geschichte des arabischen Schrifttums* (GAS) offers the most complete inventory of the Arabic manuscripts ascribed to Hermes, the list itself being an unmatched contribution because it contains detailed basic information indispensable for future work on the primary sources of this subject. The late David Pingree wrote a number of astonishingly learned studies of astrological texts and related works often touching upon the Hermetica, and these provide many important contributions to the subject undertaken here—indeed, they are one of the bases of this book—although Pingree's theories about the alleged connection between the Ṣābians of Ḥarrān and the Hermetica remain to be verified. Although I propose a number of corrections to Pingree's conclusions in what follows, I could not have proceeded to my own conclusions without his earlier studies. Charles Burnett's research on the reception of Arabic Hermetica and the Arabic occult sciences in Latin has also been foundational and has shed light on many problems. Paul Kunitzsch showed that an extant Arabic Hermetic text was translated from Middle Persian, the only case in which the proximate origin of an Arabic Hermeticum is known for certain, and he has edited this text in its Arabic and Latin versions. Manfred Ullmann's valuable bibliographical study of natural and occult sciences in Arabic is a great resource for scholarship on the subject, and in 1994 he edited an Arabic Hermetic text previously unknown, Hermes' book on snake poisons. Ullmann also discovered the only known case of a whole Arabic Hermetic text the Greek antecedent of which survives, the *Kyranis*. Isabel Toral-Niehoff has now published an edition and study of this work. Ursula Weisser's edition and study of the *Sirr al-ḫalīqa* of Pseudo-Apollonius crucially help to establish a chronology of some of the early Arabic Hermetica, including the Emerald Tablet. Thanks to these scholars, and the works of pioneering researchers of earlier generations such as J. Ruska, P. Kraus, and M. Plessner, the Arabic Hermetica are not completely uncharted territory.

Hermes in Sasanian Iran

In the early third century, a noble family of Persia overthrew the Parthian king Artabanus IV and established their Sasanid dynasty firmly over all of Iran and Mesopotamia. The new King of Kings, Ardašīr I, and his son and successor, Šāpūr I, ruled an Iranian empire that would remain a lasting rival of the Romans and Byzantines (ca 224–642 CE).[1] Like their Roman counterparts, the Persian emperors dominated a vast territory including many different ethnic and religious groups having their own languages and traditions.[2] But although the two literatures of the Roman ruling class, Greek and Latin, were substantially preserved by later generations according to their own peculiar historical circumstances, the Persian literature of the same period is almost entirely lost. Middle Persian, the language of the Sasanian court and administration of government, as well as their Magian (Zoroastrian) religion, was displaced by Arabic after the Arab conquest and colonization of Iran in the seventh and eighth centuries.[3] Arabic, the prestigious language of the new rulers and of their new religion, Islam, superseded written Iranian languages almost entirely. Education and literacy in Middle Persian and other Iranian languages became practically obsolete for Iranians who converted to Islam. The children of converts learned Arabic, the language of their scripture, as their own literary medium. Nevertheless, as the number of Iranian Magians

1. Dignas and Winter 2007.

2. Detailed comparisons of the Roman and Persian empires, and a history of the rivalry between the two, are found in Dignas and Winter 2007 and Howard-Johnston 1995. For a different approach to Sasanian Iran, see Pourshariati 2008.

3. Middle Persian is an Iranian language, an older relative of Modern Persian, Farsi. It was used as the language of court, of the Magian (Mazdaean, Zoroastrian) religion, and for other official purposes in the Sasanian Empire and is preserved in Magian texts from that period to the present. It was also an important ecclesiastical language of the Manichaeans in Iran and Central Asia, and was used to an unknown degree by Christians of Iranian background. This much is known from the surviving remains, but there is no doubt that other forms of literature in Middle Persian are now lost. The language was written in various cognate scripts derived from the Aramaic script, the best known of which is called Pahlavi; Middle Persian is itself is often called Pahlavi or Pehlevi.

diminished, some Middle Persian texts deemed worthy of preservation were translated into Arabic. Middle Persian works do survive in their original form in continuous manuscript tradition, but only when preserved by priests in the Magian communities, an ever-shrinking minority who maintained the old religion of their parents and the memory of ancient Iran's Persian kings and mythology.[4] The priests retained relatively few nonreligious texts in Middle Persian,[5] but a substantial quantity of originally Middle Persian texts survives to this day in Arabic translation, some of which are not yet recognized as translated.

As this chapter will show, the lost Middle Persian literature of Sasanian Iran also included works of Hermes Trismegistus.[6] Some of these, in Arabic translation, became a part of the Arabic Hermetica. Although this is not well known, research has sporadically suggested the existence of lost Iranian Hermetica for almost a century.[7] Paul Kunitzsch and David Pingree proved beyond doubt that at least a couple of astrological works attributed to Hermes circulated in Middle Persian language.[8] Charles Burnett, referring to the work of Reitzenstein and Pingree, concluded after studying some astrological doctrines found in Arabic Hermetic texts that "we have reasonable evidence that at least part of the Arabic Hermetica derived from Persian and ultimately Indian sources."[9] No one, however, has made a special, concerted investigation of the evidence for pre-Islamic Middle Persian Hermetica. That is the

4. Substantial quantities of fragmentary Manichaean texts in Middle Persian have been discovered at Turfan and other Central Asian sites, but these were not part of a continuously preserved heritage as Zoroastrian Middle Persian is.

5. Cereti 2001 is presently the most thorough introduction to the Middle Persian literature of the Zoroastrians.

6. Middle Persian Hermetica are not mentioned in any survey, bibliography, or general study devoted to the Greek Hermetica. The exact identity of the type of script in which the Middle Persian Hermetica were written is in most cases impossible to determine, because no Hermetic work survives in Middle Persian form. For now the reader may refer to the descriptions of Middle Persian scripts given by Tafażżolī, "Dabīre," *EIr*, 6.540a–541b, as well as Sundermann 1989: 142–143 for an overview and further bibliography.

7. Blochet's study of the Arabic Hermetica kept in the Bibliothèque Nationale (1911–1912) led him to make hypotheses of an Iranian background that could explain their peculiar features and their mysterious names. He was the first to point to the alleged parallel between a passage of the Hermetic *Kitāb Usṭuwwaṭās* and the Zoroastrian Middle Persian cosmology known as the *Bundahišn* that later excited some further interest (idem: 62–64). Julius Ruska, in his studies of the Arabic *Book of Stones* of Ps-Aristotle (Ruska 1911: esp. 43–44) and of the Hermetic *Tabula Smaragdina* (Ruska 1926), pointed to a pre-Islamic Iranian scientific literature, in the end referring to Blochet's studies and concluding that the cooperation of classicists, Arabists, and Iranists would be required to solve some of these puzzles (idem: 45–47). Richard Reitzenstein went furthest by postulating an Iranian background to the Greek Hermetica, rather than the other way around (a view not accepted today), and pointed out more parallels between Greek and Arabic Hermetica and Zoroastrian Middle Persian literature (Reitzenstein and Schaeder 1926).

8. Kunitzsch 1968, 1970, 1981, 2001. For Sasanian astrology in general, including detailed references, as well as discussion of Middle Persian Hermetica, see Pingree 1997, especially chapter 4, "The Recovery of Sasanian Astrology," 39–50.

9. Burnett 1986: 86–87.

present task. As will become clear, establishing the existence of Hermetica in pre-Islamic Middle Persian is not a difficult matter; on the other hand, fixing the date of their translation into, or composition in, the Middle Persian language is much harder to say with any degree of certainty, and correspondingly requires much lengthier investigation having less satisfying results.

The study of the Arabic Hermetica themselves cannot proceed without taking the Middle Persian tradition fully into account for the simple fact that some of the Arabic texts derive from the Middle Persian, not Greek, Hermetica. The Sasanian tradition is thus a fundamental part of the background to the Arabic works of Hermes. Where even a few Arabic Hermetic texts are known for certain to derive from a Middle Persian antecedent, any number of others may in fact have similar origins not yet detected. An attempt to answer the question completely will require careful study of the internal evidence found in the Arabic Hermetica individually. As a preliminary to such future investigation, this chapter gathers the evidence external to the Arabic Hermetica for these little-known Middle Persian works and explores the traditions pointing to the transmission of ancient Hermetica from Greek into Middle Persian long before the existence of Arabic literature.

2.1 Evidence for Hermes in Sasanian Magical Texts

The name Hermes was invoked in Sasanian Mesopotamia as a source of occult power. A few surviving texts of Syro-Mesopotamian origin provide the attestations: two Babylonian Aramaic incantation bowls containing the same formula,[10] found at Nippur (modern Niffar) in Iraq, once part of the Sasanian Empire, and a magical amulet written in Syriac on parchment dating to Sasanian times.[11] Incantation bowls are a type of popular magical apparatus inscribed with texts in different Eastern varieties of Aramaic made from about the fourth to the seventh century, that is, under the Persian Sasanid dynasty, in Mesopotamia. Unfortunately little is known about exactly how they were used.[12] The two bowls mentioning Hermes invoke him as a magical power, so that the protective operation is performed not only in the name of four angels but also in the name of "Hermes the Great Lord," *hyrmys mry' rb'*.[13] One of these bowls was made for the benefit of "Yazīdād, son of

10. Edited in parallel in Montgomery 1913, bowl 7, line 8.

11. Gignoux 1987: 25–43. "Amulet" here is *qmʿ* rather than the better-known Syriac form *qmîʿâ*. Cf. Middle Persian *qm'hg* /kumāhag/ (Durkin-Meisterernst 2004: 205b), Mandaic *qmaha* (Drower and Macuch 1963: 413b).

12. Theories about the use of these bowls are given in Naveh and Shaked 1985: 15–16.

13. The spellings of Hermes appear as *hyrmys* on one of the bowls and as *'yrmys* on the other. Montgomery 1913: 29 noted that in the spelling system used on these bowls, *y* and *w* were often used to indicate short vowels as well as long vowels (as in Babylonian Aramaic in general). Another magical formula appearing on some other similar incantation bowls was formerly thought to include the name Hermes, but Naveh and Shaked 1985: 131–132 have shown that this should be read *'dms'* = Manichaean Middle Persian *'rms'*, Arabic and New Persian *'lm's* الماس = Greek *adámas* ἀδάμας, "diamond" or "steel."

Yazdāndukh(t)," both Middle Persian names indicating a Persian, perhaps aristocratic, recipient.[14] As for the parchment amulet, although it was written in Syriac, it was made for the protection of a certain Ḫwarrawehzād, called Yazdānzādag, daughter of Dēnag, whose name is also clearly Middle Persian.[15] The magical formulae in this text are hard to understand, full of strange terms, many of which appear to be distorted names of ancient significance derived from various backgrounds, such as Uriel (Hebrew) and Nabu (Akkadian and Aramaic). Although the editor of the text has made a valiant attempt to decipher this document and explain the names, the meaning of all the terms in the text cannot, at present, be explicated precisely. In any case, among the authoritative names invoked on this amulet is unmistakably the name Hermes, Lord of the Universe (*hrms mr' kl* ܗܪܡܣ ܡܪܐ ܟܠ).[16]

The brief appearances here of the name Hermes do not make these Aramaic texts "Hermetic" in the sense used here (Hermes is not the alleged author), but they provide evidence that the magical power associated with the name Hermes was known in Sasanian Iran. Their formulae should be understood in the broader context of late antique magic, including the Greek magical papyri mentioning Hermes.[17] It is also noteworthy that the Syriac amulet mentioned above contains one of the earliest known examples of the abjad-script spelling of *hrms* for Hermes,[18] the spelling later most widespread in Arabic.[19] In the end, however, these magical amulets and bowls, so poorly understood, do little to further our knowledge of Hermes in the Sasanian Empire.

14. For the attestation of these names in Middle Persian, see Justi 1895: Yazīdād 149b; Yazdāndukht 146b.

15. The suffix *-ag*, represented in the Syriac with a *qop* [Q], is characteristic of Middle Persian, not of New Persian. This evidence almost certainly indicates that it can be dated to the Sasanian period.

16. Gignoux 1987: 32.27.

17. For a basic orientation into Greek magical texts referring to Hermes, see Fowden 1993a: 79–87 and 168–172. For the Egyptian background to the Greek magical texts, see Dieleman 2005.

18. *Abjad* is the grammatological term for the type of script that shows only consonants and some of the vowels, or in other words, a *consonantary*. This term distinguishes it from scripts like the alphabet, which renders each vowel as well as each consonant with an individual grapheme, and from other types of script. On this scientific typology of scripts see Daniels 1996.

19. The normal spelling of Hermes in Syriac is *'rmys* or *hrmys* (see, for the former, Jacob of Edessa, *Hexaemeron* 149b11–150a17, and for the latter, Bar ʿEbrāyā (Ibn al-ʿIbrī), *Syriac Chronicle*, Bodleian MS Hunt. 52 facsimile 2v.a15 and 55r.b33, trans. 5 and 153. Even Arabic authors mention that in Greek the name is *'rmys* or *hrmys*, which is indeed the way that the name would normally be transcribed into Arabic from Greek. *Hrms* appears interchangeably with *hrmys* in the third-century work *Book of the Laws of the Countries* (e.g., Drijvers 1965: 48.12 and 50.8). *Hrms* is also the spelling of the name in the texts translated from Middle Persian mentioned below as well as in most of the Arabic Hermetica. It became the standard spelling in Arabic, but the fact that it was not regarded as the Greek way of citing the name is indicated by the note sometimes accompanying the biographies of Hermes in Arabic that his name is *hrms* but the Greeks called him *'rmys* or *hrmys* (notably in Mubaššir ibn Fātik, *Muḫtār al-ḥikam*, 7.9–10; the whole text is given in translation in section 4.2 in this volume). It is highly likely, therefore, that even the spelling of the name *hrms* was transferred from Middle Persian (and not from Greek or from Syriac, in which the spelling *hrms* appears only rarely), as was done with the title of the *Almagest* (Kunitzsch 1975: 281). Paul Kunitzsch illustrated this type of transmission of spellings with reference to several Greek words and argued convincingly that the name *Almagest* spelled in Arabic *'lmǧsty*, without full letters for the Greek short vowels, must be a transliteration from Middle Persian (Kunitzsch 1974: 115–125; 1975: 280–281) and not from Greek or Syriac.

2.2 The Hermes of the Sasanian Astrologers

The present concern is rather with the book traditions of pre-Islamic Iran. Nallino, Kunitzsch, and Pingree have already proved that a number of scientific books were translated from Greek into Middle Persian in the Sasanian Persian Empire and then were later translated from Middle Persian into Arabic in the early ʿAbbāsid period (second half of the eighth century to the early ninth century).[20] All the Middle Persian versions of these texts have been lost, along with most of the literature of that language, but some survive in Arabic translation. Their survival in Arabic is, in fact, the only reason we know of the Middle Persian stage. Of these few works, three are especially important here: two directly attributed to Hermes, and one including a reference to a separate, apparently unknown, book of Hermes. In the search for texts in Middle Persian attributed to Hermes, these texts suffice to prove that some form of Sasanian Middle Persian Hermetic literature did exist.

One of these texts is the *Kitāb Asrār an-nuǧūm* of Hermes (*The Secrets of the Stars*, and in the Latin translation of the Arabic, *Liber de stellis beibeniis*),[21] a work explaining the characteristics of persons born under the influence of the brightest fixed stars. In 1970 Kunitzsch showed the Arabic to be a translation from Middle Persian, retaining many of the Middle Persian words in Arabic transliteration. Later, Pingree demonstrated that Rhetorius, a Byzantine scholar who flourished around 600 CE, writing in Greek, shared the same (presumably Hermetic) source as the Middle Persian text for his chapter *On the Powers of the Fixed Stars*.[22] By studying the celestial

20. Nallino 1922; Kunitzsch 1968, 1970, 1975, 1981; Pingree 1962, 1963a, 1973, 1975, and especially 1989a. See also Panaino 2001. The works of this kind identified so far are five in number: Cassianus Bassus Scholasticus' *Geoponica*, Dorotheus' *Carmen Astrologicum*, Vettius Valens' *Anthologiae*, Teucer of Babylon's *Paranatellonta*, and Hermes Trismegistus' "*de Stellis Beibeniis*." A possible sixth is the astrological work attributed to Zoroaster, *Kitāb al-Mawālīd* (claimed as from Greek by Kunitzsch 1993). Nöldeke 1890 also showed that the Syriac version of the historical novel of Pseudo-Callisthenes, the "Alexander Romance," had been translated from a Middle Persian version of the Greek. I do not accept the arguments of Ciancaglini 1998 and 2001 objecting to Nöldeke on this issue; see van Bladel 2007a: 61–64.

21. The word *beibeniis* is the Medieval Latin transcription (with Latin inflectional ending) of the Arabic transcription (*al-biyābānīya*) of the Middle Persian adjective *awiyābānīg*, "non-wandering," or "fixed (stars)," rendering the Greek adjective ἀπλανής (Henning 1942a: 232n3). As noted by Kunitzsch 1993: 244–245, formerly this word has been misunderstood by al-Bīrūnī and some modern scholars following him to mean "(stars) of the desert."

22. It is significant that the long portion immediately preceding this chapter in Rhetorius' work, dealing with the astrological "doctrine of the twelve places," is attributed in one class of the manuscripts to Hermes Trismegistus. (See F. Cumont in CCAG 8.4.116–117, text edited at 126–182; this will be book V, chapter 57 of the forthcoming edition of Pingree and Heilen: Rhetorius 2009.) It seems likely that the source shared by Rhetorius and the Middle Persian translator included both of these together, along with the attribution of at least the first of the two parts to Hermes, after which Rhetorius did not indicate that the section on "the doctrine of the twelve places" was also attributed to Hermes in the shared source; alternatively, the Middle Persian translator mistakenly attributed the second part of the shared source to Hermes because it immediately followed the first part bearing that attribution. On Rhetorius see Pingree 1977 and Pingree 2001: 6–13.

coordinates given for these stars, and noting the difference in their position from that given by Ptolemy's *Almagest* in the second century, Pingree argued that this Hermetic source must have been written about 505 CE, calculating this date according to Ptolemy's own guideline of a one degree increase in longitude every 100 years. The same change in coordinates of these stars, shared by the Greek of Rhetorius and the Middle Persian-Arabic version of this Hermetic work, could mean only that the two works share a common source.[23] Therefore the date ca 500, if Pingree's assumption is correct, gives a *terminus post quem* for this Hermetic text in its Middle Persian form.

The second text gives similar evidence for a Middle Persian Hermetic literature, citing even name of the translator. An Arabic manuscript now in Berlin is entitled *The Book of Hermes on Revolutions of the Years of the Nativities, as ʿUmar ibn al-Farruḫān aṭ-Ṭabarī Translated It from Persian into Arabic, it being The Rod of Gold*.[24] ʿUmar ibn al-Farruḫān aṭ-Ṭabarī (d. 816) was a court astrologer for the caliph al-Manṣūr (r. 754–775).[25] His father's non-Muslim Iranian name, his epithet indicating his family origin in Ṭabaristān, his being one of the earliest known Arabic astrologers, and his reported translation of another Middle Persian work (immediately below) all make his translation of this work from Middle Persian into Arabic totally plausible. Although we lack for now a thorough study of the text, it is nevertheless clear that this is another example of an astrological work of Hermes once written in Middle Persian.

Third is the *Carmen Astrologicum* of Dorotheus (fl. first century CE). This work survives almost complete in the Arabic translation of the Middle Persian recension of the mostly lost Greek original. The Arabic translation from Middle Persian was made, significantly, by the same ʿUmar ibn al-Farruḫān. In the Arabic, the text contains references both to a Hermes Trismegistus, king of Egypt (*al-mušarraf al-maḥmūd bi-ṯalāṯ ṭabāʾiʿ Hirmis Malik Miṣr*), and to a Babylonian Hermes, the addressee of the book, who is described as the son of Dorotheus the King of Egypt.[26]

23. Pingree 1989a: 233–234. See also Kunitzsch 2001: 13, citing the anonymous astrologer of 379 CE for Ptolemy's guideline (CCAG 5.1, 198.5–7).

24. Staatsbibliothek Preußischer Kulturbesitz, MS or. oct. 3256 (described by Schoeler 1990: 175–177). Arabic title: *Kitāb Hirmis fī taḥwīl sinī l-mawālīd bi-mā fassarahū ʿUmar ibn al-Farruḫān aṭ-Ṭabarī min al-Fārisīya ilā l-ʿarabīya wa-huwa Qaḍīb aḏ-ḏahab*. This work is mentioned by an-Nadīm (*Fihrist* 327.3–4; ed. Flügel 267.15) as *Kitāb al-Maktūm fī asrār an-nuǧūm wa-yusammā Qaḍīb aḏ-ḏahab* (*The Hidden Book on the secrets of the stars, called "The Golden Bough"*); apparently it is the same as that which he lists as a work of ʿUmar ibn al-Farruḫān under the title *Kitāb Taḥwīl sinī l-mawālīd* (*Fihrist* 333.1). It is also named by Ṣāʿid al-Andalusī (ed. Cheikho 19), and was apparently cited in Latin by Hermann of Carinthia in his *De essentiis*, written in 1143 at Béziers (Burnett 1981: 167n6), as Hermes' *Aurea Virga*, although this will have to be verified in a study of the still unedited Arabic text.

25. Ullmann 1972: 306–307.

26. Dorotheus 1.1–2, 2.20. Pingree 1999 shows evidence of a second Arabic translation of Dorotheus from the same Middle Persian version; he attributes the translation tentatively to Māšāʾallāh. According to Pingree 1975: 7, Dorotheus is represented as King of Egypt also in another Arabic translation of a Middle Persian work, the still unedited *Kitāb al-Mawālīd wa-aḥkāmihā* of Zarādušt (Zoroaster), a work of genethlialogy dependent on Dorotheus.

The first reference, to Hermes King of Egypt, says that one should consult Hermes' book to learn the places of the planets. This implies that a Hermetic astronomical work, perhaps tables of planetary motions, was known in Middle Persian—unless of course the reference is an addition in the Arabic. The other, perhaps surprising, reference, to the *Babylonian* Hermes is explained in another text, examined below, written by the Iranian astrologer Ibn Nawbaḫt, a younger colleague of ʿUmar ibn al-Farruḫān. For now, the point is that an astrological book translated from Middle Persian into Arabic refers the reader to a specific work of Hermes.

Besides these texts, some of the other earliest Arabic astrological writings, also written by Iranian astrologers, cite Hermes as an established authority. One such astrologer was Zādānfarrūḫ al-Andarzġar, of late Sasanian or early Islamic times; the surviving fragments of his writing, in Arabic translation, refer to an astrological work of Hermes.[27] The early ʿAbbāsid court astrologer Māšā'allāh (fl. 762–ca 815), an associate of ʿUmar ibn al-Farruḫān aṭ-Ṭabarī, cites Hermes as an authority in his own Arabic works.[28] He may well have known works of Hermes in the Middle Persian original; being a learned Iranian Jew, Māšā'allāh could have had access to Aramaic and Persian sources as well as Arabic.[29] Al-Fazārī, another contemporary and fellow astrologer at the court of the caliph al-Manṣūr in Baġdād, is also known to have used a work or works attributed to Hermes: one fragment of al-Fazārī's writing, reported by al-Bīrūnī (d. ca 1050),[30] credits Hermes with establishing a measure of the circumference of the earth, and another fragment of al-Fazārī, reported again by al-Bīrūnī, via Yāqūt (d. 1229), credits Hermes with the doctrine of the division of the inhabited world into seven climes.[31] Because relatively few Middle Persian texts are known to have been translated into Arabic by the time in which Māšā'allāh and al-Fazārī were working, and because they were undoubtedly working

27. Edited and translated by Burnett and al-Hamdi 1991–1992.

28. Hermes is mentioned in the beginning of Ibn Hibintā's (ninth cent.) summary of Māšā'allāh's astrological history *Fī l-qirānāt wa-l-adyān wa-l-milal* (On Conjunctions, Religions, and Nations), ed. and trans. E. S. Kennedy and D. Pingree 1971, f 214v.13. "Hermes and others" are cited together as authorities on conjunctions and world-year transfers. It is unlikely that this is an interpolation of Ibn Hibintā; the phrase "Hermes and others" cited as authorities seems to be rather from the original text.

29. Pingree, "Māshā'allāh," *DSB* 9.159–162; Pingree 1973: 37b. Māšā'allāh also cites Dorotheus in a different translation from the Middle Persian (not that of ʿUmar ibn al Farruḫān), perhaps indicating that he read it in the Middle Persian original himself and rendered it, or parts of it, again (Pingree 1999).

30. D. J. Boilot, "al-Bīrūnī," *EI*², 1.1236a–1238a.

31. Pingree 1970, 115–116 (citing al-Bīrūnī, *Taḥdīd*, 211–212 and Yāqūt, *Buldān*, 1.27.15–17). The second report includes a comment of al-Bīrūnī relevant to our understanding of the Arabic Hermetica: "Abū-Rayḥān al-Bīrūnī said, 'Hermes held this division [*scil.* of the climata] that Muḥammad ibn Ibrāhīm al-Fazārī relied on in his *zīǧ* [i.e. astronomical almanac]. Since Hermes was one of the ancients, it appears that no other (division) was used in his time; however, the subjects of mathematics and astronomy [rather than geography] are more particular to Hermes.'" For a detailed investigation of the history of the doctrine of the seven climata, see Honigmann 1929, to be supplemented by Neugebauer 1975, index under "climata: seven climata." Išox, a medieval Armenian author, says in his book on nature (*Girkʿ i veray Bnutʿean*) that Hermes and Aristotle divide the world into seven climates (Thomson 1992: 203).

in an Iranian tradition of astrology, there is every reason to think that these eighth-century figures were using not Arabic, but Middle Persian Hermetica—as ʿUmar ibn al-Farruḫān aṭ-Ṭabarī certainly was—even though the manuscript traditions preserve only the Arabic writings of these authors themselves.[32] Yet another court astrologer working among these Iranians, the Syrian Theophilus of Edessa (d. 785), refers in one of his Greek writings to works of astrology that the Persians had translated into their language from Greek.[33] Although he does not name books of Hermes among them, it is clear that he is aware of astrological works that had undergone a similar transmission into Middle Persian. It appears that all the astrologers in the court of al-Manṣūr—who knew each other personally, of course—must have been familiar with astrological Hermetica, in Middle Persian or in new Arabic translations. These Hermetica were thus among the earliest known scientific texts in Arabic and were studied by scholars enjoying imperial patronage.

Given the evidence, one can definitely conclude that there was an unknown number of Hermetica in Middle Persian concerned with a variety of subjects in astrology. Already two specific Arabic Hermetic works definitely derive from Middle Persian (the *Kitāb Asrār an-nuǧūm* and the *Qaḍīb aḏ-ḏahab* mentioned above), and it seems certain that these were not the only ones written in that language. Meanwhile, the origins of most of the Arabic Hermetica and their relationships with the surviving Greek Hermetica have remained largely obscure.[34] Perhaps, then, more of the Arabic texts are translations not from Greek, but, unexpectedly, from Middle Persian. Research in this direction may shed more light on a scarcely charted Iranian path of transmission of relatively early Greek material into relatively early Arabic, a path altogether separate from that of the majority of books of Greek philosophy and medicine translated into Arabic. Still an important question remains: how, and when, did there come to be a Middle Persian Hermetic literature in the first place?

2.3 The Later Testimonies of Sasanian Translation from Greek into Middle Persian

Ibn Nawbaḫt was a member of the second generation of ʿAbbāsid court astrologers, and he is explicitly described as a translator of books from Persian into Arabic.[35] His

32. Pingree 1989a: 229b–230a has argued that Māšā'allāh was indeed reading the original Middle Persian of Dorotheus at least.

33. See note 50.

34. The first of two exceptions is the *Kyranis* text identified by Ullmann 1975, edited by Toral-Niehoff 2004. This text has a Greek antecedent that survives in Greek as well as a medieval Latin translation; its path of transmission into Arabic is still uncertain. The second exception is a short paragraph in the *Kitāb al-Muǧtanā* of Ibn Durayd identified by Franz Rosenthal as identical to a Hermetic passage found in the fifth-century Greek *Stobaean Anthology*. On this see section 5.3 in this volume.

35. Louis Massignon, "Nawbakht," *EI*², 7.1043b–1044a. The testimony that he translated books is in the *Fihrist*: 333.14.

father (Nawbaḫt) was one of the same group of astrologers mentioned in the previous section: ʿUmar ibn al-Farruḫān, Māšāʾallāh, al-Fazārī, and Theophilus. The son took his father's place before the end of al-Manṣūr's reign in 775, and he continued in the service of the caliph ar-Rašīd (r. 786–809), working in ar-Rašīd's library.[36] Ibn Nawbaḫt offers the most important testimony about exactly when works of Hermes were translated from Greek into Middle Persian. According to him, this happened already in the third century, by the order of the Sasanian emperor Šāpūr I (r. ca 240–271). The specific claim, made in a passage also constituting the earliest known history of science in Arabic, deserves careful investigation. The history appeared in Ibn Nawbaḫt's now lost *Kitāb an-Nhmṭ'n*, surviving excerpts of which, reported in the *Fihrist* of an-Nadīm (wr. 987), are summarized briefly here.[37] Ibn Nawbaḫt's history is based almost entirely on older Iranian traditions.

His story begins during the reign of Dahāk son of Kay (Arabic aḍ-Ḍaḥḥāk ibn Qayy), a demonic tyrant of Iranian mythology whose dominion on the earth lasted for one thousand years, the second millennium of the world.[38] He gathered scholars together in ʿIrāq in twelve citadels corresponding to the twelve signs of the zodiac. These scholars ruled the people and were obeyed by them until a certain prophet appeared. We are not given information about the prophet in the account. The people, upon hearing the message of this prophet, rejected the scholars, who then went off to different countries where they became kings. Hermes, who left Babylon to become king of Egypt, was one of these sages. He then taught the Egyptians the ancient sciences. The text says the following: "Each one of them went to a country to inhabit it, to be in it, and to become chief of its people. There was among them a scholar named Hermes. He was among them one of the most perfect in intellect, most accurate in knowledge, and most subtle in investigation. He went down to the land of Egypt and ruled its people as king, civilized it, improved the conditions of its inhabitants, and revealed his knowledge there."[39] As the story goes, time passed until eventually Alexander the Macedonian conquered the Persian Empire of Dārā ibn Dārā (i.e., Darius) and brought the learning of the ancients to an end by

36. He succeeded his father as a court astrologer to al-Manṣūr (r. 754–775) (Ibn al-Qifṭī, 409.4–6). The *Fihrist*, 333.12–16 (Flügel 274.8–9), reports that he was in the *Ḫizānat al-ḥikma* for ar-Rašīd, who ruled 786–809. On this library, see D. Gutas and K. van Bladel, "Bayt al-ḥikma," *Encyclopaedia of Islam Three*, forthcoming.

37. *Fihrist* 299.9–301.2 (Flügel 238.9–239.31). For more notes on this passage, see also chapter 4 below on the biographies of Hermes in Arabic. The name *an-Nhmṭ'n* (with short vowels unknown) has never been satisfactorily explained, though presumably it masks a distorted Middle Persian or other Iranian word.

38. On Dahāk see CHI 3(1) 426–427; on his place in Persian legend, idem, 383–387. Note that Dahāk's activity is located in Babylon, in other words, Iraq, in the ninth-century Pahlavi books (for references see CHI 3(1) 426).

39. *Fihrist* 300.6–8 (Flügel 239.2–5): *fa-amma kullu ʿālimin minhum baldatan li-yaskunahā wa-yakūna fīhā wa-yatara''asa ʿalā ahlihā wa-kāna fīhim ʿālimun yuqālu lahū Hirmisu wa-kāna min akmalihim ʿaqlan wa-aṣwabihim ʿilman wa-alṭafihim naẓaran fa-saqaṭa ilā arḍi Miṣra fa-malaka ahlahā wa-ʿammara arḍahā wa-aṣlaḥa aḥwāla sukkānihā wa-aẓhara ʿilmahū fīhā.*

destroying all the written texts, except for a number of Persian books that he plundered and had translated into Greek and Egyptian (*ar-Rūmī wa-l-Qibṭī*), while some of the ancient sciences also survived scattered in India and China. After Alexander's destruction, a period of ignorance fell over Iraq (i.e., the region that was to become the administrative center of the Sasanian Empire). This came to an end when Ardašīr, the founder of the Sasanid dynasty (r. ca 224–ca 241), began to bring together the remains of the ancient Iranian writings. His son and successor, Šāpūr I, continued the effort. Some of the authors whose works were translated under Šāpūr are named: among others, Dorotheus, Ptolemy, and "Hermes the Babylonian, who had been king over Egypt" (*Hirmisu l-Bābilīyu llaḏī kāna malikan ʿalā Miṣr*). All these books that originally derived from Babylon (*allatī kāna aṣluhā bi-Bābil*) were studied and taught in Persian. Later, the Sasanid emperor Ḫusrō I (r. 531–579) again renewed interest in the sciences. Thus ends Ibn Nawbaḫt's history, the earliest extant history of science in Arabic. The translation of works of Hermes of Egypt into Middle Persian is specifically said to have been effected by the early Sasanid Šāpūr I, along with astrological works of other Greek authors.

In this account, the Egyptian Hermes is originally from Babylon. This explains the Babylonian epithet of Hermes in the Arabic translation of the Middle Persian version of Dorotheus, discussed above: it appears to reflect an idea in the Middle Persian reception of the Hermetica claiming that they were effectively of Babylonian origin. It should be remembered that Babylon (i.e., Mesopotamia) in this story is meant to indicate a territory that was part of the Persian Empire. (The Sasanian capital, Ctesiphon, was near ancient Babylon.) This becomes clear when Ibn Nawbaḫt says that Darius refused to pay the tribute imposed by Alexander on the people of Babylon and the Kingdom of Persia.[40] It is well known that Mesopotamia was considered an important part of Iran in later Sasanian times.[41] In this account, ancient Babylonian science—including the teachings of Hermes—is conceived as part of the ancient heritage of Iran, proper to the Persians.

As it is related, there are three episodes in this history. First there is the very ancient period during which the sciences are gathered in Babylon under Dahāk but then scattered, partly by the agency of Hermes who went to Egypt and preserved them there. The next episode is that of Alexander's theft and destruction of Persian knowledge. The last episode is the Persian recovery of the scattered, stolen sciences in more recent times. It will be useful to recognize that these three components have been put together to form a continuous narrative of the history of Iranian science, but that there is no evidence that they all belonged together originally.

40. *Fihrist* 300.10 (Flügel 239.7).

41. Ibn Ḫurradāḏbih, 5.18–19: *ṯumma abtadiʾu bi-ḏikri s-Sawādi iḏ kānat mulūku l-Fursi tusammīhi dil-i Ērānšahr ay qalbu l-ʿIrāq*, "Next, I begin by reporting on the Sawād [= lower Mesopotamia] since the kings of the Persians used to call it [Persian words given:] 'the heart of Iran,' that is, the heart of Iraq."

This eighth-century account of Ibn Nawbaḫt is undoubtedly based on a legend or group of legends that existed in Middle Persian, now surviving mostly in abbreviated form. Versions of this history of science became widely known in early Islamic Iran and Iraq, and it was also referred to, in forms usually much less detailed, by a number of Iranian and other authors in Arabic, Middle Persian, and Greek in the early ʿAbbāsid period (very roughly 750–900). Below is a list of testimonies to this history, with notes explaining what portion of the fuller version of the account each testimony conveys.[42] The language in which the account survives is indicated in parentheses, MP standing for Middle Persian. As one can see, the second component in Ibn Nawbaḫt's history, the destruction and theft of Iranian science by Alexander, was particularly widespread.

1. *Dēnkard* (a ninth-century Zoroastrian Middle Persian compilation based on sources of the Sasanian period):
   a. book III[43] (MP): King Ardašīr and Tōsar (a.k.a. Tansar), a Magian priest, credited with reviving Persian science.
   b. book IV[44] (MP): Lengthy account of the scattering of the Avesta in the wake of Alexander's destruction and theft, its restoration by Ardašīr, how Šāpūr I gathered the astrological, medical, and philosophical writings that were scattered in the Roman Empire (*Hrōm*) and India;[45] and the continued developments of philosophy and religious dialectic under Ḫusrō I.
2. The *Greater Bundahišn*[46] (MP): Alexander sent the Zand ("commentary" on the Avesta) to the Roman Empire, killed the Magi, and burned the Avesta.

42. Several of these are discussed in the context of ʿAbbāsid ideology and the Graeco-Arabic translation movement by Gutas 1998: 34–45.

43. Section 420 trans. of de Menasce (1973). Text in transcription and Italian translation by Cereti 2001: 42–45.

44. English translation of this passage from the ninth-century Middle Persian compilation in Zaehner 1961: 114–125; Romanized Middle Persian text Zaehner 1955: 31–32 (and earlier, uncorrected English translation 7–9); for comments and bibliography see Gutas 1998: 35n14. On the basis of the near-demonstrative pronoun *im* referring to Ḫusrō I ("this present" king) in the text, Bartholomae 1920: 9n2 (note continuing on the next page) thought that this passage dated to his reign, that is, to the sixth century. This view was later held by Harold Bailey 1943: 155, and especially 155n2, and by Zaehner 1955: 11–12, assessing the claim by discussing the passage in its philological and stylistic aspects, respectively. Boyce also accepted this view (1968a: 5), and so do I. English translations in Shaki 1981 and Shaked 1994: 99–103, with important bibliographic and interpretative annotations. A transcription with accompanying Italian translation is in Cereti 2001: 59–61.

45. Our interpretation of the motive behind these translations depends to a great extent on our understanding of one Middle Persian circumposition (*az ... bē*). The scientific texts Šāpūr sought for translation into Persian are called *nibēgīhā-iz ī az dēn bē*: that is, either "also writings *apart from* the religion," in other words, *nonreligious* works (thus Shaki 1981, 119), or "writings *based on* the religion" (thus Shaked 1994: 100; cf. 1980: 21–22, where both meanings of *az ... bē* are discussed).

46. Bailey, 1943: 153; Anklesaria 1956: 274–277, §14.

3. Māhānkard[47] (ca 750) (Arabic translation from MP): Alexander destroyed the original ancient Persian books after having them translated into Greek.
4. Ibn Nawbaḫt[48] (Arabic by Iranian author, cited by the *Fihrist* of an-Nadīm): His is the account summarized above. *It is the only one of these accounts to mention Hermes and Hermetic books specifically.*
5. Māšā'allāh[49] (Greek and Latin translations of Arabic by Iranian author): Some of Persian science survives in India.
6. Theophilus of Edessa[50] (d. 785) (Greek by a Syrian author): The Persians who loved wisdom translated Greek books into their own language and used Greek astrology.
7. Ibn Qutayba[51] (d. 889) (Arabic based explicitly on MP sources): Alexander conquered the kingdom of Iran and burned the books of their religion.
8. Abū ʿĪsā ibn al-Munağğim[52] (fl. ca 900) (Arabic by Iranian author): Šāpūr I received books as gifts and as plunder from the Roman emperor Gordian.
9. *Ardā Wirāz Nāmag*[53] (MP): Egypt-dwelling[54] Alexander the Roman[55] plundered the sacred Persian books.

47. He is quoted in Arabic translation by Saʿīd ibn Ḫurāsānḫwarrah in the introduction to his translation of the *Book of Nativities* of Pseudo-Zoroaster. See Krause 1936: 471–472, Afnan 1964: 77–78, and the notes by Gutas 1998: 37–38. Afnan's English translation of the unpublished prologue to the book explains that Māhānkard translated the book in the days of Abū Muslim (d. 755) from the language of the religion (*dēn-dibīrīh*), meaning from Avestan or Pāzand (Middle Persian in Avestan script), into Darī Persian (apparently Middle Persian in the common book script). Later it was translated into Arabic by Saʿīd ibn Ḫurāsānḫwarrah.

48. *Fihrist* 299.9–301.2 (Flügel 238.9–239.31).

49. The full version of this story is not found in an extant work of Māšā'allāh, but it was evidently known to him as the references to Persian learning preserved in India show, which are derived from a bibliography of his found only in translation in the Latin *Liber Aristotelis* of Hugo of Santalla and in a Byzantine Greek translation (Burnett-Pingree 1997: 6–7; see also Pingree 1989a: 228b).

50. CCAG 1.129–131. The relevant words are: οἱ δὲ κατὰ τὴν ἑῴαν ἅπασαν Περσῶν σοφίας ἐρασταὶ τὰς Ἑλληνικὰς βίβλους τῇ ἑαυτῶν μεταφράσαντες γλώττῃ... "The lovers of wisdom among the Persians throughout the East translated the Greek books into their own language..." He mentions by name specific authors of astrological works: Dorotheus, Valens, Critodemus, and Timocharis. Theophilus of Edessa was a court astrologer for al-Manṣūr and al-Mahdī and had access to Middle Persian traditions (Pingree 1989a). He probably knew Ibn Nawbaḫt personally.

51. Ibn Qutayba, *Kitāb al-Maʿārif*, 653.2–5, citing al-Aṣmaʿī (d. 828) and "books of the life-stories (*siyar*) of the Iranians" (652).

52. Stern 1972: 447–448 and 454; see more on this below.

53. Vahman 1986: text 76–77, trans. 191.

54. The epithet "Egypt-dwelling," *Muzrāyīg-mānišn*, may reflect a report derived from the Middle Persian translation of the *Alexander Romance*, which gives Alexander a Pharaonic genealogy, and which mentions Alexander's burial in Egyptian Alexandria at the end of the story (*Alexander Romance*, ed. Budge, text 252, trans. 142). On the Middle Persian antecedent to the Syriac Alexander Romance, see van Bladel 2007a: 61–64.

55. To inhabitants of the Sasanian Persian Empire, the Western lands were "Roman" regardless of the period.

10. *Šahristānīhā ī Ērān*[56] (MP): Alexander destroyed the Avesta, which was stored in writing at Samarqand.
11. Ḥamza al-Iṣfahānī, wr. 961[57] (Arabic on the basis of eight collated translations of the MP *Ḫwadāy Nāmag* and the account of Mūsā ibn ʿĪsā al-Kisrawī): Alexander, jealous of the unparalleled knowledge of the Persian nation, first translated what he needed from Persian, then destroyed the rest, killing the Magi, too. Although he destroyed their books of religion, he had translated their books dealing with philosophy, astrology, medicine, and agriculture from Persian into Greek and Egyptian (*al-Qibṭīya*) and sent them to Alexandria.
12. *Tansar-Nāma*[58] (New Persian translation from Arabic translation of a late Sasanian MP work): Alexander destroyed the Avesta.
13. *Dēn-vijirgard*[59] (New Persian in Pahlavi script): Alexander destroyed the *Nasks* (chapters of the Avesta) except for those concerned with medicine and the stars, which he had translated into Greek ("Roman"). This is a very late text (fifteenth–eighteenth century?) but contains this interesting tradition specifying the *types* of Mazdaean books taken by Alexander. It may be based on a much older source, or sources, because it lists all twenty-one canonical *Nasks*.

This list is far from complete. Versions of these stories about ancient Iranian science became common knowledge, appearing in many other Arabic works, early and later, including aṭ-Ṭabarī,[60] al-Hāšimī,[61] Ibn Ḥazm,[62] aṯ-Ṯaʿlabī,[63] Ibn Abī Uṣaybiʿa,[64] Ibn

56. Bailey 1943: 153–154; Markwart 1931: 9; Daryaee 2002: 13 and 17.

57. Ḥamza, 1921/1922: 18 (intro.), 30–31 and 32 (*faṣl* 4, *bāb* 1, under "Alexander" and "Ardašīr").

58. Bailey 1943: 156–157; Boyce, 1968b: 37.

59. West 1892: 446–447.

60. aṭ-Ṭabarī, *Taʾrīḫ*, 1.701.9–10, refers to Alexander's destruction of Persian books and official records.

61. Al-Hāšimī (107v) relates that, according to "some Persians," Būḏāsaf was from ancient Babylon, and he fled from Ṭahmurūṯ (i.e., Ṭahmūraṯ) to India, where he taught them a religion (*aẓhara lahum āyātin wa-šarāʾiʿa*) and discovered that they kept astronomical records. Būḏāsaf is the Iranian (and Arabic) adaptation of the name of the Bodhisattva. According to the Iranian history of science under examination here, Indian doctrines derive in part from Iran, and in particular from Babylon (presumably Babylon under Dahāk). Būḏāsaf was therefore the Indian counterpart to Hermes, who went to Egypt. Both figures were used to explain the far geographical reach of the religion of idolaters, later designated "Ṣābians." (On Ṣābianism, see the chapter 3.)

62. Ibn Ḥazm 1.138 (in the section *al-Kalām ʿalā l-Yahūd wa-ʿalā man ankara t-taṯlīṯ min an-Naṣārā wa-maḏhab aṣ-Ṣābiʾīn wa-ʿalā man aqarra bi-nubūwat Zarādušt*). See also Pines 1990: 43–44.

63. Aṯ-Ṯaʿlabī 2002, 609, claims two sources: (1) unnamed historians who say that Alexander killed the magi and burned their books and (2) al-Murtaḍā (evidently the Imāmī Abū l-Qāsim ʿAlī ibn al-Ḥusayn), who says that the ancient Iranian books were of gold and therefore were melted down for cash.

64. 1.9.7–10. He cites al-Mubaššir ibn Fātik's *Kitāb Muḫtār al-ḥikam* for this information. Interestingly, Ibn Abī Uṣaybiʿa's history of medicine describes Alexander's plunder of the Zoroastrian books in connection with the origins of medicine. The implication of this story, then, is that *medicine* came originally from Iran.

Ḫaldūn,[65] and others.[66] However, the ones given above are only the Middle Persian sources and translated Middle Persian sources, or are taken from authors who definitely had access to Middle Persian sources directly or in immediate translation.[67] Of these, only two of the older accounts (1b and 4), dating from the sixth and the eighth centuries, state *both* that the *earliest* Sasanid emperors translated books from Greek *and* that this restored what Alexander had allegedly stolen from Iran. The other accounts give either other versions of a similar story or merely references to this story, mentioning Alexander's plunder, or the Sasanian reception, but not both in the same passage. Nevertheless, it is clear that a complete story of Alexander's theft and the Sasanian restoration must have been well known.

Besides Ibn Nawbaḫt (4), the *Dēnkard* (1b) also designates Šāpūr I specifically as a proponent of the translation of learned books from foreign languages and ascribes to him the wish to assess some of the teachings that were newly translated into his language to determine whether they were compatible with (or, should be added to) the Avesta,[68] the ancient scripture of the Mazdayasnian religion. This passage in the *Dēnkard* can be dated with certainty to the time of Ḫusrō I, the sixth century.[69]

Abū ʿĪsā ibn al-Munağğim (8) wrote a "slender volume" (*muğallad laṭīf*)[70] of chronology, fragments of which are related by the *Muntaḫab Ṣiwān al-ḥikma* (an epitome made between 1191 and 1241 from a late tenth-century history of philosophy) and the historian Abū l-Fidā' (d. 1321). One of the excerpts of this work says that the Persian emperors took books from the territories of the Greeks and Romans

65. Ibn Ḫaldūn (trans. Rosenthal) 1958: 3.113–114.

66. The basic idea of this story was apparently even taken over by an Armenian tradition. The eighth-century Armenian historian known as Moses of Chorene adapted an account from an unknown Mar Abas, stating that Alexander ordered the translation into Greek of Chaldean records stored at Nineveh. Mar Abas is supposed to have found the Greek record there, taken from ancient historical records about Armenia, and to have brought them back to Armenia in Syriac and Greek. Then the Armenian king Vałaršak received them and inscribed part of the record on a stele in his palace at Nisibis. Another, similar Armenian account, the "Primary History," based partly on Mar Abas, lacks any mention of Alexander the Great. Moses of Chorene, who claimed falsely to be writing in the fifth century, is well known as a fabricator of stories, and we may suspect that the legend—explaining how ancient Armenian history can be recovered through a Greek text thanks to the actions of Alexander the Great—is just a part of his program to give Armenia a glorious past and to promote the Bagratid family. See Robert Thomson's introduction to his annotated English translation of Moses Khorenatsʿi 1978, especially 54–55, 60–61, and 82–85. (Thomson also provides a translation of the "Primary History," just mentioned, as an appendix.) In any case, we may date his adoption of the story of Alexander's translations to the late eighth century, the period in which Ibn Nawbaḫt also wrote his Iranocentric version of the legend based on earlier sources.

67. A few examples of this story were discussed by Joseph Habby 1998, with the suggestion that the story that Alexander burned Persian books is connected with the story of the burning of the library at Alexandria.

68. The precise purpose lies with interpretation of the Middle Persian phrase *abāg Abistāg abāz handāxt*, "compared (it) with the Avesta" or "added (it) to the Avesta." For a discussion and history of the interpretation, see Shaki 1981: 121–122.

69. See note 44.

70. Abū l-Fidā' 1831, 2.11. Cf. Stern 1972: 439.

as plunder, and also that the Roman emperor Gordian III (r. 225–244) gave Šāpūr books on astrology, geometry, mathematics, music, medicine, and mechanics as gifts (*ahdāhā*).[71] Thanks to a posthumously published study of S. M. Stern, we know that Abū ʿĪsā ibn al-Munaǧǧim was very well acquainted with ancient historical sources from which he could have derived the report, including Christian chronographic traditions translated from Greek or Syriac.[72] Abū ʿĪsā's testimony would corroborate Ibn Nawbaḫt's statement about Šāpūr's interest in foreign books *if* it were independent from it. Some indications suggest that Abū ʿĪsā's source is the lost chronicle of Theophilus of Edessa (6 above), a colleague of Ibn Nawbaḫt, and who wrote in Greek that the Persians who loved wisdom translated Greek works into their own language.[73] In that case, it is easy to imagine that the two astrologers in the same court shared a common source and therefore do not constitute independent witnesses to the tradition. Nevertheless, it is surprising that Abū ʿĪsā (whatever his source) gives such specific information about the agents on both sides, not only Šāpūr but also Gordian, and also the means of exchange: sometimes books were received as gifts from Gordian, and sometimes books were taken as plunder in war. Stern explains, with some reservations, how a chronicler could have invented this: "One could well imagine that the author knew of the Zoroastrian story and found that one of the Roman emperors contemporary with Shāpūr was Gordian, and thus concluded that it must have been he who gave to Shāpūr the books which the latter 'collected,' according to the story. To be sure, if so, one would have expected rather the name of Valerian."[74] I accept Stern's conclusion that there are too many lingering questions about the source of information to accept the report about Gordian and Šāpūr as an unimpeachable witness, its specificity of detail notwithstanding.

Of all the evidence reviewed above, only Ibn Nawbaḫt states specifically that *books of Hermes* were among those translated under Šāpūr. Moreover, no other account refers to Hermes' departure from Babylon and his kingship in Egypt, unless we include the reference to the Babylonian Hermes found in the Arabic version of Dorotheus. It seems therefore that this first episode in Ibn Nawbaḫt's history of science was not originally part of the narrative connected with Alexander, but is a later addition with the purpose of proving that the astrological teachings of the Egyptian Hermes derive from Iranian Babylon. When named as an author whose works were translated into Persian under Šāpūr, Hermes is linked to the story of the destruction and restoration of the sciences in Iran. This must have allowed Iranian astrologers to think that the astrological writings of Hermes, which they knew in Middle Persian

71. Presented in Arabic and in translation by Stern 1972: 454 and 462; texts published in *Muntaḫab* ed. Dunlop 212–215, ed. Badawī 95.7–13, and abridged in Abū l-Fidā' 1831: 82.12–14.

72. Stern 1972: 447–448.

73. Stern 1972: 466n53 and Walzer 1950: 185–186 point out that Abū ʿĪsā shares certain passages with Agapius of Manbiǧ, whose main source appears to have been Theophilus (see the discussion in section 5.1 in this volume).

74. Stern 1972: 448.

and used for their work, were not foreign sciences, despite their apparently Egyptian origin, but rather they were actually from Babylon in the time of Dahāk, and connected with ancient Iranian history. The account also seems to reflect an understanding that the works of Hermes were allegedly much older than Alexander; therefore a prior episode to the story was required in which the sciences of Iran were diffused to Egypt even earlier than Alexander.

Ibn Nawbaḫt mentions Ptolemy, Dorotheus, and Hermes by name as authors whose works were translated into Persian, and it is likely, given his professional expertise as court astrologer and his work as a translator of Persian books into Arabic in the court of the first ʿAbbāsid caliphs, that he knew of works by these authors in their Middle Persian form.[75] As already discussed, studies have shown that different works of these three authors were in fact translated from Middle Persian into Arabic. One might imagine that such Middle Persian translations would have some indication of the source from which they were translated, or would declare the patron supporting the translations (such as Šāpūr), but all those Middle Persian works are lost along with any such possible evidence. What remains uncertain is not whether Greek Hermetica were translated into Middle Persian, but whether they were translated as early as the third century, because all the evidence for Šāpūr's involvement in translations from Greek into Middle Persian comes from three or more centuries later. There is no definite proof that any translations were made so early.

Because Ḫusrō I was also certainly a sponsor of translations of non-Iranian scientific works into Persian,[76] a skeptic might suggest that Ḫusrō's supporters invented this history in order to justify their adoption of foreign sciences in the sixth century, and that Šāpūr's interest in the sciences was fabricated to give Ḫusrō a venerable precedent. There are possible objections to this skeptical argument. The account of the *Dēnkard* (1b), written in the time of Ḫusrō himself, is unlikely to have been believable to its audience if there were not already by that time known Middle Persian translations of Greek works of considerable age. The sixth-century *Dēnkard* account states that copies of the works translated under Šāpūr were placed in some kind of special treasury or library, implying that they were to be found in such a place in the time of Ḫusrō.[77] Also, if one supposes that the legend is a pure fabrication of later Sasanian times, then one also has to explain why its original, pro-Iranian author was

75. *Fihrist* 300.25–27 (Flügel 284.9–10).

76. A brief survey of evidence is supplied by Dignas and Winter 2007: 263–265. See also Gutas 1983.

77. The reading of the name of this treasury is disputed: it is the *ganj ī špyk'n*, the last word subject to many proposals for emendations. Shaki (1981: 115n2) suggests emending the word to read *šāhīgān*; the phrase then meaning "royal treasury" or "treasury of the palace" (MacKenzie 1971: 79 and Durkin-Meisterernst 2004: 314b); and Shaked 1994: 100n3 suggests that we read the word as apparently written: *šabīgān*, "night(-quarters)," the idea being that these books in the treasury are kept in a place convenient to be brought forth to be read to kings at night. Durkin-Meisterernst 2004: 315b has rather *šabānīg* for "of the night" in Middle Persian. The term remains obscure for now.

willing to accept a severe rupture in the history of the ancient sciences of Iran, along with Iranian dependence on other peoples for what belonged originally to them. If any fiction was acceptable, this one is far from an ideal account. The concession, implicit in the story, that Iran needed other countries for its own sciences reduces the likelihood that the account is completely fictional simply because a better fiction could have been concocted, in which Iran had always been independent. However, all of these interpretations of the limited evidence operate from likelihoods, not sure testimonies. It remains that three centuries separate Šāpūr I and Ḫusrō I, the intervening period allowing for many events in Persian intellectual history about which we know exceedingly little. The present task is not just to test the hypothesis of a third-century reception, but also to see whether the reception of specifically Hermetic works was or could have been effected at this early time. More evidence is required.

2.4 Some Supposedly Sasanian Horoscopes

Several eminent Iranists have already accepted the claim of the reports given above, that the Iranian reception of foreign science did indeed occur in the third century.[78] Other scholars have asked for more convincing evidence for a transmission of Greek works into Middle Persian at that early time.[79] David Pingree thought he had found a testimony in favor of the earlier date in two horoscopes present in the Arabic version of Dorotheus' *Carmen Astrologicum*, a translation from Middle Persian. He dated these horoscopes to October 10, 281, and February 26, 381, and claimed that they "were added to the text as illustrations by the Sassanian scholars."[80] If this were all correct, it would clearly point to a third-century Middle Persian translation from the Greek original, after which Persian horoscopes were added to the text. James Holden has argued, however, that the earlier of the two horoscopes is better dated much earlier, to October 4, 44, "which is consistent with the dates in Dorotheus' other example charts." This also falls within the lifetime of the first-century Dorotheus. As Holden puts it, Pingree overlooked the more likely early date "since he wishes to ascribe it to the Persians."[81] In the case of the later horoscope, dated by Pingree to February 26, 381, the birth date of the native (the individual about whom the horoscope was made) was in Mihrmāh, the Iranian "month of Mihr." The Persian name might indicate an Iranian context for the horoscope and argues that Dorotheus already existed in Middle Persian by that time. Then again, the date could have been

78. For example, see Henning 1942a: 245, followed by Boyce 1968b: 37.

79. For example, see Gignoux 2001.

80. Pingree 1989a: 229b; cf. Pingree's *praefatio* to his edition of Dorotheus, XV. Pingree has argued through other, unpublished, evidence that the horoscope supposedly of 281 was cited by a Middle Persian commentary on Dorotheus used by Māšā'allāh in the Middle Persian original. The evidence is too complicated to repeat here without lengthy digression. For the full argument, see Pingree 1989a: 229b–230b.

81. Holden 1996: 34n83.

converted into Persian terms. Moreover, another problem is that it is dated in the 96th year of an era appearing in the much-distorted Arabic as that of "Darīnūs." Pingree has interpreted this as the era of Diocletian, which is doubtful.[82]

If Pingree's argument were correct, this would constitute the only known directly datable indication of a scientific translation from Greek into Middle Persian earlier than the sixth century (and the court of Ḫusrō I). One element in Ibn Nawbaḫt's history of science in the Sasanian period would thus approach confirmation, and one would be that much more willing to assign translations of works of other authors he mentioned together with Dorotheus, including Hermes, to the third century. Unfortunately, the evidence does not seem as strong as that.

So far, it has been shown that there is a very widely attested Eastern tradition dating back to Sasanian times, at least to the sixth century,[83] claiming that the Sasanids made an effort to adopt Greek science, and that this was linked to an ideology of restoration of lost Iranian science. The horoscopes adduced by Pingree seem less reliable. It is necessary to look for other signs of translation activity in the court of Šāpūr I and the existence of such an ideology in his time, and particularly for evidence pointing to translations of Hermetica. If even some of the Arabic Hermetica do date from the time of Šāpūr I—again, the third century CE—then they may shed light on the earliest communities that used Hermetica.[84] For the third century seems to have been the acme of interest in the Hermetica in the Roman Empire, when many believe that the authors of the Hermetica were flourishing, and certainly the period during which authors outside of Egypt began to notice the Hermetic texts. Moreover, if any of the extant Arabic Hermetica can be shown to have been translated from Greek into Middle Persian at such an early stage, then these Arabic Hermetica will in that case be significant not only for their intrinsic value and their place within the history of the Hermetica, but also for our understanding of the religious and intellectual movements of antiquity, of the history of Roman thought and science, and of Sasanian intellectual history. They will represent a chain of transmission from Greek into Arabic that passed outside of the transformation of the Roman Empire into a society of contending Christian orthodoxies. This last point may be the most important of all, for the manuscript traditions by which we know Greek and Roman antiquity are mostly formed by the selecting and rejecting hands of later copyists with their own preferences.[85] Do any of the Arabic Hermetica represent a literature abandoned in Greek but transmitted, in altered form, by this Iranian route?[86]

82. Pingree 1973: 35b. The horoscope roughly matches the position of the planets at this date.

83. See the notes on the *Dēnkard* (1.b given above).

84. On the milieu of the early Hermetica, see Fowden 1993a: 155–195.

85. On this point with regard to the Hermetica, see Fowden 1993a: 8–9.

86. Martin Plessner posed essentially the same question, though without reference to Iran, as one of the fundamental questions pertaining to the Hermetica waiting to be solved (1954a: 45): "To which degree does Hermetic literature in the Arabic language enlarge and complete our knowledge of the Hermetic heritage in the classical languages?"

2.5 Hermetica in the Time of Šāpūr I

There is more to be said about the era of Šāpūr suggesting that it was a likely time for the transmission of Hermetica into Middle Persian.[87] It is to this evidence that I turn now, but emphasizing from the outset that the result is a suggestion, not a sure proof. Five centuries passed between Šāpūr I and the earliest Arabic account of his alleged support for translations. The intervening period is generally known today as late antiquity, during which the chief religions of Western Eurasia were organized systematically with canons of scriptures, new religious hierarchies, and their own law codes. By the end of this period, much of the process of selection for preservation of the ancient Greek texts that would be passed down to us had been completed.

In order to test the hypothesis of a third-century transmission of Hermetic texts or of Greek literature in general into Middle Persian, one must search for evidence not in later accounts but in the third century itself. If Šāpūr or members of his court

87. Because so little Middle Persian literature survives, some will perhaps even question even the existence of written literature of this kind in the Middle Persian language as early as the third century. Doubts along these lines seem excessively skeptical to me. Four pieces of evidence point to the existence of a lively culture of scribes capable of reading, writing, and enjoying Middle Persian written works in this period (a, b, c, and d as follows):

a. several of the Middle Persian inscriptions surviving from the third century include references to scribes (including the "signatures" of the scribes of some of the inscriptions), references to individuals bearing the title *dabīrbed* (chief scribe), and mention of written records (see Ahmad Tafażżolī, "Dabīr i. In the pre-Islamic Period," *EIr*, 6.540a-541b, for references, and a full index to the occurrence of the words for scribe and writing in the inscriptions in Gignoux 1972 under the Middle Persian entries *dpyr/dpywr* 'scribe' and *npšty* 'écrit');
b. Middle Persian graffiti and a Middle Persian letter on parchment have survived at Dura-Europos (captured by the Sasanians ca 257), written by scribes who were clearly able to produce writing readily and easily (see D. N. MacKenzie, "Dura Europos. ii. The inscriptions," *EIr*, 7.593b–594a, for overview and references);
c. And not least, Mani wrote his Middle Persian *Šābuhragān* (discussed below) for Šāpūr, and there can be no doubt that he expected someone to be able to read it.
d. Experts on the Avesta generally agree that the system of letters in which the Avesta comes down to us was not created until the fourth or fifth century. However, the earlier existence of a written Avesta, in Arsacid times as claimed by Zoroastrian book tradition, has long been debated (see J. Kellens, "Avesta," *EIr*, 3.35a–44a, especially 3.35b–36b; also K. Hoffman, "Avestan Language i. The Avestan Script," *EIr*, 3.47a–51a, especially 3.50b). It seems especially difficult to me to disregard completely Mani's alleged statement, preserved by his students in the Coptic translation of the originally Aramaic *Kephalaia*, that Zarathushtra's disciples wrote his words down "in (books) they are reading today" (J. Kellens, op. cit., 36a; cf. Henning 1942b: 47). The relevant passage is:... ⲍⲁⲣⲁⲇⲏⲥ ⲥⲁϩ ϫⲱⲙⲉ ⲁⲗⲗⲁ ⲛⲉ[ϥ] / [ⲙⲁⲑⲏⲧⲏⲥ ⲉⲧⲁⲩⲉⲓ ⲙ]ⲛ̄ⲛ̄ⲥⲱϥ ⲁⲩⲣⲡⲙⲉⲩⲉ ⲁⲩⲥϩⲉⲓ ⲙⲛⲛ.../ [...] ⲉⲧⲟⲩⲱϣ ⲙ̄ⲙⲁⲩ ⲙ̄ⲡⲟⲟⲩⲉ ⲙⲡⲣ... (Polotsky and Böhlig, 1940: 7.31–33) = "but [.../...] Zarathushtra (did not) write books. Rather, hi[s / disciples who came a]fter him, they remembered and they wrote [.../...] that they read today [...]." (Gardner 1995: 13, slightly modified). Even though this probably does not point to a written rendering of the Avesta itself, but to other texts, it certainly must be regarded as evidence for books among Iranian magi in the third century (Mani lived 216–276), perhaps attributed to Zoroaster. Then again, the passage just cited from the *Kephalaia* may be a later invention attributed to Mani.

For a few more traces of literacy in the Sasanian Empire of the third century, see also Huyse 2008: 145–146.

sponsored such translations, the texts that were translated from Greek naturally were chosen from what was available at the time. Greek literature in the third century was full of new religious and cosmological ideas, "Gnostic" speculations, alchemical mysteries, and oracles, among much else. Astrology was thriving, and Ptolemy's works, which would become some of the most influential in all of history, written just a few decades before, had already begun to acquire authority. The Christian church, a collection of sometimes persecuted communities with widely varying interpretations of the message of Jesus and other ancient teachers, had no group easily capable of enforcing an orthodoxy over the others. Paganism still flourished. It was the era in which Manichaeism was born in the Persian Empire and began to spread rapidly with Šāpūr's own consent. The third century was also the time in which the movement known today as Neoplatonism began, thanks to Plotinus and his pupil, the Hellenized Aramaean Porphyry. In this time the Neoplatonic school tradition of Alexandria, familiar to historians of Arabic philosophy as the source of the Arabic medical and philosophical curricula, did not yet exist. The revival of the study of Aristotle was only just underway,[88] while other philosophical traditions such as Stoicism and Cynicism still found adherents. Hermetic literature in Greek also came into wide circulation in this century, as is known from the Greek testimonia, and it included all the types of scientific learning typical of the period, including alchemy, astrology, cosmology, philosophy of the soul, and theology.

There is good evidence that Šāpūr was open-minded and receptive to new ideas. The prophet Mani, whose new religion was intended to supersede the others, and who wrote his canonical scriptures in Aramaic, also dedicated a paracanonical work to Šāpūr that he wrote in (or had immediately translated into) the emperor's native Middle Persian. Naming it *Šābuhragān* for its recipient, Mani boldly stated in its opening, "The religion I have chosen is greater and better[89] than the other religions of the ancients in ten respects."[90] This indicates that Mani considered Šāpūr to be open to his teaching, and perhaps that the latter had asked Mani to state why his religion was worth considering. It is interesting to linger a moment on some of what follows in the preserved text. The first of the ten reasons given by Mani was that his religion was to be universal: it was to be rendered in all languages, and not restricted to one language and one people as he considered the

88. Alexander of Aphrodisias, who helped give new life to Aristotelian studies, flourished around 200.

89. Because Zoroastrianism was known to its adherents as the Good Religion, *Weh Dēn*, it seems that Mani was consciously competing by calling his religion "better," *wehdar*.

90. *Dēn īg man wizīd az abārīgān dēn ī pēšēnagān pad dah xīr frāy ud wehdar ast.* This and the following excerpts are given in the transcription of Boyce 1975: 29–30. The translations are my own. Cf. the publication of the text by Andreas and Henning 1933: 295. It is clear that this was from the opening of the book from the testimony of al-Bīrūnī, *Āṯār*, 208.9ff, which gives an Arabic paraphrase of this same passage.

other religions to be.[91] However we interpret this statement in its specifics, it shows both an ambient interest in translation and a recognition of the importance of translations. Mani's second reason was that, although the other religions had lost their original messages after the demise of their founders, his was organized so as to preserve his teaching perfectly.[92] Clearly this indicates a concern, perhaps one shared by Šāpūr, about the authenticity of the ancient Iranian traditions that had come down to his time.

One could perhaps situate the motive for the translations of Greek books into Middle Persian against the background of such questions. According to the accounts reviewed above, the Greek books translated into Persian were considered to have been the remnants of the ancient wisdom of Iran. Šāpūr's Persian translations only "restored" them to their original people, in this ideology. Is it is plausible then that the Persian emperors, and Šāpūr in particular, really believed that they were recovering their own people's heritage, the authentic truth of the ancient Iranians? There is evidence suggesting that they would have believed this, and that this was more likely to happen in the third century CE than at any other time. There were books in Greek, mostly lost today but newly in circulation in the third century, attributed to Zoroaster and other ancient Iranian sages. These are the sorts of works that could be expected to attract the attention of a Persian emperor and, perhaps, stimulate royal support for translation from Greek. Šāpūr's contemporary, the philosopher Porphyry, wrote,

> At that time[93] there were many Christians and others, including the sectarians that had been led out of the ancient philosophy, students of Adelphius and Aquilinus, who had gotten hold of most of the works of Alexander of Libya, Philocomus, Demostratus, and Lydus, and who published the revelations of Zoroaster, Zostrianus, Nicotheus, Allogenes, Messus, and others of that kind. They deceived many, being themselves deceived, claiming that Plato had not come close to the depths of intelligible being.
>
> Therefore, making many refutations at our discussion meetings, he [Plotinus] wrote a book that we have entitled *Against the Gnostics*,[94] and he

91. *Yek, ku dēn ī ahēnagān pad yek šahr ud yek izwān būd. Ēg dēn ī man ād ku pad harw šahr ud pad wisp izwān paydāg buwād ud pad šahrān dūrān kēšīhād.* "First, because the religions of those of old were in one country and one tongue. But my religion is such that it shall appear in every land and in all languages, and shall be taught in distant lands."

92. *Dudīg, ku dēn ī pēšēn, andom dāš sārārān pākān andar būd hēnd. ud ce'ōn sārārān ahrāft hēnd, ēgišān dēn wihurīd, ud pad andarz ud kirdagān sust būd hēnd, ud pad* [break in text]... [*ēg dēn ī man pad nibēgān*] *zīndagān, pad hammōzagān, ispasagān, wizīdagān ud niyōšagān, ud pad wihīh ud kirdagān, dā ō abdom pattayād.* "Second, because the religions of the ancients, as long as the leaders in it were pure,..., and as the leaders were raised up (i.e., ascended to heaven), then their religion was confused, and in commandments and deeds they became lax, and in [break in text]... [Then my religion in] living [books], in teachers, bishops, Elect, and Hearers will last until the end."

93. The reference is to the time of the Emperor Gallienus (cf. *Vita Plotini* 12), who ruled 253–268. Šāpūr ruled until 270 or 272. Plotinus died in 270.

94. The section of Plotinus' *Enneads* that received the title *Against the Gnostics* is numbered 2.9.

> left to us the task of judging the rest of it. Amelius went on as far as forty books writing against the book of Zostrianus. I, Porphyry, have made numerous refutations of the book of Zoroaster demonstrating that *it is both falsely attributed and recent, and fabricated by those who adhere to the sect to create the belief that the doctrines they had chosen to put first in rank were those of the ancient Zoroaster.*[95]

Thus we have the unimpeachable testimony of Porphyry that pseudo-Persian, indeed pseudo-Zoroastrian works were in circulation in Greek in the mid-third century, during the reign of Šāpūr, and that he, his teacher, and his colleagues were aware that they were recent compositions.[96] Porphyry grouped these pseudo-Zoroastrian works together with works generally called Gnostic today. This is the kind of material one would expect to attract the attention of a Persian emperor concerned with the authentic traditions of ancient Iran: books in Greek claiming to be derived from ancient Iran.

How close Plotinus was to the intellectual currents of the east in that period may be judged by an earlier section of his biography. In Egypt Plotinus had met his teacher, the influential Ammonius. "From that day he stayed with Ammonius continually. He acquired such a disposition in philosophy that he became eager to try to acquire even the (philosophy) practiced among the Persians and the (philosophy) prospering among the Indians. When the Emperor Gordian was about to go against the Persians, (Plotinus) joined the army and went along, now thirty-eight years old—for he had studied eleven whole years under Ammonius. When Gordian perished in Mesopotamia, Plotinus barely escaped and reached Antioch safely."[97] This passage tells us that after his education in Egypt, Plotinus was in the army of the very emperor who is reported by Abū ʿĪsā ibn al-Munaǧǧim's source to have provided Šāpūr with books of science as a gift. The phrase τῆς παρὰ τοῖς Πέρσαις (scil. φιλοσοφίας)... καὶ τῆς παρὰ τοῖς Ἰνδοῖς κατορθουμένης, "the philosophy practiced among the Persians and the philosophy prospering among the Indians," must refer to intellectual currents understood to be happening in the east at that time.[98] It is,

95. *Vita Plotini*, 16 (my translation; emphasis added).

96. Zoroaster is reported by Arnobius, *Adversus nationes* I.52, to have been the *nepos* (nephew or grandson) of Zostrianus. As far as I know, this is the only clue to Zostrianus' supposed identity and Persian background. The book of Allogenes, addressed to "Messos," survives in Coptic translation, to be mentioned below.

97. *Vita Plotini* 3 (my translation).

98. We learn how near to each other in history and in geography Plotinus and Mani were from Alexander of Lycopolis, who reported that Mani accompanied Šāpūr on one of his campaigns against the Romans. His wording makes it almost certain that it was the campaign against Valerian (256–260), and not the campaign against Gordian III (242–244), and in any case one would expect the Prophet of Light to have accompanied the emperor at the later of the two times, after his influence at court had grown. Alexander of Lycopolis, *Contra Manichaei opiniones* 4.19–22: "Manichaeus himself is said to have lived during the reign of Valerianus and to have accompanied Šāpūr the Persian king during his military campaigns; but having offended him in some way, he was put to death" (Alexander of

unfortunately, difficult to know what he was referring to by this Persian philosophy.

The king of kings Šāpūr certainly could have known that books attributed to Zoroaster, Zostrianus, Hystaspes, and other ancient Iranians (which Porphyry recognized to be contemporary forgeries) were in circulation. It seems possible, then, that Šāpūr might have been one of the many who were fooled by the teachers that Plotinus and his disciples so disdained. Or could it be that these texts were actually generated to fill a demand in the Persian court? Could the Persian emperor even have commissioned some such works? This speculation is justified because of the remarkable coincidence that Porphyry was condemning newly composed pseudo-Persian books at precisely the same time that the later Arabic and Middle Persian traditions say that Greek texts of supposedly Iranian origin were "restored" into the Persian language under Šāpūr by translation from Greek into Middle Persian.

An event mentioned in Šāpūr's trilingual inscription (known as ŠKZ, for its author and location: *Šābuhr à la Kaʿba de Zoroastre*),[99] has significance for the discussion of Greek books moving into Persia. It describes how he defeated Gordian III, claimed half a million denars as tribute from Gordian's immediate successor, Philip the Arab, and led many citizens of the Roman Empire into captivity, resettling them in the Persian Empire. The Greek part of the inscription states, "and we led in captivity people from the nation of the Romans from those of non-Iran and—in our nation of the Iranians in Persis, Parthia, Huzene (= Susiana/Khuzistan/Ahwāz), Assyria and in the other nations, according to province, where the possessions of ours and our father and grandfathers and ancestors were—there we settled them."[100] When a large number of Roman citizens, many of whom certainly spoke Greek, were integrated into the Persian Empire, they must have brought with them their ideas and books. It is hard to believe that Šāpūr, an emperor who took care that his inscription be written also in Greek, would have been totally unaware of the

---

Lycopolis, trans. van der Horst and Mansfield 1974: 52, slightly modified). αὐτὸς (scil. Mani) δὲ ἐπὶ Οὐαλεριανοῦ μὲν γεγονέναι λέγεται, [lacuna here? see edition for variant conjunctions] συστρατεῦσαι Σαπώρῳ τῷ Πέρσῃ, προσκρούσας δέ τι τούτῳ ἀπολωλέναι. Though the two did not meet, the historical fact remains that Plotinus marched with an army against Šāpūr I, Mani marched with Šāpūr I, and the arch-dualist probably missed the founding figure of Neoplatonism by just twelve years. For discussion see Honigmann and Maricq 1953: 22–23, and the commentary to the French translation of the text by A. Villey 1985: 104–107. For Alexander of Lycopolis as a witness of Manichaeism, see *DPA* 1.142–144 and G. Widengren, "Alexander of Lycopolis," *EIr*, 1.830a–b. These events also help to illustrate how permeable the borders between the empires and their cultures must have been.

99. Gignoux 1972: 8.

100. The translation is my own. Greek text from Huyse 1999: 1.43, §30; cf. Honigmann and Maricq 1953: 15.34–35 and Back 1978: 324–326). Καὶ ἀνθρώπους τοὺς ἀπὸ τοῦ ἔθνους Ῥωμαίων ἀπὸ τῶν Ἀναριανῶν ἐν ἁρπαγῇ ἠγάγαμεν καὶ ἐν τῷ ἡμετέρῳ ἔθνει τῷ Ἀριανῶν ἐν Περσίδει καὶ ἐν Παρθίᾳ καὶ ἐν Οὐζηνῇ καὶ ἐν τῇ Ἀσσυρίᾳ καὶ ἐν τοῖς ἑτέροις ἔθνεσιν κατ' ἐπάρχιον ὅπου ἡμῶν τε καὶ τοῦ πατρὸς ἡμῶν καὶ πάππων καὶ προγόνων ἡμῶν κτίσματα ἦν, ἐκεῖ αὐτοὺς ἀπεκθίσαμεν. The Greek version of the text shows an almost perfect word-for-word correspondence with the surviving Parthian version, whereas the Middle Persian text is very badly damaged.

Greek books claiming to be the authentic teachings of Zoroaster, the ancient teacher of his religion,[101] and other legendary Iranians.

The spurious Greek books of Zoroaster and the ancient Persians referred to by Porphyry were mostly lost long ago and have been known only from a number of tantalizing citations in texts by other authors. However, the discovery of the Coptic codices at Nagʿ Ḥammādī in 1945 brought to light another important body of evidence relevant here. Among these famous Coptic texts copied in the fourth century, the contents of which were entirely translations from Greek, are books attributed to some of the very same pseudo-authors named by Porphyry: a work of Zostrianus, including the phrase "Words of Zoroaster" in the colophon,[102] and a work of Allogenes, addressed to a Messos, evidently the same as Porphyry's Allogenes and Messus (Mesos). Among the texts discovered there are also Coptic translations of the Greek Hermetica (and a Coptic excerpt from Plato's *Republic*), perhaps showing that the audiences of the so-called Gnostic texts and the audience of the Hermetic texts overlapped, so to speak. Could it have been, then, that Šāpūr commissioned the translation of the supposed books of ancient Persian sages from Greek into Middle Persian and received Hermetic texts along with them?

Far from Egypt another piece of relevant evidence was uncovered at Turfan, in the deserts of Chinese Turkestan. One of the Manichaean Middle Persian fragments found there seems to name Nicotheus (*Nkty'wys*), a figure mentioned by Porphyry in the passage cited above, along with Shem, Sem, Enosh and Enoch (*Hwnwḫ*) as prophets.[103] This is not insignificant for the discussion of the transmission of Greek books of the third century into Middle Persian, because it shows that Nicotheus, one of the authors of the books associated by Porphyry with Zoroaster, was known in Sasanian Iran and was possibly known to Mani himself in the third century. It is also noteworthy in this connection that at some time between the two military campaigns mentioned above, Mani sent his apostle Patteg on a missionary journey to Egypt. While visiting Alexandria,[104] Patteg must have learned something about the intellectual climate of Egypt. Could he have returned with information about this ancient prophet Nicotheus, or with a book of Hermes?

To the several speculations introduced above, one may add that the prolific Syrian poet Ephraem (d. 373), who spent his life on the Roman side of the Persian-Roman frontier in northern Mesopotamia, stated in his book against Mani that the Manichaeans regarded Hermes as one of the earlier teachers of Mani's message, along with Plato and Jesus. Ephraem took pains to argue that Hermes taught nothing resembling the teachings of the Manichaeans and he paraphrased part of a Hermetic

101. Ἐγὼ Μασδαάσνης, "I the Mazda-worshipping," are the first words of his Greek inscription just mentioned, declaring his religion (Huyse 1999: 1, 22; Honigmann and Maricq 1953: 11.1).

102. Barry, Funk, Poirier, and Turner 2000.

103. Henning, 1934: 28 (Turfan Manichaean manuscript M299a recto, line 5). If it were spelled **Nkty'ws*, without the second *y*, the spelling would closely represent the Greek pronunciation.

104. Gardner and Lieu 2004: 111.

book in order to do so.[105] If Ephraem was correct, then by the fourth century at least some Manichaeans regarded Hermes as one of their own predecessors. Beyond being an important indication that Hermetic material of some kind was known in Aramaean Syria as early as the fourth century, this is another possible point of connection between Sasanian Iran and the Hermetica. It also shows that the syncretizing Manichaeans were the first group known to treat Hermes as one in the line of the prophets. Hermes kept this position in Muslim doctrine, as chapters 4 and 5 will show. Yet no surviving Manichaean text appears to mention Hermes by name.[106]

Although our knowledge of the so-called Gnostics described by Porphyry comes in large part through the writings of Christian heresiographers who opposed them, there is one figure whose works survive and who seems to have been one of this group, or, at least, who read their books approvingly: the alchemist Zosimus of Panopolis, who flourished around 300 CE. In his best known surviving treatise, *On the Letter Omega*, not only did Zosimus refer to the Nicotheus mentioned by Porphyry, but he also discussed, together in one passage, Zoroaster and Hermes as ancient sages of authority whose views on fate he compared. He even mentioned "the holy books of Hermes," with which he seems to have been acquainted.[107] Zosimus cannot have been the only philosopher concerned with all these trends of thought that were later abandoned by subsequent generations. The third century, around Zosimus' lifetime, is the period in which Hermes and Ostanes began to be associated in the extant Greek literature. Cyprian, bishop of Carthage (fl. ca 248–258), in his treatise *Quod idola dii non sint*,[108] cited the ancient Persian sage Ostanes and Hermes Trismegistus at the same time for their views on God: that God cannot be perceived and that God cannot be comprehended, respectively. Synesius of Cyrene (fourth cent.), who corresponded with the martyr of paganism Hypatia (whose father, it is not irrelevant to mention, is said to have taught and interpreted the works of Hermes Trismegistus),[109] considered Ammon, Zoroaster, and Hermes together as among the greatest philosophers.[110] The references to Hermes and Zoroaster in such works illustrate for us one aspect of the intellectual milieu of the third century, when Hermetic texts were said by our Arabic source to have been transmitted to the Persian Empire. There are only circumstantial indicators of a possibility, not a proof.

105. The relevant passages are presented in the English translation of F. C. Burkitt in Scott 4.161–165, with notes discussing the possible source or sources of Ephraem.

106. It is not inconceivable that Manichaeans, preceding the later Muslims, identified Hermes with Enoch, who was an important figure for Mani, and was mentioned in the list of prophets cited above from a Turfan manuscript fragment. On the identification of Hermes with Enoch, see section 5.1, this volume.

107. Zosimus I.15.146.

108. *Quod idola dii non sint* 6; Scott 4.5–6.

109. Malalas XIII.343 (Dindorf): Ἐπὶ δὲ τῆς βασιλείας αὐτοῦ (scil. Gratian) Θέων σοφώτατος φιλόσοφος ἐδίδασκε καὶ ἑρμήνευεν τὰ ἀτρονομικὰ καὶ τὰ Ἑρμοῦ τοῦ Τρισμεγίστου συγγράματα καὶ τὰ Ὀρφέως.

110. Synesius, *Dion* 10. He also included Antony as a great philosopher. Cf. Fowden 1993a: 179.

2.6 Ostanes and the Magi

A water-damaged Syriac manuscript in the library of Cambridge (Mm.6.29) contains a number of alchemical texts hinting at Middle Persian translations of Hermetica and other works from Roman Egypt.[111] Among the texts in this manuscript are books surviving in Syriac, but lost in Greek, of the same Zosimus,[112] as well as works of other early Greek alchemists including pseudo-Democritus.[113] The Syriac text has never been published, but a French translation was made of the whole manuscript by Duval in volume 2 of Berthelot's *La Chimie au moyen âge*, now more than a century old.[114] In these, Zosimus demonstrates some concrete knowledge of Hermetic texts. However, the concern here is with another, more mysterious Syriac text bound together with the works of Zosimus, purporting to be the correspondence between a learned Egyptian and a learned Persian.[115]

In this correspondence, the Egyptian philosopher Pebechius[116] informs the Persian magus Osron (*'wsrwn*)[117] that he has found in Egypt the divine books of Ostanes, which had been hidden, written in *Persian* letters. Pebechius says that he is very eager to know what the books of Ostanes that he has found contain, so he begs Osron "to send him the Persian letters," apparently meaning the equivalents of the letters in the Persian writing system. Pebechius also stresses the need for quick action so that he can understand these texts before he dies, for he is old and infirm. Osron

111. Hunter 2002 begins with a valuable description and history of the manuscript.

112. The authenticity of at least one of the Syriac translations of Zosimus is confirmed by George Syncellus, 4.4–14, who quoted a passage from one of these books in the Greek which is reproduced in the surviving Syriac. Cf. Berthelot 2.238.

113. Bolus of Mendes, an Egyptian alchemist of the Ptolemaic period, is usually considered responsible for the legends of Democritus and the alchemical writings falsely attributed to that name. This will be discussed further below.

114. Because the manuscript is badly damaged by moisture, and the ink is partly washed out on many pages, Duval's translation is defective and also incomplete. The edition of the *Mémoires Authentiques* of Zosimus by M. Mertens (1995) takes note that a study of the Syriac Zosimus "pourrait... se révéler riche d'enseignements" (p. lxxviii).

115. Berthelot 2.309–312. Ruska 1926: 42ff., noted the importance of this Syriac text but his discussion of its significance was inconclusive.

116. Pebechius is a typical Egyptian name naturalized in Greek. Zosimus of Panopolis cites an authority named Pebechius in his works, perhaps supposed to be the same figure as the one appearing in the Syriac text examined here. Sezgin (GAS 4.51–54) dates Pebechius to the fourth century, following Lippmann (1918: 1.66; also 1918: 1.94–95 for an implicit identification of the figures bearing the name Pebechius/Pibechius). However, this dating is open to question because Zosimus, who flourished at the turn of the fourth century, cites a Pebechius by name. It is also possible that there is more than one Pebechius involved in alchemical tradition. In any case, Lippmann's (and Sezgin's) dating of the Pebechius of this Syriac text to the fourth century cannot be accepted without further evidence.

117. This seems to be a rendering of the Middle Persian *āsrōn*, "priest." The rounding of the initial vowel may perhaps be explained as characteristic of the well-known Western Aramaic pronunciation: "The *ā* is retained with the East-Syrians, but has become *ō* with the West-Syrians," Th. Nöldeke 1904: 31 §44. (See, however, Nöldeke's handwritten note provided in the augmented 2001 English edition, p. 342, suggesting that this sound change may not have taken place until the seventh century, though the evidence given is not conclusive.)

responds, delighted at the discovery, and sends him the letters. This seems to be enough for Pebechius to decipher the books of Ostanes that he discovered, for the next message summarizes them for Osron. He explains the subjects of Ostanes' wisdom with great awe. It is supposed to include the whole of all the sciences. He states that the disciples of Ostanes explained the writings on the seven holy tablets of Hermes that had been hidden by a king behind seven doors. The text in the manuscript is damaged here, but something can be discerned about an image of a dragon eating its tail and other symbolic figures, and then someone ordering them not to open the doors for anyone unworthy. Thus in this account, ancient wisdom of Hermes, recovered and taught by Persian Ostanes in Egypt, is restored to the Magi. The parallels with the story related by Ibn Nawbaẖt, about the loss and recovery of ancient Persian science, discussed above, are clear.

Identifying this Ostanes may help in locating the provenance and date of these Syriac letters. Ostanes is a normal Persian name (with a Greek ending) attested commonly enough already in the Achaemenid period (559–330 BCE). It occurs in cuneiform documents from Mesopotamia and Persia and in Egyptian Aramaic papyri.[118] The early historian Xanthus of Lydia (fl. late fifth century BCE) mentioned Ostaneses (Ὀστάναι, in the plural) in a succession of Persian magi who lived after the time of Xerxes (r. 486–465 BCE).[119] Xanthus was a near contemporary of these magi. But the legendary magus named Ostanes is known primarily from later Greek and Latin histories of science and works on natural properties, including the remains of ancient Greek alchemical literature.[120] Pliny the Elder (d. 79 CE) is one of our best sources on this Ostanes because he went out of his way in his *Historia Naturalis* to refute the *magicae vanitates*, "useless magian arts" and the views of the magi who, he says, invented them.[121] Pliny reported that Ostanes accompanied Xerxes on his great campaign against the Greeks and, furthermore, that he incited

118. In Greek sources the Iranian names Ὀστάνης/Ὀσθάνης, Οὔστανος, and Αὐστάνης appear, representing Iranian **Avastāna-*, to be distinguished from Ὑστάνης, Ἱστάνης, and Βισθάνης, representing Iranian **Vištāna-* and possibly also **Huštāna-*. Both **Vištāna* (*wštn*) and **Aw(a)stāna* (*'wstn*) are attested, in 441 BCE and 408 BCE, respectively, in the Aramaic papyri of Elephantine in Upper Egypt, where there was a garrison under the Achaemenids (Cowley 1923: text 14.13, text 30.18 and 31.18). On these names and their Akkadian, Elamite, and Greek representations, see Tavernier 2007: 128, 211 and 354; Hinz 1975: 128 and 267, Kornfeld 1978. 98 and 106; Dandamaev 1992: 139–141; Eilers 1940: 34n2.

119. This information from his lost history of Lydia is cited by Diogenes Laertius 1.2. Ostanes is given along with other names, all in the plural, for the magi at this time. Was Ostanes the name of a class of magi as some have thought on the basis of Diogenes, or was it perhaps a hereditary name among magi? For a discussion of Xanthus of Lydia, see Kingsley 1995.

120. A special study of this Ostanes is found in Bidez and Cumont 1938: 1.165–212. Much of the following evidence is drawn from their useful collection of references. For a nearly exhaustive collection of testimonia, see idem, 2.267–356. Beck 1991: 553–564 presents a detailed discussion of Ostanes, concentrating on the pseudepigraphic nature of the works associated with his name. There is a chapter devoted to Ostanes in Lindsay 1970: 131–158, based largely on Bidez and Cumont's work; this is quite rich in valuable information, but some of it is irrelevant and it is not organized for easy use by a historian. A letter from Pitys to Ostanes is edited in *Papyri Graecae Magicae* 1.IV.2008–2137.

121. *Naturalis Historia* XXX.1.

many Greeks to learn his magical art, presumably through his works.[122] In this case, the magical art must have been understood more in its original meaning of "the practice of the magi," the magi being the priesthood of the Iranian religion called today Mazdaeanism, Magianism, or Zoroastrianism.[123] Indeed Zoroaster, said Pliny, was without a doubt the founder of this false art,[124] though Ostanes was guilty of introducing it to the Greeks. But he also referred to at least one more Ostanes who later worked in the same tradition.[125] Although it is certain that a quantity of pseudo-Iranian literature existed in Greek, one must be cautious about rejecting it automatically as entirely spurious, because exceedingly little is known about the philosophy of the Magians before and even during the Sasanid dynasty.[126] But we see at least that already in the first century Roman intellectuals had heard of the works of ancient Iranian sages circulating in Greek.

Among the reputed students of this Ostanes' "magic" was the influential atomist Democritus, who is said to have studied with Magi and Chaldaeans in the court of Xerxes himself, where Ostanes was.[127] Democritus is cited by a certain Synesius (perhaps Synesius of Cyrene of the fourth century CE)[128] in a letter, preserved in the oldest Greek alchemical corpora, addressed to a Dioscorus, priest of the Serapeion at Alexandria. In this letter Synesius tells also of Ostanes. He says that Democritus came to Egypt where "he was initiated into the mysteries by the great Ostanes in the temple of Memphis along with all the priests of Egypt."[129] According to the

122. Pliny the Elder, *Naturalis Historia* xxx.8: *Primus, quod exstet, ut equidem invenio, commentatus est de ea* [scil. *arte magica*] *Osthanes Xerxen regem Persarum bello quod is Graecia intulit comitatus, ac velut semina artis portentosae sparsit obiter infecto quacumque ommeaverant mundo. Diligentiores paulo ante hunc ponunt Zoroastrem Proconnensium. Quod certum est, hic maxime Osthanes ad rabiem, non aviditatem modo scientiae eius Graecorum populos egit.*

123. One should not assume that this is the same "Zoroastrianism" as later promoted by the Sasanians. It was not until Kirdēr's inscriptions under Bahram II (r. 276–293) that one finds the earliest evidence for "an attempt to create a mass constituency for Magianism and an affirmation of the Mazdaean identity by opposition to other religions in the late 3rd century," M. Morony, "Sasanids," *EI*[2], 9.70b–83b.

124. Pliny the Elder, *Naturalis Historia*, xxx.3: *Sine dubia illic* [scil. *magica*] *orta in Perside a Zoroastre, ut inter auctores convenit.* "Without doubt magic [or, the magian art] arose in Persia with Zoroaster, as it is agreed among the authorities" (trans. adapted from Pliny the Elder 1956: 279, trans. by W. H. S. Jones, Loeb Classical Library 393).

125. *Nat. Hist.*, xxx.11: *Non levem et Alexandri Magni temporibus auctoritatem addidit professioni secundus Osthanes comitatu eius exornatus, planeque, quod nemo dubitet, orbem terrarium peragravit.* "In the time too of Alexander the Great, no slight addition was made to the influence of the profession by a second Osthanes, who, honoured by his attendance on Alexander, travelled certainly without the slightest doubt all over the world" (translation by W. H. S. Jones, Loeb Classical Library, 1963: 285).

126. Beck 1991 rejects all of the allegedly Iranian writings found in ancient Greek as spurious, but given our lack of knowledge about Iranian philosophy in this period, we have no way of judging their authenticity one way or the other.

127. The distinction between Magi, Persian priests, and Chaldaeans, Babylonian priests and astrologers, is not always clear in Greek and Latin literature. Cf. Kingsley 1995: 199–201.

128. PW 4A, 1366, argues that it is not the same Synesius.

129. ὃς ἐλθὼν ἐν *Αἰγύπτῳ* ἐμυσταγωγήθη παρὰ τοῦ μεγάλου Ὀστάνου ἐν τῷ ἱερῷ τῆς Μέμφεως σὺν καὶ πᾶσι τοῖς ἱερεῦσιν *Αἰγύπτου*, Berthelot and Ruelle 1888: 2.57 (reprinted in Bidez and Cumont 1938: 2.313.7–9).

abbreviated version of the same story related by George Syncellus, a Greek chronicler of the early ninth century CE, Ostanes was sent by the Persian kings to preside over (ἄρχειν) the priests of Egypt.[130] In their study of Ostanes, Bidez and Cumont regarded this as a fiction, a "fabuleux récit," because of the apparent discrepancy between the story of Ostanes accompanying Xerxes and the story of his Egyptian mystagoguery.[131]

A further complicating problem about Ostanes is the question of the authenticity of the works of his alleged disciple, Democritus. Some scholars, both ancient and modern, have argued that the alchemical works attributed to Democritus were not really written by him, but rather by an Egyptian, Bolus of Mendes.[132] The most important testimony on this comes from the *Res Rustica* of Columella (first cent. CE), referring in passing to the opinion of "the celebrated writer of Egyptian race, Bolus of Mendes, whose contrivances, which in Greek are called Hand-wrought Products and are published under the pseudonym of Democritus."[133] But it is also important to bear in mind that Pliny, who knew Columella's work, considered the magical writings of Democritus to be true works of the famous philosopher and not pseudepigraphs: "That these [works of Democritus] were accepted by any human beings and transmitted by memory is the most extraordinary phenomenon in history; so utterly are they lacking in credibility and decency that those who like the other works of Democritus deny that the magical books are his. But it is all to no purpose, for it is certain that Democritus especially instilled into men's minds the sweets of magic."[134] Yet among modern scholars it is Columella's report, helped by other important testimonies, that has been judged superior. In Greek tradition, the Byzantine encyclopedia *Suda* refers to "Bolus, Democritus" and other fragments mention "Bolus the Democritan." To judge by citations, this Bolus lived around 200 BCE, and he apparently used works attributed to Zoroaster and Ostanes.[135] This last fact, combined with the reports that say he took on the name Democritus, makes it very likely that it is the Egyptian Bolus who is responsible for transmitting or possibly inventing the legend of Ostanes, teacher of Democritus. In the Democritan

130. Syncellus 297.23–25 (§471).

131. Bidez and Cumont 1938: 1.169.

132. On Bolus and his links with alchemical tradition and Democritus, see Kingsley 1994: 5–9, with many further references. Kingsley's main concern there is to be to prove, against other scholars, that Bolus, an Egyptian, was connected with early alchemy.

133. Columella, *Res Rustica*, VII.v.17. English translation by E. S. Forster and E. H. Heffner, Loeb Classical Library 273, modified following Kingsley 1994: 8: *Sed Aegyptiae gentis auctor memorabilis Bolus Mendesius, cuius commenta, quae appellantur Graece* χειρόκμητα, *sub nomine Domocriti falso produntur*... "Hand-wrought products," χειρόκμητα, is an alchemical term.

134. Pliny, *Nat. Hist.* 30.9–10. English translation by W. H. S. Jones, Loeb Classical Library, 1963, 285.... *quae recepta ab ullis hominum atque transisse per memoriam aeque ac nihil in vita mirandum est. In tantum fides istis fasque omne deest, adeo ut qui in cetera in viro probant haec opera eius esse infitientur. Sed frustra, hunc enim maxime adfixisse animis eam dulcedinem constat.*

135. The testimonies on Bolus and arguments for dating are very neatly summarized by Robert Halleux, 1981: 62–67; also Letrouit 1995: 16.

Physika kai Mystika, usually attributed today to Bolus,[136] Democritus relates in the first person that Ostanes died "before our teaching was completed,"[137] and that they therefore summoned Ostanes from Hades in order to obtain the rest of the master's teachings.[138] Subsequently Ostanes' hidden books, the books of his ancestors, were revealed inside a pillar (στήλη) in the temple.[139] Ostanes brought the books out of the pillar himself for the group to examine. Again, Bolus is apparently either the author of this elaborate legend or the one who transmitted it. Assuming that the date of Bolus ca 200 BCE is correct, the idea of a Persian magus recovering his ancestral science from Egypt becomes quite old.

Further testimony for the argument that Bolus transmitted or created the legend of Ostanes comes from the *Oration against the Hellenes* by Tatian, a Mesopotamian of the second century (and therefore born under Parthian rule), who received a good Hellenic education and lived in the Roman Empire, where he converted to Christianity, eventually returning to his Assyrian homeland, where he died. In his tirade against the pagan Hellenes he says "Concerning the *Sympathies and Antipathies according to Democritus*," that "the one vaunting the Mage Ostanes will be given over as food for the eternal fire on the day of consummation."[140] An *Alphabetical Treatise on Sympathies and Antipathies of Stones* is attributed to Bolus of Mendes by the Byzantine encyclopedia *Suda*.[141] Probably this is the work referred to by Tatian.

It is not impossible that a figure like Ostanes could have been sent to Egypt and have participated in the Egyptian priesthood under the Achaemenids.[142] It should

136. On this attribution see Bidez and Cumont 1938: 1.198–207; Festugière 1.224–238; Beck 1991: 562–563.

137. μηδέπω ἡμῶν τελειωθέντων, Berthelot and Ruelle 1888: 2.42; also Bidez and Cumont 1938: 2.317.

138. The whole account is given with commentary and references by Bidez and Cumont 1938: 2.317–320. An English translation with some comments appears in Lindsay 1970: 102–103.

139. Ὁ δὲ Ὀστάνης ἔφασκεν ἐν αὐτῇ τὰς πατρῴας τεθησαυρίσθαι βίβλους, Bidez and Cumont 1938: 2.318.6–7. The translation of Lindsay 1970: 103 is mistaken: "However [the son] Ostanes told us that it was in this column his father's books had been placed" (bracketed expression added by Lindsay). The interpretation involving a second Ostanes (Ostanes the Younger, it would seem), who had put his books in the pillar for safe keeping, is not necessary. Rather it appears that the same (unique) Ostanes, already summoned from the dead, said that the books of his ancestors (not of his father) were stored in that pillar.

140. Tatian, *Oratio contra Graecos* 17. Περὶ γὰρ τῶν κατὰ τὸν Δημόκριτον ξυμπαθειῶν τε καὶ ἀντιπαθειῶν . . . ὁ τὸν μάγον Ὀστάνην καυχώμενος ἐν ἡμέρᾳ συντελείας πυρὸς αἰωνίου βορᾷ παραδοθήσεται. The whole passage is given with notes by Bidez and Cumont 1938: 2.293–296.

141. The *Suda* lists two different figures named Bolus, one "Bolus Democritus" and the other "Bolus of Mendes, Pythagorean," but these have been regarded as the same figure since the argument of M. Wellmann, 1921: 16.

142. Moreover, the hieroglyphic inscriptions of the Achaemenids and their officials (Posener 1936) demonstrate that Persian officials resided in Egypt and took Egyptian names and titles, and that the Achaemenid kings supported the building of temples to Egyptian gods and made offerings to them. The Ḫwārizmian (!) whose name appears in the Aramaic papyri from Elephantine in Egypt of this period (Cowley 1923: 15) was interpreted by A. D. H. Bivar as possibly "a pensioned survivor from Xerxes' Grand Army" (*CHI* 3.1.181).

be remembered that Xerxes, like his father Darius and his grandfather Cambyses, ruled Egypt, where the Persian emperors styled themselves as pharaohs.[143] Xerxes, who burned Greek temples (at the urging of the magi, it was said),[144] also built temples in Egypt. That learned men crossed the cultural borders between Persia and Egypt in this period is proven by the example of Udjahor-resnet, an Egyptian priest and physician who served under Cambyses and Darius (Xerxes' grandfather and father), and whose autobiographical inscription in hieroglyphic Egyptian outlines his career of service, which included his accompanying the Persian armies to the emperor's palace at Susa, in distant Elam, later Ḫūzistān, at the borders of Persia proper.[145] Although there is no reason to doubt the historical existence of such an Ostanes just because he was Persian and not Greek, it is a different question whether the books attributed to him are really his. Fortunately, the unknowable answer is practically irrelevant for the investigation here. The consensus of scholars is that the work describing Ostanes' alchemical activities in Egypt was at least transmitted by Bolus of Mendes. It is possible that Bolus was citing earlier material attributed to Ostanes, just as it is also possible that he invented the story entirely.[146]

In brief, the story of the Persian Ostanes in Egypt, and therefore an account about Persian science in Egypt, was in circulation, probably mediated by Bolus, long before the third century CE. By the time of Šāpūr I, this legend was well known to physical theorists, alchemists, and dabblers in the properties of natural substances. The correspondence between Pebechius the Egyptian and Osron the Magus surviving in Syriac, summarized above, surely must be related to this story, for it describes the discovery of the Persian Ostanes' revelation in Egypt of the wisdom of Hermes. Thus the legend of Ostanes, at least in one of its forms, which survives in Syriac (and in Arabic, as will be shown), described him as an interpreter of the inscriptions of the Egyptian Hermes.

That the Syriac text under examination is bound together with books of Zosimus and other similar works, all dealing with the same topic, alchemy, suggests perhaps that they may have been transmitted together from the time of Zosimus. The name Pebechius is found elsewhere in the same manuscript: in his *On the Letter Theta* (surviving only in this Syriac manuscript), Zosimus cites Pebechius as an authority.[147] It is therefore possible that the correspondence between Pebechius and Osron originated either in the same period, ca 300, or earlier. These texts may be documents illustrating the historical context of Šāpūr's alleged concern for translating Greek books. The Syriac letters clearly connect Hermes to Persian science by making Ostanes, a legendary

143. See the collection of hieroglyphic Egyptian inscriptions (with translations and commentary) of the Achaemenid period in G. Posener 1936. For an overview of early Achaemenid rule in Egypt, and the status of language in the Persian Empire, see Briant 2002: 172–184 and 507–510.

144. Cicero, *De legibus* II.26.

145. Anthony Spalinger, "Udjahorresnet," in *Lexicon der Ägyptologie*, 6.822–824.

146. For discussion of this question and scholars' views on it, see Beck 1991: 560–561.

147. Berthelot 2.245.

ancient Persian sage who lived before Alexander, the interpreter of the inscriptions on the pillars of Egypt and of Hermes' science. Notwithstanding the uncertainties, the fictional connection between a Persian magus and an Egyptian scholar in their discussions of Hermes and Ostanes, ancient sages of their respective countries, is striking. Moreover, the survival of these texts in Syriac brings them into proximity with the Sasanian Empire and indicates that they were certainly transmitted in the domain of the subsequent Arab Empire. There is also still more evidence to be considered: texts in Arabic associated with the name Ostanes and also with Hermes.

2.7 Ostanes and Hermes in Arabic

Two sections of an Arabic manuscript containing works ascribed to Ostanes were published by Berthelot in *La Chimie au moyen âge*, vol. 3.[148] Both are written in a very awkward style of Arabic that shows almost without doubt that they were translations. At least, one may judge that they could not have been originally composed by native speakers of Arabic. In the first of the two books, Ostanes has a conversation with Aristotle about the properties of special kinds of matter; clearly we are dealing with a pseudepigraphon. In the second work, part of which was published by Berthelot and which is more important for the present discussion, Ostanes describes a bizarre dream that he had after being wearied with concern for "this work" (i.e., alchemy). In the dream, evidently of symbolic character, a visitor comes, tells him to get up from his bed, and shows him seven doors. Behind these doors, the visitor explains, are the treasure-houses of science (*ḫazā'in al-ʿilm*), to be understood as libraries. But Ostanes needs the key, and these must be obtained from a ferocious self-devouring beast with the wings of a vulture, the head of an elephant, and the tail of a dragon who has just appeared. Encouraged by the visitor, he tells the beast, in the name of God, to hand over the key. Then he takes the key and finds behind the doors an inscription written in seven languages; the first of these languages is Egyptian, the second Persian, the third Indian (meaning Sanskrit), and the rest too damaged to read. He relates all of it that was legible. The Egyptian text gives a metaphor for the body, spirit, and soul, which are likened to the lamp, the oil, and the wick respectively. It states that the first elements (*ṭabā'iʿ*) are fire and water, whose mixing produced bodies, plants, and stones.[149] Then comes the Persian text of the dream-inscription, which is of importance for this investigation:

> Then there was after it Persian writing in which was much learning and knowledge, and this is what I say to you now; it is the writing that was clear to me and the knowledge that I acquired. It said:

148. Berthelot 3.79–88 (Arabic text).

149. On the uses of the term *ṭabīʿa* in alchemy, see S. Nomanul Haq and D. Pingree, "Ṭabīʿa 3. Ṭabāʾiʿ in the alchemical tradition," *EI*², 10.27a–28a.

"As for Egypt, it has excellence over the towns and the countries. That is because God the Exalted has given its people wisdom and knowledge in all things. And as for Persia, the people of Egypt and the people of all the horizons need it [i.e., Persia]. Nothing of their works is sound except by that which has come from Persia. Do you not see that each of the philosophers whose concern was in this knowledge had sent to the people of Persia and took them as brothers, and asked them to send to him that which comes from their land and is not found in other lands? Have you not heard a certain philosopher having written to the magi, people of Persia, that 'I have come upon one of the books of the first sages in the pen of Persia that I cannot read. Send to me one of their sages to read to me my book which I have found. For you are doing me a favor, if you do that, and I will thank you forever, as long as I remain. So send quickly to me the one whom I have asked from you before my life ceases and I become a corpse needing no knowledge.' And the sages of the people of Persia wrote to him that, 'When your letter reached us, we were very happy with what you wrote to us. We have sent quickly to you that sage that you asked for to read to you your book and explain to you what is hidden from you of it, for we considered that it is a necessary duty for us to you. Consider, when you have finished your book, as it should be (*ka-llaḏī yanbaġī*), and *send quickly to us a copy of that book, because our ancient fathers were the ones who established that book.* Therefore let us have a share in it with you, for it also thus should be. Peace.'"

This is everything that I read of the Persian writing that was on the tablet.[150]

In the Indian portion of the inscription, the Indians claim that their country is "the most potent of all countries" (*wa-arḍunā ašaddu l-araḍīna kullihā quwwatan*) because of its proximity to the sun. However, "if it were not for what we need from Persia, we would complete all our work with that which comes out of our land and our sea."[151] Then the Indians relate how they sent the urine of a male, white elephant from their land to a man who requested it, and how the man was very impressed with its healing properties. Thus ends the legible portion of the tablet, according to the narrative. Ostanes barely escapes the treasure-houses of science before the time at which the doors are locked, and meets an old man, who exhorts him to find knowledge in the strange and terrifying beast, who now also appears again. The text ends with the old man giving Ostanes mysterious instructions. Ostanes reports that

150. Berthelot 3.85.10–86.8 (Arabic text). I have tried to retain the awkward style of the Arabic in this English rendering. The dream is described in French by Blochet 1911–1912: 272–275 (repeated in Bidet and Cumont 1938: 2.347–352), and briefly in German by Ullmann 1972: 185.

151. Berthelot 3.84.13–15. *law-lā mā naḥtāǧu ilayhi min arḍi fārisa la-akmalnā l-ʿamala kullahū bi-llaḏī yaḫruǧu min arḍinā wa-baḥrinā.*

he followed them to the completion of "the entire work" "as Hermes had described"![152] Thus it seems that Ostanes' dream-guide was none other than Hermes.

This dream of Ostanes is important here because of the close resemblance of the account on the Persian inscription within the dream to the Syriac correspondence between Pebechius and Osron, as well as to the later Arabic accounts of Middle Persian translations of Greek and Egyptian books reviewed above. The similarity with the Syriac text, including the specific details of seven doors, seven tablets of writing, and a self-devouring monster, is too great to be accidental. The dream related by Ostanes in the Arabic text, describing the exchange of ancient science between Egypt and Persia, and mentioning the name Hermes, also represents the same ideology described by Ibn Nawbaḫt and the *Dēnkard*, to the effect that ancient Egyptian science was originally Persian. These points, combined with the absence of any reference in the dream-inscription to the Arabic language, as well as the comments indicating the dependence of Egypt and India on Persia, lead me to surmise that they originated in pre-Islamic times and in the Sasanian Empire. Moreover, listing the views of representatives of different nations is a common and well-attested device in Middle Persian literature that went on to flourish in Arabic translation and adaptation.[153] The idea of the inscriptions representing different nations also points to a Sasanian background of the story found in the correspondence between Pebechius and Osron surviving in Syriac. Was this Arabic book of Ostanes translated from Middle Persian? Even if so, the text or the idea behind it must still also somehow have its ultimate roots in Greek, for the very form of the name Ostanes in Arabic reveals by its characteristic Greek ending (*-s*) the Greek derivation of his legend. Given the legend of Ostanes that circulated in Greek, through the reports on Democritus outlined just above, one could expect no other ultimate source.

It still remains to be determined where the Arabic legend of Ostanes and his teacher Hermes originated. The same manuscript contains another work of Ostanes called "The Twelve Chapters of Ostanes the Wise on Knowing the Venerable Stone [i.e., the Philosopher's Stone]." The text begins with an explicit description of its alleged transmission into Arabic, but unfortunately it is very difficult to believe. It claims that there was a great scholar bearing the Muslim Arabic name Abū Ḫālid Šaddād ibn al-Yazīd, who lived in Ujjain (in western India) and had mastered alchemy and other occult arts, studying them in more than seventy languages; he had translated the art of Ostanes into Greek from "his [Ostanes'?] language" and taught it to someone named ʿUbaydallāh ibn Aḥmad the Indian (al-Hindī); the latter allegedly translated the work from Greek into Persian; he then taught it to Ǧaʿfar

152. Berthelot 3.88.3–4. *fa-faʿaltu llaḏī amaranī bihī fa-adraktu l-ʿamala kullahū tāmman ka-llaḏī waṣafahū Hirmis.*

153. Shaked 1984: 41–49, van Bladel 2004: 159–160.

ibn ʿUmar the Persian (al-Fārisī), who translated it from Persian into "the idiom of most remote of Ḫurāsān (Eastern Iran)"; then this Ǧaʿfar taught it to Abū Bakr Yaḥyā ibn Ḫālid of Ḫurāsān, who translated it into Arabic.[154]

The city Ujjain, where the first teacher in the account worked, was the location of the meridian at 0° longitude in Indian science. In Arabic geography, a recipient of this tradition, Ujjain, under the name Uzzayn (usually misread as Arīn), became known consequently as the center of the world. In other words, the book is supposed to have been transmitted by a brilliant scholar who lived in the center of the world. Manfred Ullmann rightly considers the tale "phantastisch und fiktiv."[155] But even if the ultimate origin of this work of Ostanes, in India, is fictional, the text may be evidence for a Middle Persian Ostanes. It is interesting that the text of Ostanes' dream inscription, reviewed above, with its claim of the superiority and originality of Iranian science, was bound together with another text allegedly translated through two Iranian languages.

To sum up, the legend of the Persian Ostanes as revealer of Hermetic wisdom in Egypt circulated together with the works of a third-century alchemist, Zosimus, who cited both Hermes and Ostanes in his own works. There is direct evidence for the existence of the legend in Syriac, and a strong suggestion for its circulation in Middle Persian, transmitted into Arabic.[156] Again the problem is to provide a date for the texts, but no solution is evident. What is clear is that Ostanes, the Persian sage of Egypt, and his association with Hermes were known by the third century, and that the teaching of Ostanes seems to have been transmitted into the Sasanian Empire, but at an unknown time. It is possible, though not provable, that the reputation of Hermes as an ancient sage, and even works of Hermes, had reached the Sasanian court because of his connection with "their own" Persian Ostanes, who was reputed to have transmitted the teachings of Hermes.

154. The introduction is cited by Berthelot 3.13–14 (in a totally corrupt form) and Blochet 1911–1912: 271n1, on the basis of Paris MS 2605. The passage is also summarized by Sezgin, GAS 4.52–53. I have examined the work under the name of Ostanes' *Kitāb al-Ǧāmiʿ* in the UCLA Biomedical Library, MS Ar. 29, 79b and British Library Add. 22,756, 84b–85a: *qāla Abū Bakr Yaḥyā Ibn Ḫālid al-Ġassānī ṯumma l-Ḫurāsānī aqraʾanī Ǧaʿfar ibn ʿUmar al-Fārisī qāla aqraʾanī ʿUbaydullāh ibn Aḥmad al-Hindī qūlu šāhadtu maǧlisa Abī Ḫālid Šaddād Ibn al-Yazīd bi-madīnat Arīn*... (The high qualifications of Abū Ḫālid in all the sciences are described.)... *wa-kāna qad naqalahū* (scil. the book of Ostanes) *min lisānihī ilā l-lisān al-yūnānī fa-naqaltuhū anā min al-Yūnānī ilā l-Fārisī ṯumma naqalahū Ibn ʿUmar min al-lisān al-Fārisī ilā l-lisān al-ʿAǧamī ʿalā lafẓ aqṣā Ḫurāsān qāla Abū Bakr fa-naqaltuhū anā min lafẓ Ḫ urāsān ilā lafẓ al-ʿarabī ʿalā luġati l-ʿIrāq.*

155. Ullmann 1972, 184. One still wonders, though, why an author wanting to give his book a fictive ancient authority did not make the more likely claim that the text was translated directly from Greek. Moreover, we apparently have a tantalizing reference to an Eastern Iranian language in the phrase "the pronunciation of furthest Ḫurāsān." Bactrian and Sogdian, languages that had literatures of their own even under early Arab rule, are likely candidates.

156. Ostanes was also known to early alchemists writing original works in Arabic. For example, he is mentioned by Ǧābir ibn Ḥayyān in his *Kitāb Sirr al-asrār* (Kraus 1943: 1.136, no. 1072).

2.8 The Hymn of the Pearl and the Iranian History of Science

The Hymn of the Pearl, an important piece of so-called Gnostic literature, survives in Greek and Syriac as a part of the apocryphal *Acts of Thomas*. Related in the first person, the hymn tells the tale of the journey of a young prince, son of the King of Kings from the East, into Egypt, sent on a mission by the great king to recover a pearl. Once there, the prince is seduced by the pleasures of Egypt and forgets his mission. Only when revived by a messenger from his father does he awaken and return with the pearl (won from a dragon) to the splendor of his home in the East.

The tale served as a metaphor for the descent of the spirit into the material world, its torpor of ignorance there, its awakening by the gospel, and its ascent to reunion with God. After the foregoing discussion, however, one may suspect that the hymn also alludes to the idea of the recovery of lost Persian knowledge from Egypt. For if the hymn was composed in a culture aware of the "Iranocentric" history of science, its metaphorical significance would be placed in the more mundane context of this history. The pearl stands for knowledge recovered in Egypt and brought back to the east. The question of dating the hymn therefore becomes important, but again there is no precise answer.[157] It has been suggested that it was composed during the last decades of the Parthian kingdom (ca 200–224 CE), because of a reference to kings of Parthia and Parthian words in the hymn. One scholar considers that the Parthian background may be part of the "mythological setting" of the tale.[158] Because the Parthian aristocracy did not disappear during Sasanian times, the work could also have been composed later.[159] In any case, the book of the *Acts of Thomas*, in which the hymn occurs, is generally thought by modern scholars to have been composed—either in Greek or in Syriac, the two languages in which it survives—in the vicinity of Edessa in Syria in the first half of the third century. This would place the hymn in the very beginning of the reign of Ardašīr, Šāpūr's father. Although the precise date is unknown, this hymn can be interpreted to suggest that a legend of Persian knowledge lost in Egypt was known popularly in the early third century.

2.9 The Role of Alexander in the Iranian History of Science

In the accounts dating from the sixth century onward, reviewed above, Alexander is supposed to have plundered the ancient wisdom of Persia and to have sent it back to his home, the Roman Empire. Why is Alexander, who, in reality, long antedated

157. Klijn 2003: 182–195 provides a recent bibliography on this hymn, on which many scholars have commented.

158. Klijn 2003: 193.

159. Pourshariati 2008 discusses at length the Parthians in Sasanian times.

the Roman Empire, continually associated with the Rome in the legends examined here? There is, of course, the simple fact that the Roman Empire was heir of Greek and Macedonian rule in the Eastern Mediterranean, and that for those east of the Roman Empire, the "Roman" language was Greek. Similarly, as is well known, later Byzantine Greeks are always known as *Rūm* in Arabic sources. However, the association between Alexander and the Roman emperors was part of Roman imperial propaganda precisely at the beginning of the third century, the very period during which the Sasanids came to power and are supposed to have initiated a policy of restoring science from Greek books stolen by Alexander.

The Roman emperor Antoninus Caracalla (r. 211–217) is particularly famous for having been extremely fond of Alexander the Great (φιλαλεξανδρότατος)[160] and for having self-consciously imitated the legend of Alexander. His *vita* in the *Historia Augusta* tells us that he thought himself worthy to be compared with Alexander the Great.[161] Dio's history describes Caracalla as obsessed with Alexander, setting up numerous images of him, organizing a phalanx composed only of Macedonians and calling it "Alexander's Phalanx," dubbing Alexander "the Eastern Augustus," and even claiming himself to be Alexander returned to life.[162] Herodian relates the story of Caracalla's special visit to Alexander's tomb in Egyptian Alexandria and his veneration of the site there, which was apparently deemed ridiculous by the Alexandrians. Herodian identifies the mockery of Caracalla as one of the main reasons for his rage and slaughter of numerous inhabitants of that city during that visit.[163] Alexander's imitator must have made himself infamous in Iran by spending part of his reign in Mesopotamia, apparently making forays into the Parthian Empire and, no doubt, trying to follow the eastern footsteps of the Macedonian conqueror. He was finally assassinated.[164]

After the brief reigns of Macrinus (r. 217–218) and Elagabalus (r. 218–222), Alexander Severus reigned (r. 222–235), having adopted the names both of Alexander the Macedonian and of Severus, Caracalla's father. A legend developed that he was so named because he was born in a temple dedicated to Alexander the Great.[165] It is said that the Senate even offered him the title "the Great," but he refused it.[166] When Ardašīr overthrew Artabanus the Parthian and established Sasanian rule, it was this Alexander against whom he made war. The first Roman enemy of the Sasanid emperors was thus an Alexander seen as prefigured by Alexander the Macedonian. This would have been a convenient coincidence for Ardašīr, if he had wanted to

160. Cassius Dio LXXXVIII.9.1.

161. *Historia Augusta, Vita Antonini Caracalli*, II:...*se Alexandro Magno Macedoni aequandum putabat*....*Alexandrum Magnum eiusque gesta in ore semper habuit.*

162. Dio, LXXVIII.7.

163. Herodian IV.9.

164. Idem, IV.10–13.

165. *Historia Augusta, Vita Severi Alexandri* V.1.

166. Idem, V.6–12.

present himself as the future restorer of the ancient Persian Empire overthrown by Alexander. In effect, he could have claimed to be taking revenge against Alexander the Macedonian by attacking the Roman Alexander. The two main Roman historians for these events, Herodian and Cassius Dio, writing in Greek, portray Ardašīr as threatening just this.[167] Dio says that Ardašīr threatened to take Syria and Mesopotamia on the claim that "they belonged to him from his forefathers, since as the Persians of old held (the lands) up to the Hellenic Sea."[168] Herodian goes further in saying that Ardašīr intended to restore the empire of Cyrus and Darius.[169]

There has been an ongoing debate in modern scholarship over how much Ardašīr really knew about the ancient Achaemenid dynasty of Persia: whether to trust Dio and Herodian at all on this matter, and how much.[170] What is at stake in the question is a narrative about cultural continuity of the history of ancient Iran. Did the Sasanids really see themselves as restorers of the Achaemenian Empire? The answers to the problem have mostly been based on the interpretation of these two Greek sources. There is, in fact, no evidence to suggest that the Sasanids had much accurate information about the Achaemenian Empire. David Frendo argues forcefully that Herodian and Dio had "recourse to a highly literary and allusive form of historical geography" in describing Ardašīr's claims. Specifically, he argues that they were using Herodotus' descriptions of the extent of the Achaemenid Persian Empire to inform their understanding of contemporary events, but that this did not accurately represent Ardašīr's policy.[171] Huyse has made a similar case.[172] An awareness of ancestral Persian land possessions of some kind is evident in the trilingual royal inscription (ŠKZ) of Ardašīr's son Šāpūr,[173] mentioned above, but were the lands mentioned there merely ancestral family estates, or were they actually supposed to

167. Herodian VI.2.2, VI.4.5; Dio LXXX.3–4 (Dio was a military commander under Alexander Severus and thus a knowledgeable contemporary of these events).

168. Dio LXXX.4 ὡς καὶ προσήκοντά οἱ ἐκ προγόνων ὅσα ποτὲ οἱ πάλαι Πέρσαι μέχρι τῆς Ἑλληνικῆς θαλλάσης ἔσχον; he says nothing about the Achaemenids as such.

169. Herodian VI.2.2.

170. Notable and more recent publications treating this problem include Yar-Shater 1971, Wiesehöfer 1986, Shayegan 2000, Rubin 2002, Shahid 2004, Frendo 2006, Dignas and Winter 2007: 53–62. One may pursue these publications (particularly Shayegan 2000: 101–102n40) for more bibliography.

171. Frendo 2006 doubts even Dio's account of Ardašīr's alleged irredentism. Like Yar-Shater 1971 (who does not appear to doubt Dio's account), Frendo sees Ardašīr's policy as a continuation of Arsacid policy. Shayegan 2000: 101 sees Dio's interpretation as a Roman literary *topos*.

172. Huyse 2002, which also brings up several of the following points about a vague reminiscence of ancient Persia.

173. In Dio's representation (80.4), Ardašīr claims the Roman territories "on the grounds that they belong to him from his forefathers" (ὡς καὶ προσήκοντά οἱ ἐκ προγόνων). This superficially resembles that of the inscription of Ardašīr's son, Šāpūr, referring to the territories that were possessions of his forefathers (see note 100: ὅπου ἡμῶν τε καὶ τοῦ πατρὸς ἡμῶν καὶ πάππων καὶ προγόνων ἡμῶν κτίσματα ἦν).

be the Achaemenid provinces?[174] One cannot assume that the alleged irredentism of Ardašīr referred to a specifically *Achaemenian* past, reflecting detailed historical knowledge of the Achaemenian Empire. It is likely, however, that Ardašīr knew of a vaguer, but nevertheless real, traditional lore about a former Persian glory which he imagined that he could restore. Ardašīr need not have had any of the information supplied by Herodotus to know that a Persian dynasty, one whose very name was perhaps forgotten in Iran, had controlled a great empire long ago.[175] Similarly, the Sasanids are not likely to have had specifically Achaemenid territories in mind. Yet the carved reliefs depicting the investiture of Ardašīr are at Naqš-i Rağab, just a few kilometers from the ruined Achaemenid palace at Persepolis, and at Naqš-i Rustam, on the very cliff face into which the tombs of the Achaemenids were carved, sites at which the subsequent Sasanian emperors also added their own reliefs and inscriptions. This indicates that Achaemenian monuments were regarded as a physical memorial of past Persian glories, even if an accurate historical record about them had been utterly lost. Moreover, it is difficult to assume that Ardašīr and his contemporaries were completely ignorant of the meaning behind the vaunted claims of and about Caracalla and Alexander Severus, the new Alexanders. Surely, if the *imitatio Alexandri* was to have a role in Roman relations with their eastern neighbors, someone would have explained to the Parthian and Persian courts just what a new Alexander meant. This would not require the listener to have any accurate conception of the Achaemenian history beyond a remembered Persian glory of a bygone age ended by Alexander. Just such a glorious age, before Alexander, is described in the Iranian traditional stories of the Kayānian kings, most famously recorded in Firdawsī's *Šāhnāma*. Numerous Arabic works (many of which are listed in section 2.3) tell on the basis of translations from Middle Persian that Ardašīr's goal was precisely to restore the unified empire of his ancient ancestor Dārā. As Dignas and Winter point out, the agreement between the contemporary Greek sources and the Arabic and Middle Persian tradition, attested much later, speaks for itself.[176]

Unfortunately, due to the paucity, subjectivity, and ambiguity of the sources, speculative debates predominate in the discussion of nearly every aspect of the culture of the early Sasanian Empire. For the same reasons, it has been difficult in this chapter to say with certainty whether works of Hermes, or a legend about Hermes, were known in the Sasanian Empire as early as the third century. If, however, the

174. For arguments that the Sasanians were indeed trying to restore the Achaemenids' empire, see Fowden 1993b: 26–30, Shahid 2004: 240–241, and Dignas and Winter 2007: 56–62. Against, see most recently Huyse 2002 and Frendo 2006.

175. Similarly Wiesehöfer: "The Sasanian claims were probably not so precisely formulated and historically based as Herodian, in particular, suggests; but these reports show that the objective of Sasanian foreign policy was well understood by their Roman adversaries…" in "Ardašīr I (?–242 CE), the founder of the Sasanian empire," *EIr* 2.371–376. Yar-Shater 1971 also thinks that the Persian claim was not about a specifically Achaemenid past; he doubts Herodian (p. 525), but not, apparently, Dio.

176. Dignas and Winter 2007: 55–56.

early Sasanian emperors did claim that lands were wrongly taken from the ancient Persians by Alexander the Great, then the Persian propaganda was particularly appropriate to match and oppose the Roman propaganda, and indeed was probably specifically designed to counter it, using the Roman ideology as a means for their own justification. Therefore, if it is accepted that the "Iranocentric" history of science, as found in the later Middle Persian and Arabic sources reviewed above, had its genesis in this early period, the third century, one may suppose that the appropriation of Greek science by the Persian rulers can be understood not just as a measure taken to promote learning, but also as a political device. According to this ideology, the Persians were not merely restoring the ancient Persian Empire, but they were also taking revenge against Alexander, now self-consciously represented by the Roman emperors themselves, who revered Alexander as their forebear and their model, but who, for the Persians, was the thief of their ancient sciences and wisdom. In any case, the passage from the *Dēnkard*, discussed above, shows that this was a historical narrative present in the Persian court in the time of Ḫusrō I (r. 531–578), by which time there can be no doubt that translations had been made from Greek into Middle Persian. If, however, the ideology existed already in the third century, then the legend about the Iranian restoration of texts stolen by Alexander supports the Greek sources stating that Ardašīr made truly irredentist claims against the Romans. One must avoid a circular argument: the two sets of evidence cannot both be evidence for the other. Nevertheless, there is a striking coincidence between the third-century Greek sources, describing the Sasanid aim to restore pre-Alexandrian Persian territory, and the sixth-century and later Middle Persian and Arabic sources, describing the Sasanid aim, in the third century, to restore pre-Alexandrian Persian science.

2.10 Conclusions

First and foremost, the preceding discussion shows that there was definitely a Middle Persian Hermetic literature: astrological texts in Middle Persian attributed to Hermes. The doubtful and more complicated matter, based on the scarce information available, is *when* Greek Hermetica were translated into Middle Persian. The evidence suggests that, already in Šāpūr's time, Iranian sciences were alleged to have been recovered in Egypt and that forged works attributed to Zoroaster, Ostanes, and other Iranian sages had entered general circulation and attracted the attention of philosophers. Hermes is connected with the accounts linking Egypt and Iran through the legend of Ostanes, a sage of Achaemenian times. This legend first appears, as far as one can now tell, in the works of pseudo-Democritus, usually identified today with a poorly known Egyptian figure, Bolos of Mendes (third century BCE). There is also evidence for an exchange of ideas between the Greek language and its Eastern neighbors in the first century of Sasanid rule in the activity of Manichaeans and, for example, in the interests of Plotinus in "Eastern philosophy."

The references in Greek sources to a supposedly Iranian wisdom, hidden in Egypt, may give support to the numerous Middle Persian and Arabic reports dating from the sixth century onward, claiming that Persian sciences—also connected with the name Hermes by Ibn Nawbaḫt, one of the most detailed sources—were "restored" to the land of their origin by translation under Šāpūr I. Unless these correspondences are fortuitous, one must consider them to refer, however distantly, to the same events: the Sasanian Persian translation of Greek works in the third century CE. Given the foregoing, it is possible—though by no means absolutely proven—that some of the works attributed to Hermes were among those translated into Middle Persian in this early phase. The preceding discussion also illustrates some of the historical circumstances in which this transmission might have taken place. As stated at the outset, this has been an investigation of the evidence external to the Arabic Hermetica. The study of the Arabic Hermetica themselves might bring more evidence.

These findings show that one must take care not automatically to dismiss as late forgeries, without further investigation, all the Arabic literature that makes claims to have been written not only by Hermes, but also Ostanes, Zoroaster, and the other more mysterious figures of Greek and Iranian background.[177] Rather, it seems that there is a potentially substantial number of works transmitted into Arabic from Greek through Middle Persian, and that these may offer an important and somewhat unexpected new piece of the history of Graeco-Arabic translations. Each text will have to be analyzed on its own in order to try to determine the extent of this. Reliable criteria will need to be sought by which an Arabic work can be shown to have been translated from Middle Persian.[178]

The special importance of the Middle Persian mediation of Greek texts into Arabic, of which some of the Arabic Hermetic literature seems to be but a part, is that it originated in Greek in a relatively early period, in a context totally different from the transmission into Arabic of the texts of the Greek philosophers and physicians, which was accomplished mainly by Aramaean and Arab Christians centuries later. While much in the dating of the Sasanian reception of Hermetica remains hypothetical, the problematic evidence requires us to be satisfied for now with likely hypotheses. What is certain is that the earliest appearance of Hermes in Arabic, in the eighth century, is from the Middle Persian tradition, not the Greek.

177. This is a position advocated by Sezgin, GAS 3.xvi, harshly criticized by Plessner 1972 at least for being a premature conclusion, but then strengthened by Kunitzsch 1975 in his survey of the works known to have been translated from Middle Persian into Greek. My position remains that we must test the hypothesis in every instance. Although the most general outlines of the Middle Persian transmission is clear, there is still a huge amount of material to be studied in manuscripts that will show how much or how little trace of this transmission survives, and possibly its general significance in the history of science and the cultures of the ancient world.

178. Nallino 1922 is still the most important contribution to recognizing Arabic translations from Middle Persian.

3

Hermes and the Ṣābians of Ḥarrān

In 391, emperor Theodosius I decreed a law effectively making Christianity the sole legitimate religion of the Roman Empire. The public practices of pagan worship that had once been at the center of civic life—sacrifices, the adoration of statues of the gods, and related public celebrations—were now officially prohibited.[1] In the following years, ancient temples were destroyed or converted into churches, and those who persisted in performing the old kind of religion were pressured by new laws gradually depriving them of ordinary rights. Over several generations, the efforts by the Roman government, combined with mounting social pressure from the growing Christian majority, were largely successful. Paganism entered a long twilight, its last famous Hellenic representatives being the philosophers of the sixth century. By 529 and the reign of Justinian, laws existed mandating baptism for all; those who failed to comply were to be deprived of all property. The actual practice of pagan worship now warranted the death penalty. Although the implementation of such laws evidently varied from case to case, they represented a real and fatally effective threat to pagan practice in the remaining, Byzantine portion of the Roman Empire.[2]

By the sixth century, pagans in the Byzantine Empire were relatively few in number, but the local worship of ancestral gods did endure longer in certain outposts where such laws could not practically be enforced, perhaps more so than is generally

1. There are several problems inherent in a historian's use of the word *pagan*. When I employ it here as a convention, I am following as much as possible the meanings of the term and its equivalents in the languages of the sources that I am discussing. I will discuss these terms and their use in Arabic in due course.

2. The brief telling of this story drastically simplifies a complicated social history. For introductions in English, see Lane Fox 1986, MacMullen 1984 and 1997, and Chuvin 1990 (to be corrected on the matter of Ḥarrān, pp. 139–141, by this chapter). Trombley 1993–1995 can be consulted for a closely source-based look at different regions; on the laws attributed to Justinian, idem 2.81–97.

thought.[3] One of the longest-lived survivals was in the Syrian city Ḥarrān (Latin Carrhae). The Ḥarrānian pagans became notorious for their stubborn resistance to Christianity, withstanding Christian persecutions until the seventh century, when the Muslim Arabs came to power over them. Under the Arab caliphs, they succeeded in negotiating a marginally legitimate status for their religion. Some of them even became renowned scholars and courtiers, allowed to retain traditions of sacrifice and the veneration of the planets while attending the caliph himself. The most famous of these is the Ḥarrānian Ṯābit ibn Qurra (d. 901), who migrated to Baġdād, where he became an eminent mathematician and translator of Greek and Syriac books into Arabic. Many of his descendants, preserving their ancestral religion in some form for at least a century longer, became physicians, historians, and astronomers associated with the caliphs' court.[4] It appears that the intelligentsia of Baġdād were acquainted through this family with a concept of the ancient, pre-Christian religion that was not entirely negative. These "pagans" in the court are mentioned in Arabic sources as late as the eleventh century, after which they converted and joined the Muslim majority.[5]

The Ḥarrānian pagans, seeming so out of place in the tenth and eleventh centuries, with their religious roots reaching back to ancient Babylonian times, have held a romantic appeal for many historians. However, information about their actual beliefs and practices is scanty and based on tendentious sources, leaving room for widely varying interpretations and uncontrolled historical speculation. It is all too easy to imagine the Ḥarrānian religion as unchanging, given its apparently conservative character. We are left largely in the dark about the developments in belief and practice that must have taken place over the centuries, not only through a passive accommodation to new circumstances, but also by the initiative of its adherents. Of course, by the eighth and ninth centuries, their "paganism" must have been different from that of their ancient ancestors. But because of the problematic sources, no one has produced a really satisfactory description of just what these last pagans of Syria were doing, what exactly they believed, and how their beliefs and practices evolved. At least a few things are certain. They were recognized, and designated themselves, as a distinct sect. The Syriac word used by Ṯābit ibn Qurra for his religion was *ḥanpūtā*, in fact equivalent to the English word *paganism*;[6] Christian authors in

3. Trombley 1993–1995. In the context of another discussion, Watts 2005: 301–305 presents evidence for pagan communities in several places in the Byzantine empire of the sixth century. Morony 1984: 384–396 and Hämeen-Anttila's study of the *Nabataean Agriculture* (2006) demonstrate the persistence of polytheism of different kinds in Sasanian Mesopotamia.

4. *Fihrist* 331.23–24: "Ṯābit ibn Qurra founded the leadership (*riʾāsa*) of the pagans (*aṣ-Ṣābiʾa*) in this country [*scil.* Baġdād] and in the presence of the caliphs. Then they attained secure status, high ranks, and eminence." Cf. Ibn Abī Uṣaybiʿa 1.215.31.

5. François de Blois, "Ṣābiʾ," *EI*[2] 8.672a–675a, provides a list and genealogy of these later pagans, the Ṣābians in Baġdād. The end of the Ṣābians is discussed in section 3.5.

6. As cited by Ibn al-ʿIbrī (Bar ʿEbrāyā), Syriac Chronicle, Bodleian MS Hunt. 52 facsimile 55v.a14, trans. 153 (where Budge renders it similarly as "heathenism").

Syriac used the same term for their pagan foes. According to the near unanimity of the Syriac and Arabic reports, the religion of the Ḥarrānians entailed planet worship and involved ritual sacrifices. On the basis of the descriptions in these sources, which mention the names of some ancient Syrian gods, modern scholars have securely connected it with Babylonian and Hellenistic religion, whether one wants to consider that to be a "survival" of ancient religion or a "reimagination" of it, the product of a creative process of imitation and adaptation with each new generation.[7]

For the history of the Hermetica, the Ḥarrānian pagans have been considered especially important. Besides the Iranian astrologers discussed in the previous chapter, they are the only special group credited with possessing works attributed to Hermes and transmitting them into Arabic. By the ninth century, Ḥarrānians were claiming that Hermes was the very prophet of their religion. Yet the evidence for the Ḥarrānian transmission of Hermetica into Arabic faces a major impediment: no one has hitherto been able to show a single surviving Arabic work attributed to Hermes to be definitely Ḥarrānian in origin, meaning that it was preserved by or written by a Ḥarrānian practitioner of the ancient pagan religion of that city—whether as a part of that religion or not. The great influence that some historians suppose "Ḥarrānian Hermetism" to have had on Arabic culture and on Islam lacks sure evidence at this stage. Still the question remains: can any extant Arabic Hermetic text, or indeed any extant work at all, be shown definitely to have a Ḥarrānian provenance?

3.1 Introduction to the Problem of the "Ṣābians"

According to a well-known account, in the reign of the caliph al-Ma'mūn (813–833), the pagan Ḥarrānians are said to have adopted the name *Sābians* (Arabic *Ṣābi'a*) in order to acquire a protected status for their religion.[8] *Ṣābian* was a designation given

7. The Babylonian background has been explored by Green 1992 (the only recent attempt to tell the entire history of the Ḥarrānian religion) and Pingree 2002. Hjärpe 1972 is an indispensable critical analysis of heresiographic sources on the Ṣābians of Ḥarrān.

8. They are also said to have changed their style of clothing and long hair to conform with those around them at this time. Their claim to the name Ṣābian is from the frequently cited report of Abū Yūsuf Īšu' al-Qaṭī'ī, a Christian, related by an-Nadīm, *Fihrist* 385.3–30 (Flügel 320.10–321.11). Chwolsohn 1856: 1.140n4 argues that the precise year of the event was 830 (followed by, e.g., Hjärpe 1972: 8); Abū Yūsuf's account of it must nevertheless come from approximately a hundred years later, in the time of Sinān ibn Ṯābit ibn Qurra, since it refers to Sinān (d. 943). It is important to bear in mind that many religions were practiced at Ḥarrān, where there were, according to a witness in 772 CE, the Syriac *Chronicle of Zuqnin*, "Christians, pagans, Jews, Samaritans, worshippers of fire and sun, Magians, as well as Muslims, Ḥarrānians and Manichaeans," trans. Amir Harrak 1999: 272–274, but with the problematic rendering of *Ḥarrānāyē*, "Ḥarrānians," as Sabaeans (*sic*), corrected here. Pingree 2002: 18 cites this passage and corrects the text to say "Ḥarrānians" without note. Pagans, *ḥanpē*, in this text may mean "Hellenes" (Greek pagans), as opposed to the Ḥarrānians (Syrian pagans of local tradition) whose descendants would later take the name *Ṣābi'a*. The matter is uncertain, but the point of the sentence, that all of these different groups were in the city of Ḥarrān at this time, is clear.

in the Qur'ān (with the form *aṣ-Ṣabi'ūn*) for one of the religious groups considered licit in the community of Muḥammad, but the original, intended meaning of the Qur'ānic term Ṣābian seems to have been lost fairly early.[9] The Ḥarrānians succeeded in adopting the name, thereby preserving their religious practice from possible persecution, and Arabic sources from about the time of al-Ma'mūn generally do call the Ḥarrānian pagans Ṣābians. They were not, however, the only group that came to be identified as Ṣābian; the Mandaeans of southern Iraq adopted the name, leading to an ongoing scholarly debate over whether the Mandaeans or the Ḥarrānians were the "real" Ṣābians. By the early tenth century, any supposed "pagans" or idolaters of any period, such as the ancient Greeks, pre-Christian Romans and Egyptians, and the Buddhists, were also called Ṣābian.[10] The precise identity of the Ṣābians whom God sanctioned in the Qur'ān was and is a matter of persistent controversy for Muslim exegetes and heresiographers.

Daniel Chwolsohn laid the foundation for modern research on the Ṣābians with his great two-volume study published in 1856. Despite the problematic character of some of its conclusions, it still remains one of the best resources on the subject, particularly because it collects most of the *testimonia* about the Ṣābians available in Arabic. Chwolsohn's achievement is especially remarkable because, working before most of these texts were printed, he had to read his sources in manuscripts without indices or other modern textual apparatus. His work has to a great extent determined the two main issues under debate regarding the Ṣābians: who were the Ṣābians referred to in the Qur'ān, and what were the beliefs and practices of the Ṣābians of Ḥarrān?

There have been many attempts to answer the first question by a wide variety of approaches to numerous different sources, with equally varied answers. Modern scholars have identified the Qur'ānic Ṣābians as the Mandaeans,[11] the *ḥunafā'* understood as Gnostics,[12] Christian Sabaeans (Saba', the people of Sheba) of South Arabia,[13] the Manichaeans,[14] Elchasaites,[15] the Gnostics understood as the Archontics or Stratiotics (a Judeo-Christian sect mentioned by Epiphanius in the fourth century),[16] the *ḥunafā'* understood as "sectarians,"[17] and even just as the Ḥarrānian

9. Qur'ān: 2:62; 5:69; 22:17.

10. Already Chwolsohn (1856: 1.13) noted the generalization of the term Ṣābian for all "pagans," ancient or medieval, by the eleventh century (noted as one of Chwolsohn's major theses by Hjärpe 1972, 3). Numerous examples of this usage, which was in fact well established in the early tenth century, will appear in this chapter; see also the appendix to this chapter.

11. Chwolsohn 1856; Gündüz 1994.

12. Pedersen 1922: 390; Hjärpe 1972.

13. R. Bell 1926: 60.

14. De Blois 1995. De Blois' argument (1995: 51–52) that the real meaning of the term Ṣābi'a is really just "converts" is the most convincing etymology yet (cf. Margoliouth 1913: 519b; Genequand 1999: 126).

15. Buck 1984.

16. Tardieu 1986.

17. Genequand 1999: 123–127.

pagans.[18] I tend to believe that the real identity of the group intended in the Qur'ān cannot be known for certain given the present evidence.[19] Gündüz made a chronological survey of the Arabic sources on the Ṣābians, showing that the Mandaeans were indeed considered to be Ṣābians from at least the middle of the eighth century, but this does not prove that they were intended as the Qur'ānic Ṣābians a century earlier.[20] Thus the question has never been answered to everyone's satisfaction. The diversity of scholarly opinions indicates the pressing need to interpret the primary sources in a more critical, comprehensive, and, especially, chronologically ordered way, rather than striking upon a possible conclusion and setting out to prove it.[21] If an answer to this question can be obtained, the solution must account for all of the evidence and at least attempt to collect all the ancient and medieval sources of information about the Ṣābians in a manner showing the development of the terms over time. Fortunately, it is not necessary here to identify the Ṣābians of the Qur'ān, but only to understand the Qur'ānic origin of the term in Arabic.

The second major issue established by Chwolsohn is the question of the religion of the Ḥarrānians who came to be called Ṣābians. What was the character of their long-lasting paganism? The answer lies in the interpretation of problematic sources. In his important dissertation on the subject, Jan Hjärpe shows that much of the surviving information describing a Ḥarrānian religion of star worship is rooted in traditions of Syriac Christian heresiography having sources traceable to the tenth century. He considers these to be basically a polemical literary fiction.[22] At the same time he argues that many Arabic reports about the philosophical doctrines of the Ḥarrānian Ṣābians relate to the family of Ṯābit ibn Qurra—Ḥarrānian Ṣābians who had migrated to Baġdād—and that these Ḥarrānians, in the capital of the empire, followed a more philosophically informed religion differing from the old traditions

18. Strohmaier 1996.

19. Stroumsa 2004: 335–341 expresses a similar skepticism about the arguments of modern scholars for the identity of the Ṣābi'a.

20. Gündüz 1994: 22–52, esp. 24–25. If it is true that the compilation of the Mandaean *Ginza* took place in the first fifty years of Islamic rule (Macuch 1965: 176–185), then it seems likely that it was Muslim dominion that prompted the codification of this holy book. The creation of a canonical scripture may perhaps be regarded as an attempt to join *ahl al-kitāb* ("the people of the book") and to obtain legitimacy in the eyes of Muslims. In this case the Mandaeans might have been given the name Ṣābian very early in the history of Islam, but not from the beginning. That they had to take measures to establish a canonical scripture and thus adopt such a name would serve to indicate that they were *not* the Ṣābians of the Qur'ān. On a Mandaic colophon referring to Mandaic manuscript copying activity before and at the time of the advent of the Arab Muslims, see Buckley 1992: 40.

21. Gündüz 1994 comes the closest of all to being comprehensive, and the collection of sources and views provided there should form a basis for future investigations, though Chwolsohn's detailed notes and comments, although not all pertinent, must not be neglected.

22. Hjärpe 1972: 43–49. The sources are probably not as completely unreliable as Hjärpe claims. For example, he showed (90–92) that the descriptions of the Ḥarrānian religion by al-Mas'ūdī, ad-Dimašqī, and aš-Šahrastāni share the common source al-Ḥāriṯ ibn Sunbāṭ, a Melkite Christian of Ḥarrān; but one need not necessarily conclude, as Hjärpe seems to do, that this Christian source is so religiously polemical that it has no value whatsoever.

of Ḥarrān itself. He suggests that it may even have been a sect in schism with the Ḥarrānian religion, and that this schism was due to the "Hermetic theology" of Ṯābit and his followers.[23] In his view, most of the information available about the Ḥarrānian Sābians, where correct, actually concerns the Baġdādī Ṣābians alone, the descendants and relations of Ṯābit. Although there appears to be too little information to suppose a true "schism" of the kind Hjärpe has in mind, it is entirely plausible that the Ṣābians in Baġdād, just a handful of related families surrounded by other, much larger religious groups, would have practiced their religion differently and held some different beliefs from those back in Ḥarrān, who could presumably comfortably maintain distinctive practices more conspicuously in their own communities or neighborhoods. In Hjärpe's view, then, the inevitable conclusion is that the "traditional" religion and practices of the pagans in Ḥarrān itself become practically inaccessible to the historian. The sources are either unreliably polemical or describe a separate sect, the Baġdādī sect of Ṯābit and his progeny.

Michel Tardieu tried to overcome these difficulties posed by the sources in an influential and often-cited study.[24] He points out that there was only one Muslim Arabic author who visited Ḥarrān personally while the Ṣābians were still alive, and then reported about it: the historian al-Masʿūdī. He argues that al-Masʿūdī's information about the Ḥarrānians can be trusted because it is a firsthand account. The problem is therefore rooted in the precise interpretation of al-Masʿūdī's words. On this basis, Tardieu built an elaborate argument, trying to show that a group of the Ḥarrānians were truly Platonists, and that they maintained in the time of al-Masʿūdī a Platonic academy founded there in the sixth century by the last Hellenic

23. Hjärpe 1972: 35–39, following a similar discussion by Chwolsohn. The report about doctrinal differences between Ṯābit and the rest of the Ḥarrānians may be to some extent a later fabrication to dissociate Ṯābit's descendants to a degree from them; it is found in its fullest form in Ibn Ḫallikān's biography of Ṯābit ibn Qurra (1.313.8–14): *wa-ǧarā baynahū wa-bayna ahli maḏhabihī ašyāʾu ankarūhā ʿalyahi fī l-maḏhabi fa-rāfaʿūhu ilā raʾīsihim fa-ankara ʿalyahi maqālatahū wa-manaʿahū min duḫūli l-haykali fa-tāba wa-raǧaʿa ʿan ḏālika ṯumma ʿāda baʿda muddatin ilā tilka l-maqālati fa-manaʿūhu mina d-duḫūli ilā l-maǧmaʿi fa-ḫaraǧa min Ḥarrāna wa-nazala Kafratūṯā wa-aqāma bihā muddatan ilā an qadima Muḥammadu bnu Mūsā min bilādi r-Rūmi rāǧiʿan ilā Baġdāda fa-ǧtamaʿa bihī fa-raʾāhu fāḍilan faṣīḥan fa-staṣḥabahū ilā Baġdāda wa-anzalahū fī dārihī wa-waṣalahū bi-l-ḫalīfati fa-adḫalahū fī ǧumlati l-munaǧǧimīna fa-sakana Baġdāda wa-awlada l-awlāda wa-ʿaqibuhū bihā ilā l-āni.* "Some things transpired between [Ṯābit] and his coreligionists, for which they reproached him on the matter of the religion. They summoned him before their leader, who reproached his doctrine and forbade him to enter the temple. He repented and desisted from it, but after a while he came again to that doctrine, so they forbade him from entering the assembly. Then he left Ḥarrān and settled in Kafratūṯā [a town near Dārā]. He lived there a while until Muḥammad ibn Mūsā [a scholar influential in the caliph's court] arrived from the Byzantine lands on his way back to Baġdād, and met him there. He saw that he was an excellent man and fluent [in Arabic], so asked him to accompany him to Baġdād, and put him up in his own house. He brought him to the caliph, who admitted him into his group of astrologers. Then he stayed in Baġdād and had children. His descendants are there until now." The second part, omitting the information about Ṯābit's doctrinal differences with the Ḥarrānians, was abbreviated already in the *Fihrist* (331.22–23) of 987; because the account is evidently written after Ṯābit's death in 901, it must come from a tenth-century source.

24. Tardieu 1986.

philosophers upon their return from their stay in the Persian Empire as guests of Ḫusrō I. I do not agree with this theory, which I discuss in detail below.

Many questions about the nature of Ḥarrānian paganism under Muslim rule remain. Scholars are almost as divided about this issue as they are about the identity of the Ṣābians intended in the Qur'ān. Were the Ḥarrānian Ṣābians star worshippers,[25] Platonists or Neoplatonists,[26] Gnostics or Hermetists,[27] perhaps both Platonic and Hermetic simultaneously,[28] or something else?[29] And were there two differing sects of Ḥarrānian Ṣābians, Ḥarrānian and Baġdādī? All these questions cannot be answered here, but the knotty problem of the alleged "Hermetism" of the Ḥarrānians requires investigation in this study of the Arabic Hermes. In the introduction, I showed how the term *Hermetist* has been used in recent times very loosely to describe a number of different authors in Arabic without any special definition of the term. The same remarks apply equally here: because Hermetism as such has long been a subject of debate, we are forced to reject the term entirely for now on the grounds that it has never been adequately defined for the study of Arabic literature. Therefore, rather than seeking to label the Ḥarrānians as Hermetists or something else—for such labels explain nothing about their practices and beliefs, and no Arabic source ever calls the Ḥarrānians by such a name—I will confine myself solely to the search for *works attributed to Hermes* among the Ḥarrānians. If these Syrian pagans did indeed use books attributed to Hermes, it might be possible to call them Hermetists in some restricted sense. We might be able to seek a better term of description for the Ḥarrānians only after we have actually comprehended their beliefs, though this prerequisite in itself may be impossible, particularly so if their religion was more a matter of customary practice than philosophical doctrine.

3.2 A Ḥarrānian Platonic School?

Tardieu's hypothesis of a Platonic school at Ḥarrān lasting from the sixth to the tenth century requires close attention here because it supposes an institution of

25. Chwolsohn 1856 and many after him have taken the Arabic sources at face value and accepted the description of the Ḥarrānians as astrolaters.

26. Hjärpe 1972 argues that the Ḥarrānians who moved to Baġdād were Neoplatonists. Tardieu 1986 argues that one segment of Ḥarrānian society had been Platonists since the sixth century.

27. None of those who suggest that the Ḥarrānian Ṣābians were "Hermetists" offer a definition of the term's meaning in the context of Arabic literature or in any context. Marquet 1966: 55: "Les ḥarrāniens, en effet, sont incontestablement hermétistes, et fort peu néo-platoniciens." Hjärpe 1972 considered the original Ḥarrānian tradition to be both Hermetic and Gnostic.

28. Hadot 2007: 53–56 argues that the Platonism inherited by the Ḥarrānians had long before integrated "Hermetism" into itself.

29. Walbridge 2001: 38: "It seems that the ancient moon cult of Harran had evolved into a Hermetic sect that worshipped the planets." H. Corbin's imagination created a Ḥarrānian Ṣābianism all of his own out of sources drawn randomly from different eras (1951). Muḥammad 'Abdalḥamīd al-Ḥamd 1998 unconvincingly attempts to prove that the Iḫwān aṣ-Ṣafā' were Ḥarrānian Ṣābians.

learning wherein ancient Greek books, including Hermetica, might be imagined to have been continuously preserved. Although this hypothesis seems generally now to be regarded as controversial at best, it has been cited as authoritative by a number of scholars in the last twenty years, and, despite critical scrutiny in several publications, still very recently some continue to defend it strenuously, particularly I. Hadot.[30] The hypothetical Ḥarrānian transmission of Hermetic books is implicated in the matter insofar as the Platonists or Neoplatonists alleged to be at Ḥarrān can readily be assumed to have preserved works of Hermes.[31]

Tardieu developed his theory beginning with a critical reevaluation of Hjärpe's study.[32] Ultimately he accepts Hjärpe's distinction between two types of Ḥarrānians, one more philosophical than the other. However, instead of treating them as a traditional Ḥarrānian group and a separate offshoot sect of intellectuals at Baġdād, as Hjärpe did, Tardieu situates the split between the two putative groups within the city of Ḥarrān itself, not in two different locations. The basis for his distinction lies in claiming the existence of an old institution of instruction, in fact a Platonic academy, surviving in Ḥarrān long after the advent of Islam, in the tenth century. The appeal of the theory for a historian is that such a school would seem by its very existence to help to explain the revival of Greek philosophy in Arabic translation, as a sort of direct survival from pre-Islamic antiquity.

His attempt to prove that this school was at Ḥarrān under Muslim rule depends in fact on the interpretation of just two main pieces of evidence in Arabic.[33] The first

30. The basic arguments are laid out in Tardieu 1986, supplemented by 1987. Examples of publications accepting, repeating, or expanding Tardieu's hypothesis are Monnot 1986: 5; Hadot 1987, 1990: 280–289, 1996: 28–50, 2001: xiii–xxxiii, and 2007; Chuvin 1990: 11 and 139–141; Athanassiadi 1993: 24–29; Thiel 1999; and Vallat 2004: 19–23 (but not entirely: 19n2). Lameer 1997 and R. Lane Fox 2005 provide important correctives that I accept in most, though not all, of their details, and build upon them here. Gutas 1999: 156n2 and Genequand 1999: 114–115 and 124–125 disagree with Tardieu's hypothesis. Endress 1990: 10n24 demands more evidence for the theory. Green 1992: 168 and Brittain and Brennan 2002: 3–4 give reasons not to be convinced. Sheppard 2000: 841–842 considers the evidence insufficient. Luna 2001 criticizes Tardieu's and Thiel's arguments in detail, dealing with the misinterpretation of the Greek evidence. Stroumsa 2004: 338–340 agrees with Gutas and Lameer. Watts 2005 is devoted to a critique of and alternative to Tardieu's hypothesis about the philosophers' settling in the sixth century Ḥarrān. Hadot 2007 offers rejoinders to several of the objections, mostly just repeating the arguments of Tardieu (missing Lane Fox 2005, no doubt too recent to have been taken into account). Pingree 2002 basically accepts Tardieu's argument, rejecting Lameer's criticisms on the grounds that he "ignores the fact that he met men there who had an assembly hall (*majma*ʿ) and who informed him correctly that the Syriac inscriptions on its door-knocker were quotations from Plato" (10n10). On this *maǧma*ʿ, and criticism of Pingree's understanding of the Ḥarrānians, see below.

31. Thus Hadot 2007: 53–56 and 103 makes their supposed Neoplatonism the basis for their alleged Hermetism.

32. Tardieu 1986: 3–11. One of Tardieu's positive contributions here is to call into question the designation of Ḥarrānians as "Gnostic."

33. Tardieu's argument for the foundation of a Platonic school in Ḥarrān in the sixth century uses also Greek evidence challenged by Luna 2001 and Watts 2005; Hadot 2007 raises objections to the latter two. The present discussion is concerned primarily with the Arabic data that were Tardieu's starting point.

is part of al-Masʿūdī's rambling description, in the early tenth century, based on his own visit to the city, of the Ḥarrānian Ṣābians and their *maǧmaʿ* (gathering place).[34] The passage is this:

ولهذه[35] الطائفة المعروفة بالحرّانيّين والصابئة فلاسفة إلاّ أنّهم من حشويّة الفلاسفة وعوامّهم
<وهم> مباينون لخواصّ حكمائهم في مذاهبهم وإنما أضفناهم الى الفلاسفة إضافة نَسَب ولا
إضافة حكمة لأنّها يونانيّة وليس كلّ اليونانيّين فلاسفة إنما الفلاسفة حكماؤهم
ورأيتُ على باب مجمع الصابئة بمدينة حرّان مكتوباً على مدقّة الباب بالسريانيّة قولاً لأفلاطون
أفهمني تفسيره ملك بن عقبون وغيره منهم وهو من عرف ذاته تألّه وقد قال أفلاطون
الإنسان نبت سماوي والدليل على هذا أنّه شبيه شجرة منكوسة أصلها الى السماء وفروعها
الى الأرض

This community, known as the Ḥarrānians and the Ṣābians, has philosophers [*consists of philosophers (see discussion below)]; however, they are some of the common riffraff of the philosophers, differing from their élite sages [scil. the élite sages of the philosophers] in their doctrines.[36] I consider them philosophers only by way of lineage, not by way of wisdom, because they are Hellenes (*Yūnānīya*); not all of the Hellenes are philosophers; only their [scil. the Hellenes'] sages are the philosophers.

I saw on the door-knocker of the gathering-place of the Ṣābians (*maǧmaʿ aṣ-Ṣābiʾa*) in Ḥarrān a saying of Plato written in Syriac of which Malik ibn ʿUqbūn and others of them informed me the interpretation. It is, "He who knows his essence becomes divine."

Plato has said, "Man is a heavenly plant. The indication of it is that he is like an inverted tree: its roots are toward heaven and its branches are toward the earth." [Al-Masʿūdī then adds that there are many doctrines attributed to Plato on the soul, etc.]

Tardieu begins with a misunderstanding of the first sentence. He takes the referent of the possessive pronoun in *ḫawāṣṣ ḥukamāʾihim* (lit. "the élite of their sages," equivalent to "their élite sages") to refer back to the Ḥarrānians rather than to the philosophers, the natural referent.[37] Thus, in his interpretation, the philosophers of the Ḥarrānians differ "from *the Ḥarrānians*' élite sages in their doctrines."[38] This

34. This is found in his *Murūǧ aḏ-ḏahab* (ed. Pellat §§1394–1395 = 2.293.1–8, trans. 2.536–537) a work completed between 943 and 956 (Charles Pellat, "al-Masʿūdī," EI^2, 6.784a–789a).

35. On the reading of this word, see below.

36. The Arabic term *ḥašwīya*, which I have translated as "riffraff," was a term of derogation used by some Muslim sectarians for other Muslims who adhered to an especially literal interpretation of the Qurʾān and *ḥadīṯ*, scholars who collected religious traditions without being capable of interpreting them. The implication here is that these Ḥarrānians follow the same religion as the philosophers but are incapable of interpreting their religious traditions in a sophisticated or intelligent way. See "Ḥashwiyya," EI^2, 3.269b.

37. Pointed out first by Lameer 1997: 187–188; seconded by Lane Fox 2005: 235–236.

38. Tardieu 1986: 17–18.

leads him to posit two distinct groups among the Ḥarrānians: "low" sages and "high" philosophers, "followers of the pagan religion of the town, and the 'sages' in the strict sense, heirs of the Greek philosophers."[39]

Joep Lameer pointed out a potentially serious objection to Tardieu's reading beyond the crucial misunderstanding of the pronoun. The great majority of the manuscripts of the text (as observed by Chwolsohn) lack the possessive *li-* at the sentence's beginning. As there is no sufficiently critical edition of the text, Tardieu followed the reading he found in print. However, if one follows the better attested alternative along with Lameer, one must translate: "This community... *consists of* philosophers..." This sentence, differing by a single letter, clearly leaves no room for holding that there was more than one group.[40] Sure enough, in another passage in the same work, al-Masʿūdī repeats himself using the same words and with no ambiguity at all, referring to "the Ṣābi'ans of the Ḥarrānians, who are the common Hellenes and the riffraff of the ancient philosophers." This leaves no doubt that Lameer's reading is the correct one.[41] Even if this evidence is ignored, al-Masʿūdī's immediately subsequent words in the passage under discussion invalidate Tardieu's interpretation, showing that in his opinion there were no real philosophers in Ḥarrān: "I consider them philosophers only by way of lineage, not by way of wisdom, because they are Hellenes (*Yūnānīya*); not all of the Hellenes are philosophers; only their sages [scil. the sages of the Hellenes] are the philosophers." The word *Yūnānī* is the Arabic equivalent of Greek Ἕλλην (Hellene) in its contemporary Byzantine usage, having the same two overlapping meanings: it means either ethnically Greek but not Christian,[42] or, more generally, just pagan.[43] Insofar as pagan Greeks were largely a thing of the past by the time of Arabic literature, Yūnān is often treated as one

39. Tardieu 1986: 17–18. The same objections have already been discussed by Lameer 1997: 188. Hadot's criticism (2007: 59–60) of Lameer's argument does not address this point.

40. Lameer 1997: 187–188. In response, Hadot 2007: 59–60 makes much of rejecting Lameer's preferred reading (the single variant letter) on the grounds that it is not the quantity, but the quality, of manuscripts that should determine the archetype; a *stemma codicum* should determine which manuscripts, and which of their readings, have greater value. All this is quite true. However, she surprisingly concludes from this that Lameer's reading "has no value." In so arguing, she overlooks a fundamental consideration: lacking a critical edition, *her own preferred reading*, that of Tardieu, based on a small minority of manuscripts, *is equally arbitrary*. Strangely, she does not address any other part of Lameer's discussion, which challenged Tardieu's interpretation of the passage on other salient points.

41. *Murūǧ* (Pellat §219 = 1.109.1–2): *aṣ-Ṣābi'atu mina l-Ḥarrānīyīna wa-hum ʿawāmmu l-Yūnānīyīna wa-ḥašwīyatu l-falāsifati l-mutaqaddimīna*. A key-word here is *ʿawāmm*, which refers to uneducated masses or the uninitiated.

42. Christian intellectuals, even those monolingual in Greek, had long considered themselves as having a separate, non-Hellenic ethnicity. On the development of this narrative, see Lechner 1955 and more recently A. P. Johnson 2006.

43. The use of Ἕλλην purely to mean "pagan" in the ninth century is shown by its rendering as *umamī*, "gentile," in the Arabic translation of Paul's *Epistle to the Colossians* (Ullmann 2002–2007: 1.234). In the tenth century, the Syriac-Arabic lexicographer Bar Bahlūl (2.764) defines the *ḥanpā* or pagan as idol-worshipper, adding "in Greek (it is) Hellenios" (*Hēlêniyôs*); furthermore he glosses it in Arabic as "ignorant *ḥanīf*" (*al-ḥanīf al-ǧāhil*). Interestingly, the Ṣābian Sinān ibn Ṯābit was one of this lexicon's sources (Bar Bahlūl 2.734.3; cf. 3.xxiv), though he is not cited in this entry. Already Procopius in the sixth century clearly alludes to the generalization of the Greek word (*De bellis* 1.20):... ἐπειδή

of many bygone nations of ancient history, a nation that, as Arabic historians knew, had once produced important philosophers, but had long ago been overcome and replaced by ar-Rūm, the Roman-Byzantine Empire of the Christians. In the present case, al-Masʿūdī demonstrates his awareness of the word's ambiguity when he tries to clarify what he means, albeit in his characteristically sloppy manner: the Ḥarrānians are philosophers not in the strict sense of the word, but only in that they are *descendants of the ancient pagans* or Hellenes; they are certainly not philosophers because of *wisdom*. The distinction is not, as Tardieu would have it, between common pagans and pagan philosophers, two different groups in Ḥarrān; it is between two groups within a single historical ethnic category, the Hellenes. The Ḥarrānians are excluded from the ranks of real philosophers *despite* their Hellenic, pagan, lineage.[44]

Thinking he has discovered a special group of practitioners of a distinctively Greek philosophy in tenth-century Ḥarrān, Tardieu then turns to the second part of the passage cited and supposes that the word *maǧmaʿ* (most neutrally "gathering place") here must signify "academy," a meaning of the word not otherwise attested in premodern Arabic, and that the door-knocker inscription ascribed to Plato proves it to have been specifically a *Platonic* academy.[45] He thinks, moreover, that the reported saying derives from 1 *Alcibiades*.[46] Because 1 *Alcibiades* was an introductory

---

Ὁμηριτῶν τῶν ἐν τῇ ἀντιπέρας ἠπείρῳ ἔγνω πολλοὺς μὲν Ἰουδαίους ὄντας, πολλοὺς δὲ δόξαν τὴν παλαιὰν σέβοντας, ἣν δὴ καλοῦσιν Ἑλληνικὴν οἱ νῦν ἄνθρωποι, ἐπιβουλῇ μέτρον οὐκ ἐχούσῃ ἐς τοὺς ἐκείνῃ Χριστιανοὺς χρῆσθαι… "…since he learned that many Jews and many others venerating the ancient faith, which people today call Hellenic, of the Ḥimyarites in the opposite land [Yemen], were treating the Christians there with exceeding treachery…" Here even South Arabian Ḥimyarites follow the Hellenic, that is polytheistic, pagan, religion; this does not mean that they are ethnically Greeks.

44. Cf. Genequand 1999: 114–115. Vallat 2004: 19–23 claims that, at least in some cases, "Yūnānīyūn = Ἕλληνες = Platonists." The last part of the equation does not hold. Vallat cites very few witnesses for it, and none of them actually establishes that Yūnānīyūn ever certainly meant "Platonists," an interpretation requiring one to ignore countless other examples of the word's use in Arabic. Yet Hadot 2007: 56–69 cites Vallat at length, trying thereby to bolster Tardieu's argument by making Platonists out of the Yūnānīyūn/Hellenes described by al-Masʿūdī at Ḥarrān. This is based on ignorance of the common Arabic usage, to which Vallat in fact refers (2004: 20n3).

45. Tardieu 1986: 17–18.

46. The resemblance of the text reported by al-Masʿūdī to Plato's *1 Alcibiades* 133c is not precise (as also noted by Green 1992: 167, Lameer 1997: 189, and Lane Fox 2005: 236). Al-Masʿūdī has, "He who knows his essence becomes divine," *man ʿarafa ḏātahū taʾallaha*; Plato, *1 Alcibiades* 133b–c, writes, "Socrates: And if the soul too, my dear Alcibiades, is to know herself, she must surely look at a soul, and especially at that region of it in which occurs the virtue of a soul—wisdom, and at any other part of a soul which resembles this? Alcibiades: I agree, Socrates. Socrates: And can we speak of any part of the soul more divine than this, which is the seat of knowing and thinking? Alcibiades: We cannot. Socrates: Then this part of her resembles God, and whoever looks at this, and comes to know all that is divine, will gain thereby the best knowledge of himself." (Translation by B. Jowett, slightly modified.) This says that the one who sees the part of the soul that is wisdom, and thus knows the divine, also gains knowledge of himself. This is a somewhat different meaning from the Arabic just cited, although the pithy character of the Arabic phrase allows easy comparisons. Al-Masʿūdī mentions this saying, cited in different form, in another discussion of the Ṣābians in his *Kitāb at-Tanbīh wa-l-išrāf* (162.4): *man ʿarafa nafsahū ḥaqīqata l-maʿrifati taʾallaha.* "He who knows himself with true knowledge becomes divine."

text for late antique Platonists, he suggests that it would be symbolically appropriate for a door-knocker, the entrance of the supposed school.[47] The idea that it is an apt symbol is taken as corroboration of the view that the building is a Platonic academy.[48] However, loosely Platonic sayings of all types were commonplaces by the tenth century, and one maxim on a door obviously cannot prove the existence of an entire academy.[49] Even if al-Masʿūdī had seen an actual school (the evidence for that being nil), there would be no reason to think that it was either very old or specifically devoted to Platonic studies.[50] But Tardieu, assuming there definitely to be Platonic philosophers at Ḥarrān, argues speciously that, "Now, as the School of Ḥarrān did not fall from the sky in the first century of the *hiğra*," one must assume that this was the last resort of the Neoplatonist philosophers on their return from the Persian Empire in the sixth century, and that these philosophers are the ones who founded the school seen by al-Masʿūdī.[51] According to this assumption, the Platonic academy must have existed there continuously for more than three hundred years. Tardieu and those following him find no impediment to the argument in the total absence of evidence from the intervening period for such an institution: no authors from that school, unless we take the sixth-century philosophers as the founders, but certainly none thereafter;[52] no references to such a pagan school of

47. Tardieu 1986: 19.

48. Repeated by Hadot 2007: 52–53.

49. On a number of closely similar, yet not all strictly Platonic, sayings in Arabic, see Rosenthal 1940: 409–410 and Altmann 1963. As Tardieu (1986: 14) says, the Platonic maxim next reported by al-Masʿūdī in the same passage ("Man is a heavenly plant; the indicator of that is that he is like an upturned tree with its trunk toward heaven and its branches toward the earth") is surely derived from Plato's *Timaeus* 90a7, but we do not know at how many removes; it was common enough to be found, with different wording, in Ibn Waḥšīya's *Nabataean Agriculture* (1.360.7–9), evidently translated from Aramaic (mentioned in this context also by T. Fahd, "Ṣābi'a," *EI*², 8.675a–678b) and in an Ismāʿīlī's reminiscences of a conversation with a well-read *dāʿī* in Qayrawān in 909 (Madelung and Walker 2000: trans. 154, text 108). In any case, there is no indication that al-Masʿūdī took this second saying from Ḥarrānians, as Tardieu thinks (1986: 14); the conjunction (*wa-qad qāla Aflāṭūn...*) shows rather that this is just one of al-Masʿūdī's typical digressions (as also intuited by Lane Fox 2005: 237).

50. Cf. Lameer 1997: 188–189.

51. Tardieu 1986: 23. On this issue see Watts 2005.

52. The thirteenth-century historian Ibn al-Qifṭī (206.9) says that Simplicius "had companions and students known as his" (*wa-kāna lahū aṣḥābun wa-atbāʿun yuʿrafūna bihī*). Hadot takes this to show "L'existence d'une école renommée de mathématiques" (2007: 102) that "Simplicius directed" (1990: 302). It is too great a leap from the unproved statement that Simplicius had students to positing a "famous school of mathematics" or to the belief that Simplicius founded a Platonic academy lasting for several centuries at Ḥarrān and nowhere else. Indeed, Ibn al-Qifṭī knows almost nothing about Simplicius, saying in a short entry of just four and a half lines that he was *formerly* famous *in Greece*—not currently—as a mathematician, and naming only one of his works. In Arabic translations of Simplicius' mathematical works, Simplicius makes reference to his colleague or companion (*ṣāḥib*) whose name is rendered *'ġ'nys*, a name distorted in Arabic and from whom passages are cited (Sabra 1969: 5–7). Hadot (1990: 302, 2001: xx, 2007: 78–79) takes this individual as further evidence of an entire school, where Aganis is understood, following Tardieu's interpretation, as an Egyptian name, meaning "cup" or "jug"(!). (On the reading of the name, perhaps as Agapius, see Sabra 1968: 13n6.) Following Tardieu again in rather wild onomastic speculation, Hadot has also pointed repeatedly (1990: 293n73,

Ḥarrān anywhere in Syriac Christian literature; not even one name of any master or disciple from this Ḥarrānian school from the seventh to the tenth century; no books the Ḥarrānian Platonists produced or copied, no commentaries they authored, not even one reference to a title designated as specific to this supposed academy, over the three intervening centuries, surviving in Greek, Syriac, or Arabic.[53]

Tardieu also points to a reference to the *maǧmaʿ* of the Ḥarrānians found in another source, the Ḥarrānian calendar in the *Fihrist*. The Ḥarrānians are reported there to eat a certain annual meal in their *maǧmaʿ*, an activity perhaps to be expected in a town's gathering-hall. Yet Tardieu concludes from this rather that there must then have been *two maǧmaʿ*s, two separate institutions frequented by the two different communities he has hypothesized from his misreading of the first part of al-Masʿūdī's description. He takes these as communities "in rivalry": the old pagans of Ḥarrān, who have some kind of ordinary gathering place, and the Platonists, who have their Academy (both places being *maǧmaʿ*s).[54] This interpretation requires one to overlook the fact that in both al-Masʿūdī's report and the Ḥarrānian calendar, the *maǧmaʿ* is grammatically definite and singular. Neither text says "in one of the *maǧmaʿ*s of the Ḥarrānians." Probably their *maǧmaʿ* was just a public hall of assembly or a temple used by the pagans of Ḥarrān on such festive occasions.[55] Indeed, the passage from al-Masʿūdī mentioning the *maǧmaʿ* occurs under the rubric "Discussion of the Venerated Houses and Honored Temples of the Ṣābians," wherein the subject is just the temples of the Ṣābians. Tardieu overlooks this context.

For his second piece of evidence, Tardieu refers to a well-known account, related by al-Masʿūdī, al-Fārābī, and others, of the transfer of the instruction in the ancient sciences from Alexandria to Antioch, then to Ḥarrān, and then to Baġdād.[56] Yet this is far from a proof of a Platonic school in Ḥarrān. Dimitri Gutas has analyzed these

---

1996: 42, 2001: 22, 2007: 82–83) to a reference in the *Fihrist* to a lost work of Simplicius dedicated to a certain person whose name appears in the printed edition as *ʾṭʾwʾlys* (*Fihrist* 311.27–312.1). In order to prove that this person was Syrian—because the view that Simplicius had settled in Ḥarrān might, in this view, be furthered if Simplicius at some time dedicated a work to a Syrian—Tardieu proposed to Hadot the far-fetched interpretation of *ʾṭʾwʾlys* as a combination of the name of a god *ytʿ* or *ʾtʿ*, found in five Ṣafaitic graffiti and one Greek inscription (Ṣafaitic inscriptions, from the Syrian desert, dating from the first century BCE to the third CE), and the Greek name Οὐάλης (= Valens) to make a "nom théophore." Tardieu takes the loss of the consonant *ʿayn* in the supposed composite name as proof that the work of Simplicius was composed in Greek. This interpretation requires us to accept the existence of the unattested and highly implausible name ʿEtha-Oualês and the unlikely premise that the Arabic name should be read exactly as the editors of the *Fihrist* read it. For a more complete and sober approach to this name see Aouad 1989.

53. Cf. Brittain and Brennan 2002: 5.

54. Tardieu 1986: 17–19, referring to the text of *Fihrist* 287.25. It says *wa-yaʾkulūna fī maǧmaʿihim*, "and they eat in their *maǧmaʿ*."

55. Yaḥyā al-Anṭākī's chronicle (Kratchkowsky 1997: 150.17) refers to the confiscation in 1031 of a (not *the*) *maǧmaʿ* of the Ṣābians (*maǧmaʿan li-ṣ-Ṣābiʾa*), identified specifically as the moon temple. The event is further discussed in section 3.5. See also the passage in note 23 stating that Ṭābit ibn Qurra was forbidden from entering *the maǧmaʿ*.

56. Tardieu 1986: 24–26.

accounts in detail, showing that they were concerned not with Platonic tradition, but rather the Alexandrian medical curriculum, including a limited course in Aristotelian logic.[57] The reasons that Ḥarrān appears in this narrative at all are still not clear.[58] In any case, Tardieu interprets the reports not at face value but merely as confirmation of the "intellectual vitality" at Ḥarrān (as well as Antioch).[59] In fact, only a single anonymous Ḥarrānian philosopher is mentioned in the more detailed rendering of the account by al-Fārābī.[60] A certain Quwayrā is supposed to have studied with this Ḥarrānian (who learned something from another, unnamed man from Antioch) before going to teach in Baġdād, where the Christian philosopher Mattā ibn Yūnus (one of al-Fārābī's teachers) was one of his students. This Quwayrā is not credited with any works apart from commentaries on the first four books of the *Organon*. Even these commentaries were obsolete by the late tenth century, for an-Nadīm described them as "rejected and avoided because their style was halting and obscure."[61] A physical description, perhaps fictitious, of Quwayrā appears in a correspondence attributed to as-Saraḫsī (d. 899), probably Quwayrā's contemporary, in which Quwayrā is portrayed as a Christian monk.[62] Thus the "Ḥarrānian" tradition of philosophy in the narrative seems to be constituted of only two individuals, perhaps Christian monks, interested in logic but of no known special attainments.[63] The Ṣābians or pagans of Ḥarrān are never named in connection with the account of the transfer of instruction from Alexandria to Baġdād. Only the city of Ḥarrān is mentioned.

One of the reasons to search for an institution of philosophical instruction in Ḥarrān is the simple fact that it was the place of origin of the Ṣābian scholar Ṯābit

57. Gutas 1999. Tardieu 1986: 22 already recognized that these narratives deal with the medical curriculum, but claimed that this medical curriculum was one part of the teaching of philosophy; thus he claims the narratives really deal with the transfer of a school of philosophy. Hadot 2007: 61n64 points out that the narratives appear to describe philosophical education when related by philosophers, and medical education when related by physicians. Her implication appears to be that al-Mas'ūdī, whose narrative is the most abbreviated and least specific about the subjects of instruction, is to be accepted as most reliable. Yet it remains that only the logical *Organon* and the medical curriculum are specified in these narratives, whereas neither Plato nor any Neoplatonic philosopher is so named.

58. Gutas 1999: 187–188. Lameer 1997 suggested that Antioch and Ḥarrān appear as philosophical centers in this complex of accounts because the former was the place of "spiritual origin" of the Eastern Christians, and Ḥarrān of the Ṣābians; important translators came from both groups. This theory appears quite unlikely. The eastern Iranian city Marw also appears in the account of the transmission of the sciences, occupying approximately the same position of importance (or lack thereof) as Ḥarrān (Gutas 1999: 165), but no one today tries to locate a school of Platonic philosophy in Marw!

59. Tardieu 1986: 26.

60. The accounts of al-Mas'ūdī and of al-Fārābī derive from a common source (Gutas 1999: 187n89).

61. *Fihrist* 321.15–18 (Flügel 262.23–25).

62. The amusing correspondence is studied in English translation by Rosenthal 1943: 86–94. The description of Quwayrā as a monk is on 88, the epithet "Christian" comes on 90 and 91, and the judgment of Yāqūt on the authenticity of the correspondence is given on 93–94.

63. Tardieu 1986: 25 accepts that the Ḥarrānian in al-Fārābī's lineage must have been a Christian.

ibn Qurra. Tardieu writes, "Platonism or, if one prefers, Neoplatonism was not created at Baġdād *ex nihilo* by Ṯābit ibn Qurra. He himself was dependent on the instruction he had received in Ḥarrān."[64] The supporters of Tardieu's theory seem to wish to prove that Ṯābit, along with "Arabic Neoplatonism" as such, was the product of a direct line of Platonic teachers from Simplicius, reaching back through his teachers to Proclus and even earlier. There are two main problems with this idea. One must bear in mind first that there was a great deal of information about the philosophy of Plato and his Neoplatonic heirs in Arabic that clearly has nothing to do with any Ḥarrānian.[65] As for Ṯābit himself, only one among the nearly two hundred Arabic works attributed to him appears to deal directly with Plato;[66] of his many original translations from Greek, his revisions of the translations of others from Greek, and his epitomes of Greek works, not even one is a work of Plato, although he did undertake to translate a Neoplatonic commentary on the *Golden Verses* of Pythagoras (apparently late in life, because he died before accomplishing much of it).[67] The titles of his own research and writings indicate that Ṯābit's interests were primarily, not to say overwhelmingly, with mathematics and their applications in astronomy and astrology, and secondarily with Galenic medicine and Aristotelian physics.[68] Tellingly, Tardieu does not refer to a single specific work of Ṯābit in his reference to Ṯābit's supposed Platonism.[69] In fact, we know almost nothing about the early education of Ṯābit in or near Ḥarrān, and we do not know even when he moved to Baġdād and entered the company of the court astrologers, though it was perhaps while he was still an unformed young man.[70]

64. Tardieu 1986: 11.

65. Rosenthal 1940, Gutas 1988, Adamson 2002, Endress 2007.

66. This is his *Book on Resolving the Allegories in Plato's Republic* (*Kitāb fī ḥall rumūz Kitāb as-Siyāsa li-Aflāṭūn*) (Ibn al-Qifṭī 120.7). He also wrote a *Treatise on the Proof attributed to Socrates on the Square and Its Diagonal* (*Kitāb Risāliatihī fī al-ḥuǧǧa al-mansūba ilā Suqrāṭ fī l-murabba' wa-quṭrihī*) (*Fihrist* 331.26). While the contents of this work concern a geometrical problem discussed in Plato's *Meno*, Ṯābit apparently does not deal with the main issues of that work in his treatise: see Endress 2007: 328–329 and Gutas, forthcoming.

67. Ṯābit translated into Arabic three out of one hundred folia of a Syriac translation a *Commentary on the Golden Verses of Pythagoras* (*Fihrist* 313.2–3; ed. Flügel 252.16–18), but died leaving it unfinished. (The report says that the commentary was by Proclus, but this is probably a mistake for Hierocles; thus R. Walzer, "Buruḳlus," *EI*², 1.1399b–1.1340b.) It need hardly be said that an interest in this book does not prove that he was trained in a Neoplatonic Academy in his home of Ḥarrān.

68. This general point also made by Lane Fox 2005: 239–240. In different contexts, Hadot takes both the study of mathematics (Hadot 1990: 302) and of medicine (2007: 61n64) as supporting evidence for a school of Platonic philosophy, even when there are no connected references to Plato whatsoever.

69. He cites Wiedemann 1920–1921, where there is no reference to Ṯābit's supposed Neoplatonism. Of course, Ṯābit was probably interested somehow in Plato, as much as any of his philosophically minded contemporaries were (on which see Endress 2007).

70. Hadot twice (1996: 31, 2001: xix) demonstrates how much confusion Tardieu has inspired when she refers to "la fameuse école de Bagdad, fondée par Tabit b. Qurra vers 901 après qu'il eut quitté Harran . . . héritière des néoplatoniciens harraniens." No Arabist, as far as I am aware, has ever suggested that Ṯābit founded a *school*, let alone a famous one. Moreover, 901 is the year of Ṯābit's death!

In short, none of the Arabic evidence, upon careful scrutiny, supports the hypothesis of a Platonic school at Ḥarrān.[71] Consequently there is no reason for the desperate attempts to prove that the Hellenic philosophers of the sixth century settled at Ḥarrān and nowhere else.[72] Although current publications continue to repeat Tardieu's arguments, the history of early Arabic philosophy and the reception of Platonic or Hermetic Greek works in Arabic translation are much more complicated than the hypothesis of a single important, continuously existing, but totally unattested institution of learning will allow.

3.3 The Alleged Influence of Ḥarrānian Hermetica

A general conclusion, often an assumption, among scholars who have dealt with the Arabic Hermetica is that these texts come from the Ṣābians of Ḥarrān.[73] The

71. This is not to say that no Ḥarrānians took any interest in Plato, but that is different from claiming the existence of an institution of learning with a continuous existence over several centuries.

72. See Hadot 2007 versus Watts 2005 and Luna 2001. The argument over where the philosophers settled on their return from the Persian Empire is a separate problem based on the analysis of evidence mostly internal to the works of the philosophers of the sixth century. One must avoid a circular argument whereby a supposed tenth-century school proves its supposed sixth-century foundation and vice versa. Nevertheless it is not irrelevant to indicate here a flaw in one of the main arguments for the philosophers' relocation to Ḥarrān in the sixth century. Tardieu points out (1986: 24n105) that the philosopher Simplicius argues against certain Manichaean doctrines, which, Simplicius says, were explained to him by a Manichaean sage. Tardieu thinks (followed most recently by Hadot 2007: 68) that there were only two cities in the Byzantine Empire where Manichaeans could with certainty be found in the sixth century: Byzantium and Ḥarrān. Partly on this basis he concludes that Simplicius must have lived in Ḥarrān after his return from the Persian Empire. With the exception of Lane Fox 2005: 232–233, the published arguments, whether supporting or criticizing the theory, focus on the nature of Simplicius' possible contact with Manichaeans (Luna 2001: 490–492; Watts 2005: 292–293; Hadot 2007: 63–75) and do not address the exceedingly tenuous character of Tardieu's evidence for Manichaeans at Ḥarrān at this time in the first place, namely just two texts: (1) The *Zuqnin Chronicle* (also known as the *Chronicle of Pseudo-Dionysius of Tell-Maḥrē*), reports that, in 764–765, the governor of al-Ǧazīra punished Manichaeans who had a monastery one mile from Ḥarrān (Chabot 1927–1933: 224.1–226.3; Harrak 1999: 202–204; see also Harrak 2004); (2) According to John Kyparissiotes in the fourteenth century, anti-Manichaean synods were held by Theodore Abū Qurra, bishop of Ḥarrān, who flourished in the early ninth century (Lamoreaux 2002: 38). These testimonies cannot in any way prove the existence of Manichaeans *exclusively* at Ḥarrān two hundred years before the events in question, especially when is well known that Manichaeans flourished to a degree in the intervening Umayyad period (F. C. de Blois, "Zindīḳ, *EI*²", 11.510a–513b). The supposed exclusivity of Ḥarrān as a site for Simplicius' possible conversation with a Manichaean in the sixth century is untenable and Simplicius' reference to a Manichaean cannot be accepted as evidence for his final habitation.

73. Examples: Stapleton, Lewis, and Sherwood Taylor 1949: 69–70: "The influence of the star-worshipping Sabæans of Harrān, to whom Hermes-Thoth was the god of all civilising inventions, was widespread in the Islamic world." Fück 1951: 111 stated that "there can hardly be any doubt" that Abū Maʿšar's views on different figures named Hermes "reproduced notions prevalent in his time, i.e. about A.D. 860, amongst the so-called Ṣābians of Ḥarrān." Nasr 1967: 71: "Hermeticism was propagated by the Sabaeans [*sic*] who made known to the Islamic world the writings that were attributed to Hermes." Pingree 1968: 10 posited that Abū Sahl ibn Nawbaht's account of the Babylonian Hermes who became king of Egypt was due to direct or indirect Ḥarrānian influence (this is discussed in chapter 2).

Ḥarrānian Ṣābians have often been described as Hermetists. Although there has been no concerted study of the Arabic works of Hermes themselves to establish this idea with certainty, nevertheless several early Arabic reports appear, at least at first glance, to suggest that some Ḥarrānians did have works of Hermes at their disposal. Before scrutinizing the evidence for this in the primary sources, it is instructive to recall an ingenious account suggested by Walter Scott in the first volume of his edition of the Greek and Latin Hermetica. With an understanding of the history of Islamic philosophy that, he said, was based on the (now much-antiquated) works of T. J. de Boer and B. Carra de Vaux,[74] he indulged in what he admitted to be speculation about the Hermetica. The scene of his story is set against the rise of a supposed Islamic hostility to "freedom of thought" in which Ṣābians would not have been able to keep their beliefs and practices:

> Now the time at which the Sabians disappear at Bagdad (c. A.D. 1050) is just about the time at which documents of the *Corpus Hermeticum*, after an interval of five centuries during which nothing was heard of them in Europe, reappear at Constantinople, in the hands of Psellus. Is there not something more than chance in this? It may be that one of the Sabians of Bagdad, finding that his position under Moslem rule was becoming unendurable, migrated to Constantinople, and brought in his baggage a bundle of Greek *Hermetica*—and that our *Corpus* is that bundle. If so, the line along which the *libelli* of the *Corpus* have been transmitted runs through Harran and Bagdad. This is merely an unproved hypothesis; but it is one that agrees well with the facts known to us. The Pagans of Harran almost certainly possessed the whole collection of *Hermetica* (including many documents that are not now extant) in Greek, at the time when they adopted these writings as their Scriptures, in A.D. 830; and there can be little doubt that Thabit, who was a good Greek scholar, still had a copy of them in Greek at the end of the ninth century. During the 150 years which had since elapsed, knowledge of Greek must have almost, if not quite, died out at Bagdad, and the *Hermetica* must have been now read only, or almost only, in Syriac or Arabic translations. But a man such as the Sabian I am supposing would, even if he did not himself know the Greek language, have good reason to preserve with care, and to take with him when he migrated to a place where Greek was spoken, any portions of his Scriptures, in the original Greek, that had chanced to escape destruction and to come into his hands; and it is just such a chance collection of specimens that survive in the *Corpus*.
>
> Moreover, if we choose to indulge in further conjectures, there is nothing to prevent us from supposing that it was the arrival in Constantinople of a few such Sabian Neoplatonists from Bagdad, and the writings brought with

74. De Boer 1901, Carra de Vaux 1900 and 1902.

> them, that first started that revival of Platonic study in which Psellus took the leading part.[75]

Scott added that "[i]t is almost surprising that no extracts or quotations from the *Hermetica*...have been found in Arabic writings."[76] Unfortunately, there is no evidence for any of this. To his great credit, Scott sought very diligently to find parallels between the Greek *Corpus Hermeticum* and Arabic texts known to him in translations in European languages. The result of his investigations, his collection of Arabic testimonia on Hermes, completed by A. S. Ferguson, is still unmatched and very valuable. Scott's speculation, however, also shows how histories of Arabic philosophy such as de Boer's have misled scholars to think that there was an orthodox Islamic persecution of philosophers, which is not exactly correct.[77] Indeed, this is the very period when philosophers in the Graeco-Arabic tradition such as Ibn Sīnā, al-Bīrūnī, Ibn al-Hayṯam, and Ibn Riḍwān were flourishing under different Muslim rulers in many lands. Why would a pagan at that time think that he would be more welcome, and find greater "freedom of thought," in Byzantium of all places?

Scott's surmise also shows how ready historians have been to assume, with little evidence, that the Ḥarrānians possessed a substantial Hermetic literature. In this case, even the *Greek* Hermetica have been imagined as a part of a Ḥarrānian legacy. Although this is not impossible, there simply is no positive evidence for it. The Ḥarrānians can easily be made into secret agents or scapegoats of intellectual history (as with the Tardieu hypothesis) precisely because so little is known about them. In his valuable studies on the ninth-century astrologer Abū Maʿšar, David Pingree developed the theory that "the immediate sources of Abū Maʿshar's Aristotelianism were not the Arabic translations of the *De Caelo*, the *Physica*, and the *De generatione et corruptione*, but the purported writings of the Ḥarrānian prophets, Hermes and Agathodaimon."[78] Pingree's argument for the influence of the Ḥarrānians on Abū Maʿšar is as follows:

> Since the Ḥarrānians were interested in the laws of perceptible nature precisely because they saw the same relationships between the ethereal spheres and the sublunar world of change that Abū Maʿshar seeks to prove (as well as a further relationship between the ethereal spheres and the One which Abū Maʿshar only hints at), it is an easy step to the conclusion that this justification of astrology is, in its main outline, taken by Abū Maʿshar from the books of the Ḥarrānians, and is thus only part of a much more elaborate universal philosophy of emanation.[79]

75. Scott 1.108–109.
76. Idem, 109.
77. For fuller criticisms of this work and others like it, see Gutas 2002: 8–12.
78. Pingree, "Abū Maʿshar," *DSB* 1.33b.
79. Ibidem. Another argument of Pingree for the Ḥarrānians' influence on Abū Maʿšar is discussed critically in the appendix to this chapter.

Beginning with this assumption, Pingree then proceeded to outline what he thought Ḥarrānian cosmology was.[80] However, the influence of the Ḥarrānians on Abū Maʿšar is far from proven. First we must ask, with Hjärpe, which Ḥarrānians could have been responsible for this influence: Ḥarrānians of Ḥarrān itself or of Baġdād? Second, what Ḥarrānian writings do we possess that could demonstrate this influence? Third, where does Abū Maʿšar show his dependence on these works? The answer to the first question is likely to be Ṣābians at Baġdād, perhaps specifically Ṯābit, as Hjärpe advised, in which case their cosmology may bear no real resemblance to that of the "traditional" Ḥarrānians, and may have much more to do with the science of classical authors such as Ptolemy or other astrologers, whose works Ṯābit was translating, studying, and applying. As for the second question, there are no extant cosmological works, and indeed very few works of any kind, definitely identified as concerned with Ḥarrānian Ṣābian religion and belief, so their influence cannot yet be determined.[81] As for Abū Maʿšar's dependence on Ḥarrānian doctrines, there is very little evidence. Al-Bīrūnī tells us of an otherwise unknown work of Abū Maʿšar called *On Houses of Worship*, in which he dealt with the ancient temples of the Ḥarrānians and other pagans;[82] in fact, this appears to be a section of Abū Maʿšar's lost *Book of the Thousands*, not a work in which a Ṣābian cosmology is likely to have been described.[83] Also, a Greek translation of one of Abū Maʿšar's works cites the Ḥarrānians as an authority on prognostications based on the annual

80. He did the same in his article "Abū Maʿšar," *EIr* 1.337b–340a (in a fascicle dated 1983): "(Abū-Maʿšar's) astrological theories, though derived from . . . diverse sources, were founded upon that Neoplatonizing concept of the universe associated with the self-styled Sabeans of Ḥarrān. They hypothesized three levels of being analogous to three concentric spheres: the divine (the sphere of light), the ethereal (the celestial spheres), and the hylic (the sublunar core . . .). The sciences of astrology and theurgy are, in the system, the most useful to man . . . " But where do we find these Ḥarrānian doctrines explained by Abū Maʿšar as Ḥarrānian?

Later in the same article he adds, "Of less importance, but present, are some writings on the talismans used in theurgy and on the planetary temples in which was performed the liturgy of the Harranians [without diacritics now]. Abū Maʿšar asserted that the Sabean astrological religion was the original and true belief taught to the earliest man by God." Which writings are these, and where do they mention Ḥarrānians? It seems more likely that Pingree was thinking of Arabic books attributed to Hermes, some of which Abū Maʿšar surely read. Even then, although Pingree studied some of the Arabic Hermetica in manuscript, his case is not based on published material or indeed any cited manuscript evidence.

81. Chwolsohn's survey was limited to the titles of books of Ḥarrānians, the passages from the *Fihrist* that allegedly came from one of the books of the Ḥarrānians, and the important passage cited by Bar ʿEbrāyā (Ibn al-ʿIbrī) in his Syriac *Chronicle* (Chwolsohn 1856, vol. 2). An important contribution to the literature of Ḥarrānian Ṣābians is found in Rosenthal 1962, who presents two fragmentary apocalypses. Beyond that, nothing else is known for certain. Some of the Arabic Hermetica present a possible Ḥarrānian corpus, but whether these are actually Ḥarrānian or forgeries of works as Ṣābian (pagan, as imagined in the ninth or tenth century) remains to be seen in future studies.

82. al-Bīrūnī, *Āṯār*, 205.15–20, trans. 187.32–42. See also Pingree 2002: 26–29, arguing—I think correctly—that al-Masʿūdī repeated more information from this lost work of Abū Maʿšar.

83. As-Siġzī's *Muntaḫab* of Abū Maʿšar's *Book of the Thousands* mentions that the original work explained "the construction of the great, renowned, famous buildings" (*bināʾ al-abniya al-ʿaẓīma al-madḵūra al-mašhūra*) in terms of historical astrology (British Library Or. 1346, 81 recto 29).

heliacal rising of Sirius.[84] These references are far from sufficient to demonstrate that Abū Maʿšar adopted a Ḥarrānian cosmology. Abū Maʿšar's student Šāḏān listed a *Book of the Ḥarrānian* in a bibliography of works belonging to Abū Maʿšar. Again, there is no special reason to think that this work was devoted to explaining a supposed system of Ṣābian cosmology.[85] In short, the elaboration of a Ḥarrānian cosmology on the basis of Abū Maʿšar's writings, or those of any other, is premature.[86] Such a Ḥarrānian influence, carefully qualified, may eventually be proven in the case of figures such as Abū Maʿšar, and then this view may be vindicated, but, for the moment, the evidence for the alleged influence is too slender for such far-reaching conclusions.

3.4 The Evidence for a Ḥarrānian Ṣābian Transmission of the Hermetica

What, then, is the evidence in primary sources for the Ḥarrānian Ṣābian transmission of the Arabic Hermetica? There are several indications, but most of them do not permit us yet to identify existing Arabic Hermetica as being of Ḥarrānian provenance. What follows here is an analysis of these pieces of evidence, in chronological order, and an evaluation of their collective meaning for the question of the alleged Ḥarrānian Ṣābian Hermetica.

3.4.1 Evidence from a Syriac Antipagan Christian Polemic ca 600 CE

In 1983 Sebastian Brock published the edition and translation of a Syriac work, entitled *Prophecies of the Pagan Philosophers* (*Nbîyawwātâ d-pillôsôpê ḥanpê*), addressed to the "unbaptized Ḥarrānians," which argues that the statements of philosophers esteemed by the Ḥarrānians actually prophesied the truth of Christianity, and therefore that the Ḥarrānians should convert; for if they disbelieved these prophecies, they disbelieved their own teachers, but if they believed them, they believed the truth of Christianity.[87] Brock convincingly dated this Syriac text to the late sixth or early seventh century and suggested that it could have been a part of the brutal initiative ordered by emperor Maurice (r. 582–602) to convert the Ḥarrānians, known from a report handed down by the *Chronicle* of Michael the Elder (d. 1199) and the

84. CCAG 4.124–125. Prognostications on the rising of Sirius were a part of the folklore of most peoples of the ancient Near East and also in other parts of the world. They also formed the subject of one of the most popular Arabic texts attributed to Hermes, *Kitāb Aḥkām ʿinda ṭulūʿ aš-Šiʿrā l-yamānīya* (KŠY).

85. Rosenthal 1963: 456. Rosenthal's prudent comment was, "As usual, the ascription of a given work to a Ḥarrānian allows of no plausible identification." The great majority of Ṯābit's works have nothing to do with especially Ṣābian matters; see Gutas, forthcoming. Given the context of the report, one would expect the *Book of the Ḥarrānian* to be a work of astrology.

86. See also the appendix to this chapter.

87. Brock 1983.

Chronicle ad annum 1234.[88] It is evidence in some way for pagan Ḥarrānians' views as understood by a Christian polemicist at the time of Muḥammad and the advent of Islam, showing that the Ḥarrānians were supposed to respect the views of the Greek philosophers.

Included in its collected pagan sayings are a few passages attributed by name to Hermes Trismegistus. Brock showed that these quotations of Hermes derived from a source held in common with the sixth-century Greek chronicler Malalas of Antioch, and that this common source in turn probably derived these sayings from the extant work known as the Greek Theosophia or the "Tübingen Theosophy," which was written in the second half of the fourth century. Cyril of Alexandria also cited some of the same passages. Clearly a study of the complicated interrelationships of these sources is needed.[89] In any case, these quotations attributed to Hermes in the *Prophecies of the Pagan Philosophers* themselves do not refer directly to any text of the surviving Greek *Corpus Hermeticum*. The only clue directly indicating a known Hermetic text in this collection is a reference not to Hermes, but to Poimandres, a name otherwise known only at the beginning of the Greek *Corpus Hermeticum*, in the very title of CH 1. Under the heading Poimandres, however, the *Prophecies of the Pagan Philosophers* cites CH 13.1–2 (not CH 1). As Garth Fowden has pointed out, this is interesting evidence for the collection of the extant Greek Hermetica together, with the so-called Poimandres at the beginning.[90] Evidently the name Poimandres was already generalized to refer to the whole collection, and it was thought to be relevant for pagans in Syria around the year 600.

The existence of these few passages in a work against the pagans of Ḥarrān gives us evidence that Hermes was considered to be important for the Ḥarrānians, although only as one name of authority among the other ancient philosophers. The occurrence of the sayings of Hermes here does not make the Ḥarrānians "Hermetists," or even show that they definitely had Hermetic books among them. The text shows merely that the Ḥarrānians were regarded as being like other pagans of late antiquity, and that their particular religion was seen as part of the greater legacy of Hellenic paganism. In the same way, the pagan Greek *Anthology* of Stobaeus (early fifth century) could cite Hermetic works extensively, among citations of ancient Hellenic philosophers, without being described as "Hermetic" in character. The *Prophecies of the Pagan Philosophers* also cites Plato, Porphyry, Plotinus, Pythagoras, Orpheus, the local figure Bābā, and others. Given the character of these citations, as Edward Watts has remarked, "There is also no evident expectation that (the) Ḥarrānian audience would have any significant Platonic background" in education."[91] The text thus shows not that the Ḥarrānians definitely had a copy of the Greek

88. Brock 1983: 207–209.
89. Brock 1983: 206–207.
90. Fowden 1993a: 34.
91. Watts 2005: 313–314.

Corpus Hermeticum, but only that they were expected to view Hermetica with the same high regard as they would the works of other pagan sages.[92] It is true that they may have possessed parts of what is today the Greek *Corpus Hermeticum*, but they equally may not have.

3.4.2 *Evidence from Theodore Abū Qurra, Bishop of Ḥarrān (fl. ca 805–829)*

The next association of Hermes with the Ḥarrānians comes more than two hundred years later, long after the establishment of the Arab empire. This is a great gap of time, during which one cannot suppose that the Ḥarrānian pagans' way of life stayed the same. Little is known about Theodore Abū Qurra except that he was Chalcedonian bishop of Ḥarrān for some time until he was deposed around 813. Thereafter he remained an important theologian in Syria, taking part in interconfessional debate and writing works to prove the truth of his religion. He translated pseudo-Aristotle's *De virtutibus animae* for Ṭāhir ibn al-Ḥusayn, governor of Mesopotamia (r. 813–820), and may have engaged in a theological debate in the presence of al-Ma'mūn in 829. He is one of the first Christians to write in Arabic whose works survive.[93]

In one book he describes briefly the followers of "the religion of the ancient pagans" (*dīn al-ḥunafā' al-awwalīn*).[94] This comes first in a series of descriptions of different religions, remarkable for their even-handed treatment, lacking overtly polemical terms. Because Theodore was at one time the bishop of Ḥarrān, he can be assumed to possess some true information about the pagans there: "They claimed that they worship the seven planets—the sun, the moon, Saturn, Mars, Jupiter, Mercury, Venus—and the twelve zodiacal houses, because they are the ones that create and govern this creation and give good fortune and prosperity in the lower world (*ad-dunyā*), and ill fortune and suffering. They said that their prophet in that is Hermes the Sage (*al-ḥakīm*)."[95] Although this description does not mention the city of Ḥarrān by name, its similarity to later descriptions of the Ḥarrānians, combined with the fact that Theodore was once bishop of Ḥarrān, can leave hardly any doubt about the identity of these people. The passage is therefore the earliest known description of the religion of the Ḥarrānian pagans in Arabic, antedating their adoption of the name *Ṣābi'a* as well as the arrival of the influential

92. Hadot 2007: 80–81, esp. 81n130, takes a conflicting view based on the same lack of information: she seems to think that because it is not sound to guess the philosophical competence of the audience on the basis of the text's contents, this is no impediment to her greater argument that the audience in Ḥarrān *might* include real and well-educated Neoplatonic philosophers, in fact an entire academy of them.

93. For this information, see Lamoreaux 2005: xi–xviii; see also Graf 1944–1953: 2.7–26.

94. *Ḥunafā'* (sing. *ḥanīf*) is a term that evolved in complicated ways in Arabic, which I will discuss in more detail below. For now, suffice it to say that Christians used the word in Arabic as an equivalent to the related Aramaic word *ḥanpā*, meaning just "pagan," as shown by de Blois 2002: 19–20.

95. Theodore Abū Qurra 1982: 200.8–201.1; cf. Lamoreaux 2005: 1.

Ṯābit in Baġdād.[96] It is also the earliest known association in Arabic of Hermes with Ḥarrān, where Hermes is explicitly called their prophet "in that," meaning their prophet in the matter of the planets and the zodiac (*inna nabīyahum fī ḏālika Hirmisu l-ḥakīm*).

Later in the same work Theodore claims, as a part of his argument, that every religious community has received a messenger, and that one of these messengers must be the true messenger. For his argument, the pagans needed a prophet to give their religion a status comparable to that of Christianity, Judaism, Manichaeism, and Islam. One might therefore suppose that it was Christians like Theodore who imagined Hermes to be the "prophet" of the pagans in order to make all the religions systematically similar enough so that his argument would work neatly. On the other hand, one might also imagine pagans in Ḥarrān claiming defensively, against members of other religions, that they, too, had a messenger, and resorting to Hermes as the oldest and first. As I will show in chapter 4, some late antique Christian chronicles placed Hermes in the primordial times before the Flood, making him earlier even than Moses. The claim to chronological priority might have had its appeal to pagans of Ḥarrān, who could have made Hermes into their prophet for just that reason.

In Theodore's view, apostles and prophets brought books,[97] so he probably considered one or more books of Hermes as the source of their teaching. Theodore says specifically that Hermes was their prophet "in that." "That" might refer to their worship of the planets, or rather merely to the *justification* of their worship of the planets, the view that the planets bestow good and bad fortune. In the latter case, any astrological work attributed to Hermes would suffice. It is impossible to know for certain what Theodore meant exactly, but his description of the pagans is an important one. It is the first known reference in Arabic to Hermes as a prophet of the Ḥarrānians. The characterization of the Ḥarrānian pagan religion as particularly astral appears regularly in Arabic accounts from this time onward, including a later pagan's own implicit account.[98] The next time that the sources offer a description of the Ḥarrānian religion, it is under the name Ṣābian and shows signs of involvement with the currents of ancient Greek philosophy that characterized Arabic literature in the ninth century.

3.4.3 *Evidence from al-Kindī (d. ca 870), Ibn al-Qifṭī (d. 1248), and Ibn al-ʿIbrī (d. 1286)*

The most important testimony about the possible existence of books of Hermes possibly connected with the Ḥarrānians is from al-Kindī, "the first philosopher of the Arabs," related in a work by his student as-Saraḫsī (d. 899), a tutor of the caliph

96. Theodore's description of the pagans may even antedate Ṯābit's birth in or around 826.
97. Theodore Abū Qurra 1982: 211.12–212.2.
98. See the *ġazal* of Abū Isḥāq Ibrāhīm aṣ-Ṣābiʾ in section 3.5.

al-Muʿtaḍid. A long passage from as-Saraḫsī's treatise entitled *Description of the Doctrines of the Ṣābians* (*Risāla fī Waṣf maḏāhib aṣ-Ṣābi'īn*), describing the Ḥarrānian sect on the authority of al-Kindī (*ʿan al-Kindī*), survives in citation by an-Nadīm (wr. 987), who himself read the work in as-Saraḫsī's own handwriting.[99] It is important to note that this information must belong to al-Kindī's time, the middle of the ninth century, only about a generation after the Ḥarrānians are said to have adopted the name Ṣābian. The religion described is called that of the "Ḥarrā-nian Chaldaeans known as Ṣābians" (*al-Ḥarnānīya al-Kaldānīyīn al-maʿrūfīn bi-ṣ-Ṣābi'a*) and *al-ḥanīfīya*, or "paganism."[100]

Modern scholars have used the description of the Ḥarrānian religion given in this text as one of the basic sources for their beliefs and practices. As several have remarked, however, the Ḥarrānian system is described in strikingly Islamic terms.[101] For example, the Ḥarrānians pray according to a fixed number of bowings and prostrations (*rakaʿāt* and *suǧūdāt*) toward a particular direction (*qibla*) at specific times of the day reckoned by the position of the sun. They have three supererogatory prayers "corresponding to the *witr* of the Muslims," an explicit likening of the Ḥarrā-nian religion to Islam.[102] The Ḥarrānians have a fast (*ṣawm*) of thirty days, as well as special days of fast, they have a feast at the end of their fast called *ʿĪd fiṭr*, which one may suppose came at the end of their month of fasting (like the feast of the same name at the end of Ramaḍān for the Muslims), and they wash themselves after ritual pollution.[103] In all of these areas of the description, the Ḥarrānian religion seems quite like Islam. Although there are considerable differences between the two in the theoretical domain, as described by as-Saraḫsī, the numerous remarkable similarities require some explanation. It must be either that as-Saraḫsī, al-Kindī, or their source consciously portrayed the Ḥarrānian religion in Islamic terms, inventing religious practices that did not in fact exist at that time, or that the Ḥarrānians who informed al-Kindī deliberately developed practices like those of the Muslims or deliberately described them that way. Because there would be no apparent motive

99. The treatise is reported in the *Fihrist* 321.14 (Rosenthal 1943: 40). The whole passage (*Fihrist* 383.7–385.2; ed. Flügel 318.14–320.9) is given in annotated English translation by Rosenthal 1943: 41–51, in comparison with the two abridgements made by al-Muṭahhar al-Maqdisī (writing in 966). Rosenthal also noted later ("al-Sarakhsī," *EI*^{2}, 9.35a–36a) that another fragment of this is given by ʿAbdalǧabbār, *Muġnī* 5.152.

100. For more on the term *ḥanīf*, see section 5.2. Ibn al-ʿIbrī, *Ta'rīḫ*, 153, summarizes this passage from as-Saraḫsī in his historical entry on Ṯābit ibn Qurra, saying, "What I have ascertained about the doctrine of the Ṣābians is that they preach precisely what the ancient Chaldaeans preached" (*anna daʿwatahum hiya daʿwatu l-Kaldānīyīna l-qudamā' bi-ʿaynihā*). The very next report in his history concerns as-Saraḫsī himself. We may suspect therefore that as-Saraḫsī was his source, directly or indirectly, though he does not name him as the author from whom he took the previous passage.

101. Pines 1970: 787, Hjärpe 1972: 40–42, Levy-Rubin 2003: 215–216, and Lane Fox 2005: 234 (citing F. Zimmerman). Levy-Rubin points out that the Andalusian scholar Ibn Ḥazm (d. 1064) noted the similarity of the "Ṣābian" practices to those of Islam.

102. Rosenthal 1943: 43–44.

103. Rosenthal 1943: 46–47.

for Muslims to portray Ḥarrānians in such terms, whereas Ḥarrānians would surely have had reason to do so, the latter appears to be the only real possibility.[104]

The Ḥarrānian view of the world as described in this report is based entirely on Aristotelian philosophy (not, significantly, on works of Plato or any Neoplatonist). Their views on various subjects were said to correspond to those found in Aristotle's books, including the following: *Physics*, *De caelo*, *De generatione et corruptione*, *Meteorologia*, *De sensu et sensato*, *Metaphysics*, and *Posterior Analytics*. Perhaps the queries of the philosopher al-Kindī into the religion of the Ḥarrānians elicited these sorts of responses from an accommodating informant eager to please him.[105] Yet because we can be sure that the Ḥarrānian religion was assimilated to Islam, the élite religion in Baġdādī society, insofar as it was described with specifically Islamic terms, it is also likely that the Ḥarrānians claimed the most prestigious philosophical system of the ninth century, the system that so many scholars of that time, including Ṯābit of Ḥarrān himself, were engaged in recovering through translation, study, and commentary, as their own.[106] It must have been obvious, and obviously advantageous, for pagans to claim the ancient pagan philosophers as a part of their heritage.

The Ḥarrānian Ṣābians are thus portrayed in terms that would have been highly appealing to many Muslim intellectuals of the ninth century. In fact, al-Kindī's description makes them seem like nothing other than Aristotelian Muslims with special astrological interests, and this is perhaps the main point of his description. Such a picture can have derived only from a deliberate attempt to make the Ḥarrānian religion as acceptable as possible to the élite of the Caliphs' capital in that

104. Hjärpe 1972: 40–42 argued that this was not at all due to a "fraude consciente," but rather the natural change that the Ḥarrānian religion would undergo in the environment of Baġdād, dominated by Muslim culture. Although Hjärpe's arguments are strong, I find it hard to believe that such changes could have come about so quickly that already al-Kindī could report them, unless there had been a conscious effort on the part of Ḥarrānian informants to make their religion more acceptable to Muslims. Strohmaier's suggestion (1996: 54), that the similarity of the two religions might be accounted for by supposing that it was Muḥammad who imitated the practice of the Ḥarrānian pagans, cannot be accepted.

105. Rowson 1988: 242 pointed out the similarity between the list of works said here to be authoritative among the Ṣābians and a similar list at the end of the Pseudo-Aristotelian *Kitāb at-Tuffāḥa* (Latin *De pomo*) (in reality an adaptation of the *Phaedo* of Plato: Gutas 1986: 31); the latter list also included works of Aristotle and the "Book of Hermes" where knowledge of the principles, *uṣūl*, is to be sought. This led D. S. Margoliouth 1892: 190 to suppose that a Ḥarrānian Ṣābian wrote the *De pomo*. On this see Gutas 1988: 45–46n39. I discuss the *Kitāb at-Tuffāḥa*, and its references to Hermes, in chapter 5.

106. This would mean that al-Kindī's description of the Ḥarrānians could not be used as evidence for a Platonic school in Ḥarrān in the sixth century. Most recently Pingree 2002: 15 argued in favor of Tardieu's view on the matter, adding, "The Ṣābians before Thābit almost never translated Aristotle's works into either Syriac or Arabic, though they used them in composing their own treatises in those two languages." Where does this information come from? For there are no books known for certain as both Ḥarrānian and Ṣābian extant in either Syriac or Arabic, from before the time of Ṯābit, citing the works of Aristotle listed by al-Kindī.

period. The survival of the Ḥarrānian Ṣābian family of Ṯābit in the court of Baġdād for so long testifies to the degree to which they succeeded in doing so. Given that as-Saraḫsī corresponded at some point with the Ḥarrānian Ṯābit ibn Qurra, it is likely that Ṯābit himself was a source for both al-Kindī and as-Saraḫsī.[107]

Like Theodore Abū Qurra, this passage says that Hermes (*Hrmys*) was one of the Ḥarrānian Ṣābians' famed luminaries; here Hermes is not specifically a prophet, and he is listed together with Agathodaemon, a mysterious figure called *'r'ny*, and Solon.[108] But new information about an actual work of Hermes appears at the end of the extract from as-Saraḫsī in the *Fihrist*, credited explicitly to al-Kindī: "Al-Kindī said that he examined a book which that people [the Ḥarrānians] regard as authoritative. It is *Chapters of Hermes on the Doctrine of God's Unity* which he wrote for his son most expertly on God's unity.[109] Philosophers who have exerted themselves will find no alternative to them and to professing their doctrine."[110] Modern scholars have appropriately tried to guess which books these might have been. Chwolsohn suggested that this book was the same as the Syriac work mentioned by Ibn al-ʿIbrī in his Arabic chronicle (discussed below) and that it could be the Greek *Corpus Hermeticum* itself.[111] Scott suggested that these treatises could have been the Γενικοὶ λόγοι (*General Discourses*) known from internal references in the Greek Hermetica themselves or the Διεξοδικοὶ λόγοι (*Detailed Discourses*) mentioned by Cyril of Alexandria along with the Γενικοὶ λόγοι.[112] The Arabic word used, *maqālāt*, can refer to individual treatises or sections of a book, like chapters, again calling to mind the collected treatises of the

107. As-Saraḫsī and Ṯābit almost certainly knew each other in person. An-Nadim (*Fihrist* 321.12–13) attributes to as-Saraḫsī a "Book of his correspondence in answering to Ṯābit ibn Qurra the questions he was asked about" (*Kitāb Risālatihī fī ǧawāb Ṯābit ibn Qurra fīmā su'ila 'anhu*). Ibn al-Qifṭī, 118.4, mentions Ṯābit's answer to a *Kitāb* of as-Saraḫsī. It is unclear which of these is prior to the other. In another recorded anecdote, Ṯābit tutored, and became a close friend of, the future caliph al-Muʿtaḍid when the latter was under house arrest; al-Muʿtaḍid was also at some time as-Saraḫsī's student. In such circumstances the two scholars can be assumed to have known each other quite well (Ibn Abī Uṣaybiʿa 1.216.5–13; cf. Rosenthal 1943: 22–23 and 103–104n234).

108. *Fihrist* 383.12–13. For more on Urānī, see section 5.3.

109. *At-tawḥīd*, strictly meaning monotheism or unity of God, is often a synonym for (Islamic) theology (cf. Gardet, "ʿIlm al-Kalām," *EI*², 3.1141b–1150b). It is possible that the title of the work is to be construed *Chapters of Hermes on (Monotheistic) Theology Which He Wrote for His Son Most Expertly on Theology*. Compare some of the titles of sections in the extant Greek *Corpus Hermeticum* that bear similar long descriptions in their headings.

110. *Fihrist* 385.1–2 (Flügel 320.7–9): *qāla l-Kindīyu innahū naẓara fī kitābin yuqirru bihī hā'ulā'i l-qawmu wa-huwa l-maqālātu li-Hirmisa fī t-tawḥīdi katabahū li-bnihī 'alā ġāyatin mina t-taqānati fī t-tawḥīdi lā yaǧidu l-faylasūfu iḏā at'aba nafsahū mandūḥatan 'anhu wa-l-qawli bihā*. The English translation is mine; I have consulted the previous English translations by Rosenthal 1943: 51 and Dodge 1970: 2.750, and the German translation by Chwolsohn 1856: 2.13–14.

111. Chwolsohn 1856: 2.126. Chwolsohn apparently knew the Greek *Corpus Hermeticum* by the name *Poemander*, properly the title only of the first treatise in the *Corpus*.

112. Scott 4.250n2; Cyril *Contra Julianum* 553A, 588B; CH 10.1, 10.7, 13.1; SH 3.1, 6.1; Mahé 1984 B.6. For a bibliography of views on these "General Discourses," see Copenhaver 1992: 155. For a discussion of possible uses of the *General Discourses* in their original context, see also Fowden 1993a: 97–99.

Greek *Corpus Hermeticum*.[113] Because there is no reason to doubt the testimony of al-Kindī, one can assume that in the ninth century he had obtained a work attributed to Hermes in Arabic (the only language he could read). Interestingly, al-Kindī does not say that this text is of Ḥarrānian *origin*, even though he could have just as easily said so. Instead he claims merely that they consider it authoritative (*yuqirru bihī*). Al-Kindī or one of his associates could have come across such a book anywhere in the search for ancient Greek manuscripts to be translated. In order to learn that the book was authoritative among the Ḥarrānian Ṣābians, all that al-Kindī would need to do is to ask Ṯābit or another Ḥarrānian about the ancient work of Hermes he had obtained. They would presumably have no reason to reject an ancient work of their prophet.

What is practically certain is that al-Kindī had before him either a text from the known Greek Hermetica or something related to them, a text from Hermes to his son on a theological subject.[114] In conclusion, al-Kindī's testimony alone suffices to indicate that a translation of the Greek *Corpus* or a text related to it was available in Arabic in the ninth century. Depending on how one interprets al-Kindī's words, however, the Hermetic work in question may not have had any special status in Ḥarrān. It could have been acquired anywhere, and any Ḥarrānian, considering Hermes to be his prophet, might have declared it authoritative upon al-Kindī's inquiry. Then again, it is possible to speculate that al-Kindī did indeed receive the book from a Ḥarrānian. We simply do not have any way of knowing. In any case, it is not surprising that such treatises do not survive today. The absence of other references to such works in Arabic literature probably means that these Hermetica did not find the audience that al-Kindī had thought they deserved, and that they were lost in Arabic at an early stage, along with so much other Arabic literature of the ninth century.[115] Although the existence of this testimony does not justify exaggerated talk about the influence of "Hermetism" in Arabic, it does, however, entirely warrant studies like that of Charles Genequand comparing the *Corpus Hermeticum* and al-Kindī's *Fī n-nafs*.[116]

113. In the ninth century, *maqāla* is documented as translating a variety of words for *book*, *treatise*, or *chapter*: βιβλίον, ὑπόμνημα, σύγγραμμα, and γράμμα (in the sense of chapter heading) (Ullmann 2002–2007: 1.163 and 185, S2:356 and 553).

114. A new Hermetic discourse from Hermes to his son was discovered in Coptic translation among the Nag' Ḥammādī texts (i.e., Nag' Ḥammādī VI.6, *The Ogdoad Reveals the Ennead*). This, along with the fragmentary state of the Greek *Corpus*, proves that the Greek collection is incomplete.

115. For a list of titles of lost early Arabic works (including a work attributed to Hermes) serving as examples of how "our bibliographical information of the older period—that is, the early ninth century—is still quite incomplete," see Rosenthal 1963: 454–456.

116. Genequand 1987–1988. Unfortunately, he found only one real parallel passage between the *Corpus Hermeticum* and al-Kindī's *Fī n-nafs*, and that parallel does not prove direct borrowing or influence. The shortcoming of Genequand's useful study is rooted in its attempt to connect al-Kindī's treatise on the soul mostly with general intellectual tendencies poorly defined rather than specific works (cf. his note 40, p. 11: "It is not always possible to establish clear-cut distinctions between Platonism, Hermetism, and some forms of Gnosis.") Similarities between al-Kindī's psychology and the

Much later, in the thirteenth century, the testimony of Ibn al-Qifṭī gives a similar and more detailed report on the existence of Eastern versions of the texts of the Greek *Corpus Hermeticum* or texts similar to them. He wrote, "Portions of his [i.e., the third Hermes'][117] scrolls have been translated. They are part of his chapters to his disciple Ṭāṭī by way of question and answer between the two. They lack organization and continuity because the original was decayed and divided into separate parts."[118] Tat is familiar from the Greek *Corpus Hermeticum* as the most frequently mentioned disciple of Hermes Trismegistus. In the Greek discourses, Tat addressed Hermes as father and Hermes addressed Tat as his child.[119] Evidently, then, this also is a reference to texts like the Greek *Corpus Hermeticum*, a collection of texts known to us in Greek in a somewhat tattered form that could be described accurately by Ibn al-Qifṭī's words. Ibn al-Qifṭī is probably citing an earlier author, one of many from whom he compiled his own work. When Ibn al-ʿIbrī copied this passage verbatim from Ibn al-Qifṭī in his discussion of Hermes (in this case, referring to the first Hermes), he added tantalizingly, "The manuscript is extant with us in Syriac."[120] This might be taken to imply also that Ibn al-ʿIbrī did not think it survived in Arabic.

Neither Ibn al-Qifṭī nor Ibn al-ʿIbrī mentions the Ṣābians or the Ḥarrānians in connection with this text. They bring it up only in their short biographies of Hermes. Syriac, however, was the language not only of Christians but also of the pagans of Ḥarrān and of the early Ḥarrānians of Baġdād.[121] Hermetic discourses in Syriac might be due either to the Ḥarrānian Ṣābians or to Christians; there is no way to be certain. Ibn al-ʿIbrī was able to list some of the Syriac works allegedly by Ṯābit ibn Qurra in his Syriac *Chronicle*, and even to cite a Syriac passage from one of them as an example of Ṯābit's eloquence. On the other hand, there is precedent for Christian Syriac authors taking an interest in the Hermetica for their apologetic purposes.[122] One may assume that al-Kindī's copy of the discourses of Hermes was translated from a Syriac version; if so, this would suggest that Ibn al-ʿIbrī's copy might also be of Ḥarrānian origin. Unfortunately, there is no more

---

Greek Hermetica are demonstrated, but no direct or definite connection is established through this method. Adamson 2000: 112–119 and 2002: 193–197 show numerous parallels rather between al-Kindī's psychology and the Plotinian *Theology of Aristotle*. For an early effort like that of Genequand, but involving different Arabic philosophers, see Affifi 1951.

117. On the three Hermeses, see chapter 4.

118. Ibn al-Qifṭī, 349.23–350.2: *wa-nuqilat min ṣuḥufihī nubaḏun wa-hiya min maqālātihī ilā tilmīḏihī Ṭāṭī ʿalā sabīli su'ālin wa-ǧawābin baynahumā wa-hiya ʿalā ġayri niẓāmin wa-wilā'in li-anna l-aṣla kāna bāliyan mufarraqatan.*

119. For example, in CH 10 passim.

120. Ibn al-ʿIbrī, *Ta'rīḫ*, 12.1–3, adding the words *wa-n-nusḫa mawǧūdatun ʿindanā bi-s-suryānīya.*

121. I know of no good indication of when Syriac ceased to be used by the Ḥarrānians in Baġdād.

122. For example, Jacob of Edessa in his *Hexaemeron*, who considered the words of Hermes to be in accord with Christianity, although he apparently knew Hermetic passages only through Cyril of Alexandria's *Contra Julianum* (Jacob of Edessa, *Hexaemeron*, 149b11–150a17; cf. CH 4.148–141, frag. 32).

evidence than this concerning these collections of Hermetica. Rather than speculating further, it would be wiser to accept the evidence for what it is and to keep the possibilities in mind. Even though these reports of dialogues between Hermes and Tat do exist in Arabic, there is no trace of them known today nor has any other report about them come to light. Although they surely existed, they must have been rare.

3.4.4 Evidence from Abū Maʿšar (d. 886), al-Bīrūnī (d. ca 1050), and Others

The famous astrologer Abū Maʿšar said that the Ḥarrānians claimed the wisdom (*ḥikma*) of the first Hermes as their own.[123] Abū Maʿšar's legend of the three Hermeses was the main source on Hermes for later authors and forms the special subject of chapter 4. In light of al-Kindī's contemporary report, discussed above, one may perhaps take this as a reference to Hermetic books among the Ḥarrānians. Similar references are found in many other works after Abū Maʿšar. An often-cited example appears in the masterpiece of chronological science, the *Kitāb al-Āṯār al-bāqiya* of al-Bīrūnī, which quoted Abū Maʿšar in a discussion of the Ḥarrānians.[124] Although Hermes the Egyptian is listed there as one of the prophets of the Ḥarrānians, including Agathodaemon, Valens,[125] Pythagoras, Solon,[126] and Bābā the Ḥarrānian,[127] this list does not really give us information about any *works* of Hermes suspected to have been held among the Ḥarrānians. It shows little more than the Syriac *Prophecies of the Pagan Philosophers*, discussed above in section 3.4.1, which attributed to the Ḥarrānians a high esteem for several of the same Greek sages.

3.4.5 Evidence from Ṯābit ibn Qurra (d. 901) or Sinān ibn Ṯābit (d. 942/943)

Ṯābit ibn Qurra, the father and chief of the Ḥarrānian Ṣābians of Baġdād, is recorded to have written several works in Syriac relating to his religion. This is in addition to more than one hundred and fifty scientific and philosophical works that he wrote for a more general learned audience in Arabic. Among the Syriac titles are works dealing with burial, Ṣābian doctrine, ritual purity, allegorical language, the appropriate and inappropriate animals for sacrifice, the times of worship, and the

123. *wa-huwa llaḏī taddaʿī l-Ḥarrāniyyatu ḥikmatahū*, Ibn Ǧulǧul, 5.7.

124. al-Bīrūnī, *Āṯār*, 205.20; trans. 187.44. It is likely that this "book" is actually a chapter of Abū Maʿšar's *Kitāb al-Ulūf*, which al-Bīrūnī knew and quoted in his *Āṯār*.

125. Reported in Sachau's edition (318.16) as *Wālīs*, this is Vettius Valens, the astrologer whose work was translated into Middle Persian and Arabic, where it was highly influential.

126. Aside from being one of the Seven Sages of ancient Greece, Solon figures prominently as the source of allegedly Egyptian lore, the tale of Atlantis, in Plato's *Timaeus*. This may have some bearing on the discussion below on the reports of al-Masʿūdī.

127. On Bābā see Rosenthal 1962.

orthopraxis of prayer. None of these works, doubtless part of Ṯābit's own interpretation and formulation of the Ṣābian religion, are known today.[128]

Two thirteenth-century sources, one Arabic and one Syriac, mention another work entitled *The Laws of Hermes and the Prayers that the Ṣābians Use*. Both agree in attributing it to a Ṣābian of Baġdād, but they disagree about precisely which one. Ibn al-Qifṭī says that the physician Sinān, son of Ṯābit ibn Qurra, translated into Arabic a work entitled *The Laws of Hermes and the Verses and Prayers that the Ṣābians Use in Prayer* (*Nawāmīs Hirmis wa-s-suwar wa-ṣ-ṣalawāt allatī yuṣallī bihā ṣ-Ṣābi'ūn*).[129] Because Ibn al-Qifṭī copied this information from Sinān's great-grandson, al-Muḥassin (d. 1010), who wrote lists of the works of both Ṯābit and Sinān, we can reasonably expect his information to be correct.[130] The list attributes to Sinān several more works also bearing interesting titles, including a history of the kings of the Syrians and a treatise on the historical narratives of his fathers and ancestors.[131] These, too, are otherwise unknown today.

This information about Sinān's translation of *The Laws of Hermes* would be acceptable without hesitation if it were not for a passage in Ibn al-ʿIbrī's Syriac *Chronicle*. Here Ibn al-ʿIbrī lists the titles of the Syriac works of Ṯābit. He begins with the same titles of Syriac works given by al-Muḥassin in Arabic, but adds to them several of the titles of works attributed by al-Muḥassin (as cited by Ibn al-Qifṭī) rather to Sinān. The work in question here is entitled in Syriac as *The Laws of Hermes and His Prayers that the Pagans Use in Prayer* (*Ktābā d-nāmōsē d-Hermīs wa-ṣlawwāteh hānēn da-bhēn mṣalleyn ḥanpē*), corresponding almost exactly with the Arabic title mentioned above and indubitably the same work.[132] Ibn al-ʿIbrī lists all of these as works of Ṯābit and, importantly, he presents them in just the same order as given by al-Muḥassin. He prefaces this information by saying, "He (Ṯābit) wrote in Syriac about sixteen books, the majority of which I have seen and acquired."[133] Therefore he may well be claiming to have a copy of *The Laws of Hermes* in Syriac, along with many of the others in the same list. Notably he does not say that he actually read these works, nor does he mention Sinān here at all.

Daniel Chwolsohn did not find a discrepancy between the two sources. He concluded that the Arabic title listed by Ibn al-Qifṭī refers to Sinān's translation of a work

128. Ibn al-Qifṭī 120.14–19 = Ibn Abī Uṣaybiʿa 1.220.25–29.

129. Ibn al-Qifṭī, 195.12–13. *Wa-naqala ilā l-ʿArabī Nawāmīs Hirmis*... = Ibn Abī Uṣaybiʿa, 1.224.16–17.

130. Ibn al-Qifṭī, 195.6–7 = Ibn Abī Uṣaybiʿa 1.224.8–9. See also Gutas, forthcoming, "The Works of Ṯābit b. Qurra."

131. Ibn al-Qifṭī 195.14 = Ibn Abī Uṣaybiʿa 1.220.9–10, 15–16.

132. Bar ʿEbrāyā (Ibn al-ʿIbrī), *Syriac Chronicle*, trans. Budge 152–153. Note the equivalence of Syriac *ḥanpē*, "pagans," with Arabic *Ṣābi'a*. For a similar example see Gutas 1988, esp. 37–47. Both Syriac *ḥanpē* and Arabic *Ṣābi'a* were equivalents for Greek *Héllēnes*. On the development of the Aramaic term *ḥanpā* and Arabic *ḥanīf*, see de Blois 2002: 16–25, summarized in section 5.2.

133. Bar ʿEbrāyā (Ibn al-ʿIbrī), *Syriac Chronicle*, Budge 152, Bodleian MS Hunt. 52 facsimile 55r. b.17–19: *akteb a'k 16 ktābē men hānōn da-ḥnan l-sugāhon ḥzayn wa-qnayn*. Cf. trans. Budge 152.

originally composed in Syriac by Ṯābit, attested by Ibn al-ʿIbrī.[134] It appears, however, that Ibn al-ʿIbrī, or his source, has simply added together parts of two different lists deriving from the same earlier source, al-Muḥassin. He has mistakenly listed works of Sinān, including *The Laws of Hermes*, as belonging to his father Ṯābit. This is confirmed by his relating them in the same order, something extremely unlikely if he had an independent source of information (i.e., a list not derived from al-Muḥassin). This being the case, we have to regard Ibn al-ʿIbrī's explicit statement that he knew these "works of Ṯābit" in Syriac as an exaggeration. Moreover, he, or his source, must have translated the titles of works by Sinān *from Arabic into Syriac*. The Arabic titles are original.

We are left then with no reason to doubt al-Muḥassin's report, stating that Sinān translated *The Laws of Hermes and the Prayers that the Ṣābians Use* into Arabic—but we still do not know who originally composed it. Although it is conceivable that his father, Ṯābit, was indeed the original author, and that Ibn al-ʿIbrī and Chwolsohn were fortuitously correct, we just do not know for certain. In any case, Sinān is responsible for the Arabic text: he is the interpreter of something supposedly preexisting in another language, that language surely being Syriac. One may therefore accept al-Muḥassin's report (via Ibn al-Qifṭī) as evidence of some kind of pamphlet or book of laws of Hermes among the Ṣābians of Baġdād in the first half of the tenth century.

Unfortunately, there is no clue as to its contents apart from what the title tells us. Different possibilities occur: it could have been based on older Ḥarrānian traditions or even older Hermetic works, if one wants to posit the existence of such a tradition, but it also might also be entirely Sinān's independent creation. The limited evidence does not help in deciding. Yet *The Laws of Hermes* deals with the Ṣābian religion (their laws and their prayers) and is definitely attached to a Ṣābian of Baġdād in the early tenth century. This strongly suggests that it was part of a self-conscious formulation of Ṣābianism. That it was translated into Arabic might suggest that it was intended for non-Ṣābian readers, a possibility to be addressed again later in this work.

3.4.6 *Evidence from al-Mubaššir ibn Fātik (wr. 1048/1049)*

A description of Hermes and his teachings is preserved in the collection of wise sayings made by al-Mubaššir ibn Fātik in Fāṭimid Egypt, *Kitāb Muḫtār al-ḥikam*. These passages are treated extensively in sections 5.2 and 5.3 in this volume, but a brief summary here will help to make this survey of testimonia about Ḥarrānian Hermetica complete. Al-Mubaššir's source describes Hermes as a prophet teaching pious commandments in the form of maxims, as well as an outline of rules for Hermes' religion and his wise advice. Although al-Mubaššir's treatment of Hermes and his instructions includes no direct references to Ḥarrānians or to Ṣābians in general, the religion taught by Hermes in this account is similar to as-Saraḫsī's description of the Ḥarrānian Ṣābian religion: it included feasts at astrological conjunctions and at the sun's entry into a new zodiacal sign, as well as sacrificial offerings to the planets at the

134. Chwolsohn 1856: 2.i–v, esp. v, n16.

appropriate times. Hermes is also said to have commanded them "to perform prayers that he stated for them in ways that he described." On the other hand, the religious laws of Hermes given here bear close resemblance to Islamic law: they require ritual purity, abstinence from intoxication, *ǧihād* against the enemies of the religion, alms (*az-zakāt*), and prescribe most of the punishments called *ḥadd* punishments in Islamic law. All this leads me to conclude that the "religion of Hermes" described here was developed and described well after the establishment of Islam and Islamic law. Again, this analysis will be elaborated in sections 5.2 and 5.3.

Al-Mubaššir relates the wise sayings not only of Hermes, but also those of Ṭāṭ immediately following. Here one may immediately acknowledge an ancient source for the connection between the two names, Hermes and Tat. It is interesting to note that in one branch of the stemma of the manuscripts of al-Mubaššir's text, the name Ṭāṭ is replaced by Ṣāb.[135] The published text of al-Mubaššir says, "The Lessons of Ṭāṭ, who is Ṣāb son of Idrīs—peace be upon him. The pagans (*al-ḥunafāʾ*) are named for him, so they are called the Ṣābians (*aṣ-Ṣābiʾūn*)."[136] Already in 956, al-Masʿūdī referred in passing to Ṣābī, "for whom the Ṣābians were named."[137] One section of the teachings of Hermes is made up of his testament to King Ammon, another name associated with Hermes in the ancient Greek Hermetica. The original compiler of these sayings of Hermes and Tat, some time before al-Mubaššir, must have had at least a bit of information—nothing necessarily substantial—derived ultimately from the ancient Greek Hermetica in order to make the association between the names Hermes, King Ammon, and Tat.

Other than the Ḥarrānians, there was no religious community that claimed Hermes as one of its special prophets, and for this and other reasons, to be elaborated in section 5.3, I conclude that al-Mubaššir's life and maxims of Hermes probably derive from a tenth-century Ṣābian Ḥarrānian living at the court in Baġdād. As the reader will see, I tentatively propose Sinān ibn Ṯābit or his less well known grandson, Isḥāq aṣ-Ṣābiʾ, as that author, and *The Laws of Hermes* (just discussed in section 3.4.5 above) as a candidate for al-Mubaššir's source.

3.4.7 Evidence from al-Masʿūdī (d. 956) and the Arabic Reception of Porphyry's Letter to Anebo *and the* De mysteriis Aegyptiorum

Al-Masʿūdī visited Ḥarrān personally, and therefore Tardieu and those who have followed him have regarded al-Masʿūdī as the most important source on the Ṣābians of Ḥarrān.[138] One ought to keep in mind, however, that there is no particular

135. Rosenthal's handwritten notes, 66.

136. al-Mubaššir 26.4–5: *Ādāb Ṭāṭ, huwa Ṣābu bnu Idrīsa ʿalayhi s-salāmu wa-ilayhi nusiba l-ḥunafāʾi fa-qīla lahumu ṣ-Ṣābiʾūn*. See also Ibn Abī Uṣaybiʿa, 1.215.26–27, who used al-Mubaššir's work: "It is said that the Ṣābians are named for Ṣāb, who is Ṭāṭ, son of the prophet Idrīs—peace be upon him."

137. al-Masʿūdī, *Tanbīh* 91.3–4. He mentions two genealogies that have been proposed: Ṣābī ibn Methuselah ibn Idrīs and Ṣābī ibn Mārī.

138. Tardieu 1986: 13.

reason to think that al-Masʿūdī's information actually derived entirely from his stay in Ḥarrān. Like most learned tourists, he must have acquired some of his information from books as well, and if he has mixed book-learning with his first-hand reports, one must be cautious in using his description of the Ṣābians.[139]

Along with Abū Maʿšar al-Balḫī, al-Bīrūnī, al-Mubaššir, and other authors, al-Masʿūdī names Hermes as one of the prophets of the Ṣābians, in a list with Agathodaemon, Homer, Aratus, the first and second Urānī,[140] and others.[141] He says in his *Murūǧ ad-ḏahab* that the Ṣābians claim Enoch to be Hermes, and that Hermes means ʿUṭārid (Mercury).[142] The report is mixed with traditions about the prophethood of Enoch that derive ultimately from the time of Wahb ibn Munabbih, traditions concerning the number of scrolls revealed to each prophet.[143] None of this is original.

But al-Masʿūdī is an early and important witness to the generalization of the term *Ṣābi'a* to refer to ancient pagans of all kinds. In one passage dealing with the Ḥarrānians as Ṣābians, al-Masʿūdī distinguishes the Egyptian Ṣābians as a particular group of the ancient Ṣābians, which also include the Greek pagans and Buddhists. Significantly, al-Masʿūdī says that the Ḥarrānians are the surviving remnant of the *Egyptian* Ṣābians and refers to their differing directions of prayer and dietary restrictions, particularly that the Ḥarrānians do not eat the same foods as the ancient Greeks.[144]

139. Pingree's discussion (2002: 26–29) of al-Masʿūdī's description of the temples at Ḥarrān in his *Kitāb Murūǧ ad-ḏahab* suggests that he might have been using a work of Abū Maʿšar in his description of that place. What al-Masʿūdī actually says is, "Abū Maʿšar the astrologer has made mention of the temples and great buildings whose construction took place in the world in each millennium in his book entitled *The Thousands*. The disciple of Abū Maʿšar, Ibn al-Māzyār, also reported it in his book *Selected Passages from the Thousands*. Others who came before and after them have made mention of many of the buildings and wonders in the world. I have avoided making mention of them [scil. the wonders, not the authors]." *Murūǧ* (Pellat §1419 = 2.407.9–13). This certainly does seem to indicate that al-Masʿūdī used the lost works of these two astrologers, but it also implies that he used other sources as well in the same passages.

140. This name is discussed in chapter 5.

141. al-Masʿūdī, *Tanbīh* 121.16–122.1.

142. al-Masʿūdī, *Murūǧ* (Pellat §62 = 1.42.16–43.4) under *ḏikr al-mabda' wa-ša'n al-ḫalīqa*, entry: Uḫnūḫ. ʿUṭārid is merely the Aramaic word for the planet Mercury (Hermes).

143. Discussed in section 4.6.

144. al-Masʿūdī, *Tanbīh*, 161.10–15: *wa-š-Šamanīyata wa-hum Ṣābi'atun li-Ṣīna wa-ġayrihim wa-hum ʿalā maḏāhibi Būdāsafa wa-ʿawāmmi l-Yūnānīyīna wa-tawaǧǧuhuhum fī ṣalātihim ilā l-mašriqi wa-Ṣābi'ata l-Miṣrīyīna llaḏīna baqīyatuhum fī hāḏā l-waqti Ṣābi'ū l-Ḥarrānīyīna wa-tawaǧǧuhahum fī ṣalātihim ilā t-tayammuni wa-huwa l-qiblatu wa-stidbārahumu š-šamāla wa-mtināʿahum min kaṯīrin mina l-ma'ākili llatī kāna Ṣābi'atu l-Yūnānīyīna ya'kulūnahā ka-laḥmi l-ḫinzīri wa-l-firāḫi wa-ṯ-ṯūmi wa-l-bāqillā wa-ġayri ḏalika.* "[We also mentioned] the Šamanīya [Buddhists], who are Ṣābians [= pagans] of the Chinese and others, and who follow the doctrines of Bodhisattva; and the common Hellenes [as opposed to philosophers], their orientation in their prayer toward the east; and the Egyptian Ṣābians [= pagans] whose remainder is at this time the Ṣābians [= pagans] of the Ḥarrānians, their orientation in their prayer toward the south, which is [their] *qibla*, with their backs turning toward the north, and their abstention from many of the foods that the Greek Ṣābians [= pagans] would eat such as pork, chicken, garlic, fava beans, and other foods..."

Charles Genequand has suggested that the reference to Egyptian Ṣābians reflects al-Masʿūdī's awareness of "the Egyptian origin of Hermetism."[145] It seems rather that the reference has nothing directly to do with ancient Hermetism. Instead, al-Masʿūdī provides us with its meaning when he mentions elsewhere in the same passage "the questions and answers that transpired between the Syrian Porphyry… and the Egyptian priest Anebo." This is a reference to two ancient Greek works: Porphyry's *Letter to Anebo*, which survives today only in fragments, and the work known as *De mysteriis Aegyptiorum*, written in response to it in the name of Anebo's Egyptian teacher, Abammon—who is generally accepted to be in reality the philosopher Iamblichus, writing under this guise.[146] The extant fragments of the *Letter to Anebo* deal with the subject of Egyptian religion in a reasoned critique of theurgy, posing challenging theoretical questions about a number of pagan religious practices from a philosophical point of view; the *De mysteriis* offers a thoroughgoing

---

As-Saraḫsī's description of the Ḥarrānian Ṣābians says that fava beans (*bāqillā*) and garlic were forbidden for the Ḥarrānians (*Fihrist* 384.14). The injunction against fava beans is apparently of Pythagorean origin. There is some evidence that, even in the late tenth century, Ṣābians of Ḥarrānian origin did indeed follow these dietary restrictions and take them seriously. An interesting anecdote reported by Abū Manṣūr aṯ-Ṯaʿālibī (d. 1038) in his *Kitāb Yatīmat ad-dahr*, 2.242.10–14 (in part 2, *bāb* 3, in the section *fī ḏikr Abī Isḥāq aṣ-Ṣābī wa-maḥāsin kalāmihī*) on the authority of Abū Naṣr Sahl ibn al-Marzubān (d. ca 1030) has it that Abū Isḥāq Ibrāhīm ibn Hilāl aṣ-Ṣābiʾ (925–994), a great-grandson of Ṯābit ibn Qurra, refused to eat fava beans when they were served with a meal at the house of al-Muhallabī (i.e., Abū Muḥammad al-Muhallabī, vizier of the Buyid amīr Muʿizz-ad-dawla, whose chief secretary Abū Isḥāq Ibrāhīm was). When urged to join in eating them, Abū Isḥāq aṣ-Ṣābī is reported to have said that he did not wish to disobey God's command (not to eat them), whereupon he was excused from that part of the meal.

The *Nabataean Agriculture* has one of its Aramaic sources, Yanbūšād, claiming that "Hermes (*Irmīsā*) and before him Agathodaemon (*Aġātādaymūn*) forbade fish and fava beans to the people of their country, and forbade it very strongly…" It goes on to say that although Hermes and Agathodaemon forbade this, the Egyptians used to grow fava beans anyway for medicine (Ibn Waḥšīya 1.499.12–13 and 1.500.3–4). The *Nabataean Agriculture*, exhibiting here an Aramaic form of the name Hermes, perhaps points to a wider familiarity among Mesopotamian Aramaeans with Hermes and such "pagan" dietary laws than might otherwise be expected (cf. Hämeen-Anttila 2006: 181 and 199).

145. Genequand 1999: 118.

146. al-Masʿūdī, *Tanbīh*, 161.2–3, 162.5–12: *wa-ḏakarnā fī Kitābi l-Maqālāti fī uṣūli d-diyānāti wa-Kitābi Ḫazāʾini d-dīni wa-sirri l-ʿālamīna… mā ǧarā bayna Furfuriyūsa ṣ-Ṣūrīyi ṣāḥibi Kitābi Īsāġūǧī fī l-mudḫali ilā kitābi Arisṭāṭālīs fī l-manṭiqi wa-kāna naṣrānīyan yanṣuru maḏāhiba Ṣāʾibati l-Yūnanīyīna muḫfiyan li-ḏālika wa-bayna Anābū l-kāhini l-Miṣrīyi wa-kāna yanṣuru l-falsafata l-ūlā llatī kāna ʿalayhā Fūṯāġūrasu wa-Ṯālīsu l-Milaṭīyu wa-ġayruhumā wa-hiya maḏhabu Ṣābiʾati l-Miṣrīyīna min al-masāʾili wa-l-ǧawābāti fī l-ʿulūmi l-ilāhīyati wa-ḏālika fī rasāʾila baynahum maʿrūfatin ʿinda man ʿuniya bi-ʿulūmi l-awāʾili wa-mā kānū ʿalayhi mina l-ārāʾi wa-n-niḥali.* "We recorded in *Chapters on the Principles of the Religions* and *The Treasuries of Religion and the Secret of the Ages* [two books by al-Masʿūdī not extant today]…the questions and answers on the divine sciences that transpired between Porphyry of Tyre—author of the *Eisagoge* on the introduction to Aristotle's book on logic—he was a Christian, defending the doctrines of the Greek Ṣābians [= pagans] to hide that fact—and Anebo the Egyptian priest, who was defending the first philosophy, which Pythagoras, Thales the Milesian, and others followed—that is, the doctrine of the Egyptian Ṣābians [= pagans]. This is in epistles between them [among them? NB: plural, not dual] that are well known among those who are concerned with the sciences of the ancients and with the doctrines and the sects to which they adhered." (French translation of this passage and further references in Kraus 1942–1943: 2.128n5.)

rejoinder from the adopted point of view of an Egyptian priest.[147] Therefore it seems highly likely that the reference in this passage to "Egyptian Ṣābians," of whom the Ḥarrānians are supposed to be a remaining part, and "Greek Ṣābians" (Ṣābian here clearly meaning just "pagan") has only to do with the discussions found in these "questions and answers," meaning these two works, mentioned in the very same passage by al-Masʿūdī—and does not have anything to do directly with Hermes or Hermetica. Al-Masʿūdī's identification of the Ḥarrānian pagans with the Egyptian pagans in the context of these works, which, in their discussions, explicitly connect Egyptian religion with astrolatry, is indeed striking, but there is nothing to suggest that the information here derives from the Ḥarrānians themselves.[148] More interesting is al-Masʿūdī's statement that Pythagoras, Thales, and others followed the way of the Egyptian Ṣābians. This probably echoes the *De mysteriis*, which emphasizes that sages like Pythagoras and Plato formed philosophy out of what they took from the stelae of the Egyptian Hermes.[149]

Unfortunately, there are no more than these interesting hints and questions, because we do not know al-Masʿūdī's source. Yet there is a clue. With the next stroke of his pen, al-Masʿūdī writes that many books were composed in accordance with the doctrines of the Pythagoreans and in defense of them. He adds that someone else who wrote about "that"—meaning a subject just mentioned: either the *Letter to Anebo*, the Ṣābians, or the Pythagoreans, or perhaps all of these things—was Abū Bakr Muḥammad ibn Zakarīyā' ar-Rāzī (d. 925 or 935), the famous physician of notoriously heterodox views.[150] Then al-Masʿūdī immediately digresses to explain

---

The interesting claim that Porphyry was once a Christian derives ultimately from a late antique Greek church history. Socrates of Constantinople says in his *Ecclesiastical History* (3.23, written ca 440), that Porphyry had renounced Christianity in a rage after having been beaten by Christians at Caesarea. The source for this is named as Eusebius of Caesarea. A derivative of this story is repeated in the *Tübingen Theosophy*. (On these passages, see Schott 2008, 274–276.) The account made its way into Arabic probably by way of a Syriac translation. Thus we find in a disorderly passage in the history of al-Masʿūdī's contemporary, the Christian Agapius of Manbiǧ (wr. ca 940s), perhaps derived from the lost chronicle of Theophilus of Edessa (d. 785), that Porphyry was a Christian but apostasized (Agapius 1.2.183 = *Patrologia Orientalis* 11.1, 45.6-9): "Porphyry—commentator on the Book of Aristotle, who was one of the deacons of the church of Tyre, because he was boasting proudly over all his fellow-deacons, so they fell upon him one day on the altar and struck him a painful blow, so he went out, left Christianity, opposed the Messiah, dissembled, and refuted the Gospel—said . . . " Al-Masʿūdī's report alleging that Porphyry was formerly a Christian must derive from a Christian chronographic source. What may have begun as figurative, "intellectual" blows against Porphyry have by this time been taken unambiguously as physical blows.

147. Porphyry, *Epistola ad Anebonem*, ed. A. R. Sodano, 1958.

148. The *De mysteriis* (1.17–19) explains the gods in astrological terms, and states that the visible planets are gods. This might have prompted an identification of the Ḥarrānian religion with the ancient Egyptian religion. Perhaps also related to this problem are the Arabic Hermetica that relate texts allegedly discovered at different sites of ancient Egyptian ruins. Two of these have been edited in Vereno 1992.

149. *De mysteriis* 1.1–2. Pythagoras was listed by al-Bīrūnī as one of the prophets of the Ḥarrānians. See 3.4.4.

150. L. E. Goodman, "al-Rāzī, Abū Bakr Muḥammad b. Zakariyyā'," *EI*², 8.474a–477b.

that ar-Rāzī is the one who wrote the *al-Manṣūrī fī ṭ-ṭibb* (apparently this medical work, although written just three decades before, was already well known),[151] and next, for no apparent reason, begins to give information about the *Timaeus* of Plato and its Arabic translations and commentaries, and other works on the ordering of the world.

It may be possible to explain these digressions and to shed some light thereby on the sources on ancient paganism at the disposal of Arabic scholars in the tenth century. Ar-Rāzī wrote a work dealing with Porphyry's lost *Letter to Anebo the Egyptian* in some way. According to one list of ar-Rāzī's works, he wrote a refutation of Porphyry's *Letter*; according to another list, it was a refutation of the *response* to Porphyry's *Letter*, that is, it would seem, a refutation of the *De mysteriis Aegyptiorum*.[152] The latter is more likely, as this would explain al-Masʿūdī's knowledge of the *answers* to Porphyry's questions. Al-Masʿūdī's connected reference to ar-Rāzī as someone who wrote on the subject of this letter suggests that ar-Rāzī, and perhaps specifically this refutation by ar-Rāzī, was at least one of al-Masʿūdī's sources on the subject.[153] As it also happens, the heresiographer aš-Šahrastānī claims to quote from Porphyry's *Letter to Anebo* in his *Kitāb al-Milal wān-niḥal* (*Religions and Sects*) under the heading of Porphyry's philosophy. The passages cited, however, are not translations of any of the fragments of the *Letter* extant in Greek. Rather, this Arabic citation is actually entirely the same as one of the surviving fragments of Porphyry's *Commentary on Plato's Timaeus*![154] One may suppose, then, that somehow Porphyry's *Commentary on Plato's Timaeus* became mixed up with or circulated together with his *Letter to Anebo*, and that this confusion must have happened by the time of ar-Rāzī, so that when al-Masʿūdī used ar-Rāzī's work as a source on the Ṣābians, he began also to include information about commentaries on the *Timaeus* because he associated the two with one another. Yet the contents of the real *Letter to Anebo* or the *De mysteriis* must have been known in some form in Arabic, for otherwise the reference to "Anebo the Egyptian" in the title of ar-Rāzī's work and al-Masʿūdī's discussion of Egyptian Ṣābianism could not be explained. Beyond these suggestions, the problem resists definitive conclusion for now because too many of the texts are missing. Although al-Masʿūdī can provide important clues about the development

151. Al-Masʿūdī says that ar-Rāzī's *al-Manṣūrī* was written in AH 310 (i.e., 922–923 CE). Al-Masʿūdī stated just a few pages preceding (109.8) that this work of his was written in AH 345 (i.e., 956–957 CE).

152. The first list is that reported by al-Bīrūnī, *Fihrist-i Kitābhā-yi Rāzī*, 14, no. 128: *Naqḍ Kitāb Furfūriyūs ilā Anābū al-Miṣrī* (*Refutation of Porphyry's Book to Anebo the Egyptian*). Cf. Kraus 1942–1943: 2.128n5. The second is in the *Fihrist* (358.9; ed. Flügel 300.17–18): *Kitāb fī naqḍ kitāb Anābū ilā Furfūriyūs fī šarḥ maḏāhib Arisṭālīs fī l-ʿilm al-ilāhī* (*On the Refutation of the Book of Anebo to Porphyry, on the Analysis of the Doctrines of Aristotle on the Divine Science*).

153. A future study should investigate the surviving writings of ar-Rāzī for evidence of his familiarity with the *Letter to Anebo* or the *De mysteriis*.

154. As pointed out by Jolivet and Monnot in their translation. aš-Šahrastānī, *Livre* 2.357–358n2. Cf. Porphyry, *In Platonis Timaeum*, ed. Sodano, 119–122.

of medieval Arabic interpretations of paganism, he does not offer any real evidence for works of Hermes or the Arabic understanding of "Hermetism."

3.4.8 Evidence from the Ġāyat al-ḥakīm, *the* Kitāb al-Isṭamāḫīs, *and Mūsā ibn Maymūn (d. 1204)*

The notorious eleventh-century Andalusian book of magic *Ġāyat al-ḥakīm* contains, in the seventh *faṣl* of its third *maqāla*, a long passage describing the Ṣābian ritual practices for "procuring the powers of the planets."[155] Dating the contents of the passage is hindered by its apparently compilatory nature; it has several sources.[156] The extract begins with a reference to an astrologer named "Ṭabarī" of unknown identity. Pingree discovered and published a unique Latin translation of the lost Arabic source from which the passages of this "Ṭabarī" are derived. This latter work itself claims to have been found among the books of Hermes and transmitted by an Apollonius, philosopher of Egypt.[157] Another work cited in this section of the *Ġāyat al-ḥakīm* is the physician ar-Rāzī's *Kitāb al-ʿIlm al-ilāhī* (*Theology*). Ar-Rāzī died ca 930, which would place the compilation of the contents of this section, at the earliest, in the middle of the tenth century.[158] A third work cited is a certain *Muṣḥaf al-ḥunafāʾ* (*Codex of the Pagans*) otherwise unknown.[159]

155. *istiǧlāb quwā l-kawākib* (*Ġāyat al-ḥakīm* 195.2–3). The *faṣl* occupies pages 195–228 of the Arabic edition.

156. Much of the information summarized here was brought together by Burnett 1986: 87.

157. Latin translation edited in Pingree 1992: *Hic est Liber planetarum inventus in libris Hermetis. Transtulit eum Appolonius* [sic] *philosophus Egipti*. In the article presenting this text, Pingree argues that the author of the work may be Abū Ḥafṣ ʿUmar ibn al-Farruḫān aṭ-Ṭabarī, an important early ʿAbbāsid court astrologer (fl. ca 762–812), but concluded that the matter is open to some doubt (idem, 111). In the Arabic *Ġāyat al-ḥakīm* itself (195.2) the name is given without the definite article plainly as "Ṭabarī," although the German of Ritter's and Plessner's translation gives it with the definite article, calling its absence an error (trans. 206.12n4). Pingree's edition of the Latin version of the work, based on the unique manuscript, gives the reading *Athabary* (112.15), perhaps pointing to the presence of the definite article at one time.

However, the Latin translation of the Arabic text from which this author was cited in the *Ġāyat al-ḥakīm* gives his name as *abuelabec altanarani*. Pingree recognized that these letters would most likely represent "Abū l-ʿAbbās aṭ-Ṭabarānī" (with the easy confusion of the Arabic letters *bāʾ* and *nūn*), but because no astrologer by that name is known, he made an elaborate argument (Pingree 1992: 107–111) that this author was truly Abū Ḥafṣ aṭ-Ṭabarī and that "the translator [into Latin] and the scribes were fully competent to commit the sequence of errors that would pervert the latter into its troubled Latin form" (idem, 110). I am not convinced by the argument. In a later publication, thinking that this conclusion was secure, and that the evidently distorted name did in fact refer to ʿUmar ibn al-Farruḫān aṭ-Ṭabarī, Pingree characterized the practices outlined in this passage as the earliest evidence in Arabic for Ḥarrānian ritual syncretism (Pingree 2002: 21–22). A major problem with this view is that "aṭ-Ṭabarānī's" treatise, at least as it is known in its Latin form, contains no references either to Ṣābians or to Ḥarrānians. Pingree recognized this (1992: 107), but argued that the similarities between the rituals described in this text and those later described as Ṣābian established the Ḥarrānian identity of these early ones. The possibility rather of *non-Ḥarrānian* Ṣābian practices is considered below.

158. Pingree 1992: 105–106.

159. *Ġāyat al-ḥakīm*, 205.15. Pingree 1992: 106 mistranscribes the title and translates it in a misleading fashion as "book of the true believers." Pingree 2002: 22 gives the title erroneously as *Muṣannaf al-ḥunafāʾ*, translated now strangely as *Book of the "Old Believers."* On the complicated evolution of the term *ḥunafāʾ*, see de Blois 2002, and chapter 5 in this volume.

Arabic works of Hermes extant today in manuscript, but not mentioned in the text by name, are also cited in the description of Ṣābian rituals in the *Ġāyat al-ḥakīm*. Plessner discovered correspondences between passages here and in the (still unedited) Arabic Hermetic work *Kitāb al-Usṭuwwaṭās*.[160] Another correspondence had already been discovered decades earlier, in which the *Ġāyat al-ḥakīm* gives the names of the spirits of the planets that the Ṣābians used. The nine names of Jupiter are the same as those appearing in the Arabic (also still unedited) Hermetic work *Kitāb al-Isṭamāḫīs*.[161]

At first glance, it might seem that the correspondences between passages in Arabic works of Hermes and the words of the "Ṣābian" rituals should lead us to attribute the *Kitāb al-Isṭamāḫīs* to Ḥarrānians, either those of Ḥarrān itself or, more probably, those of Baġdād.[162] If this conclusion were correct, then at last there would appear a seemingly definite link between an extant Arabic Hermetic text and Ḥarrānian Ṣābians.[163] Because the *Kitāb al-Isṭamāḫīs* is related to a number of other pseudo-Aristotelian Hermetica in Arabic, including *Kitāb al-Usṭuwwaṭās*, *Kitāb al-Hādīṭūs*, and *Kitāb al-Isṭamāṭīs*, which even share some passages among each other, the link would become a series of connections and an entire complex of Arabic Hermetica would find a home with the Ḥarrānians.[164] If one adds to this the testimony of Mūsā ibn Maymūn, who in his *Kitāb Dalālat al-ḥā'irīn* (*Guide for the Perplexed*) lists the *Kitāb al-Isṭamāḫīs* as one of the books of the Ṣābians, the link would become stronger still.[165] Finally, the appearance of a book named "The Secrets of the Ṣābians" (*Kitāb Asrār aṣ-Ṣābi'a*) in Leiden Arabic manuscript 1235, containing also the *Kitāb al-Usṭuwwaṭās*, would seem to offer further evidence of the connection.

However, a serious problem impedes these conclusions: who are the "Ṣābians" described in the *Ġāyat al-ḥakīm*? It has already been shown several times in this chapter that references to Ṣābians, especially those coming from the tenth century or later, frequently indicate nothing more specific than "pagans" as imagined by Arabic intellectuals in this time. They do not necessarily refer to Ḥarrānians in every case. But there *are* references to Ḥarrān and things Ḥarrānian toward the end of this section that would indeed seem at first to link them with the Arabic works of Hermes

160. Noted by Pingree 1992: 106: see Ritter and Plessner's 1962 German translation "Picatrix" 224n4, 227n1, 229n3, and 230n2, where references to MS Paris 2577 and the *Buch über die Pneumata* indicate the *Kitāb al-Usṭuwwaṭās*.

161. De Goeje and Dozy 1885. These findings are briefly summarized in Burnett 1986: 87.

162. This conclusion was most recently repeated by Pingree 1992 passim, and 2002: 31.

163. Though Pingree 1992: 107 and 2002: 30–31 argue fairly convincingly that these are *Ṣābian* books, questions remain as to whether they are *Ḥarrānian* in origin (rather than merely *Ṣābian* in the sense "pagan" as the idea was understood in ninth-century Baġdād) and even what these books actually say.

164. A table of correspondences between a couple of these texts was published by Burnett 2001a: 184.

165. Maimonides, *Kitāb Dalālat al-ḥā'irīn*, 588.9–10: *wa-min tilka l-kutub Kitāb al-Isṭamāḫīs al-mansūb li-Arisṭū*. Note, however, that in his discussion of the Ṣābians, Maimonides did not make particular mention of Ḥarrān or Ḥarrānian Ṣābians. He was interested rather in the primordial religion, that is, "paganism" as understood from a Jewish point of view (Stroumsa 2004: 341–352).

just mentioned. This is in fact the only part of the whole *Ġāyat al-ḥakīm* containing direct references to Ḥarrān, named here as the place of the ritual testing and initiation of young men.[166] Dayr Kādī, a location near Ḥarrān appearing in other Arabic descriptions of Ḥarrānian festivals, is named a few pages later as the site of a ritual involving a decapitated boy whose head is placed on an altar where it howls; its howls were used to predict the future of the Ṣābian people.[167] Moreover, the pagan god *Mārā Samyā* (Aramaic: the Blind Lord) is named;[168] he is known also from the *Fihrist* in a passage describing the Ḥarrānian religion.[169]

These references, however, occur toward the very end of this section of the *Ġāyat al-ḥakīm*, which, as has been shown, is a compilation from several sources. Most of the *faṣl* consists of descriptions of rituals for sacrificing to the planets, as just described, including the names of the planets given correctly according to various ancient languages, among them Persian, Greek, and even Sanskrit.[170] But neither Ṣābians nor Ḥarrānians are mentioned in the source attributed to "Ṭabarī" in its surviving Latin version. Nor do the extant Arabic Hermetic sources of this passage, the *Kitāb al-Isṭamāḫīs* and the *Kitāb al-Usṭuwwaṭās*, seem to contain direct references to Ḥarrānians (though they lack editions and almost any studies of their contents to ascertain this conclusively).[171] On the other hand, the direct references to things Ḥarrānian come only in the last four pages of the *faṣl*, after the phrase *hāḏā maḏhab aṣ-Ṣābi'īn*, "This [the preceding rituals] is the doctrine of the Ṣābians [i.e., pagans]." This brief sentence is a marker indicating that one excerpt has come to an end and another is about to begin. The subsequent passage describes the kinds of sacrifices and offerings they make and blames them for the practice of child sacrifice.[172] One may assume, therefore, that the end of this compilation on the Ṣābians,

166. *Ġāyat al-ḥakīm* 226.9.

167. Idem 228.16–17: "They bear the head (of the decapitated youth) to Dayr Kādī, and then they put it on top of an altar and it howls unbearably. They conclude from its howl whether the number of the Ṣābians will increase or decrease and whether rule will come to them or not..." On this and other stories concerning rituals involving decapitated heads, see Green 1992: 178–180. Dayr Kādī (also Kādā and Kadā) appears as the site of Ḥarrānian festivals in Bar Bahlūl's *Kitāb ad-Dalā'il* (267.11–268.4) and in an-Nadīm's *Fihrist* (ed. Flügel) 322.9–10, 325.4–5, 325.26–28 (deriving reports from Abū Sa'īd Wahb ibn Ibrāhīm the Christian and another, anonymous source). It is also mentioned by al-Bīrūnī, *Āṯār*, 320.19 and 321.1 (deriving reports from Muḥammad ibn 'Abdal'azīz al-Hāšimī).

Reeves 2004 has given the most satisfying solution yet to the problems in the numerous reports by Jews, Christians, and Muslims of members of other religions using decapitated heads in ceremonies. He shows with high likelihood that these reports are distorted accounts that originated to defame Manichaeans, who used an image of their prophet Mani's head in their annual *Bēma* festival. See also Harrak 2004.

168. *Ġāyat al-ḥakīm* 226.5.

169. *Fihrist* 389.23 refers to *ar-rabb al-a'mā al-Mirrīḫ*, the Blind Lord Mars.

170. Pingree 1989b: 10 argued that the Sanskrit names in this passage "conclusively prove(d) that the Ḥarrānians were influenced by India in their planetary worship." This seems to me extremely unlikely; see the appendix to this chapter.

171. The fullest summary of their contents is Burnett 2001a ("Liber Antimaquis").

172. *Ġāyat al-ḥakīm* 225.8–17.

with its different type of contents, comes from an altogether different source, perhaps one related to the other Arabic descriptions of the Ḥarrānian calendars, with which these last pages share some information.[173] In that case, the connection between Ḥarrān and the rest of the chapter, which, as just seen, was derived from sources making no mention of that city or its people, is effectively severed. At the very least, it makes it impossible for one to assume that the planetary rituals are connected in a common *Ḥarrānian* origin.

If the ritual offerings to the planets found here are not Ḥarrānian, what could their origin be? Indian, Greek, and Persian elements have been recognized in the Arabic Hermetica that served as sources for the planetary rituals. Charles Burnett pointed out that such a combination of pieces "fits in with the cosmopolitan culture of scholars serving the early ʿAbbāsid caliphs in Baġdād."[174] On the other hand, Pingree wrote, "It is appealing to think that the... Hermetic treatises disguised as letters from Aristotle to Alexander were forged by a Ṣābian [of Ḥarrān] after that sect had declared Hermes to be its prophet...; but, of course, they could have forged some of them before that date and still attributed their doctrines to Hermes. The contents and circumstances of the composition of these treatises are in need of a comprehensive investigation."[175] It is still uncertain whether the Ḥarrānian Ṣābians who moved to Baġdād are responsible for any of this literature. Because, as argued above, "Ṣābian" came to mean nothing more than "pagan," a member of the primordial religion as imagined by Arabic scholars of the early tenth century and later, these astral rituals may be nothing more than one author's attempt to realize or recreate what that primordial religion really was. The author need not have been from Ḥarrān in order to have an interest in the rituals of paganism. This is demonstrated by al-Kindī's *De radiis*, a work explaining ritual magic in philosophical terms; al-Kindī was no Ḥarrānian but he exhibited such an interest. Therefore it is possible to think that others, not Ḥarrānians, wrote these works in an effort to devise a pagan—"Ṣābian"—system. In other words, this "Ṣābianism" may have become for Arabic scholars nothing more than a scholarly experiment, even a fad, of the ninth and tenth centuries, joining together diverse elements in an attempt to reconstruct a rationale for the pagan thought as recently resurrected and newly important, and which had already been authorized by its Qur'ānic name. In a time and place when ancient pagans like Socrates, Hippocrates, and Aristotle now enjoyed the highest prestige as masters of truth obtained through reason, while theological and interconfessional debates were foremost in the minds of scholars in all fields, it would be surprising if no one tried to investigate the

173. Perhaps it is even to be viewed as the kind of antipagan Christian polemic that Hjärpe suspected in descriptions of Ḥarrānian religion.

174. Burnett 1986: 87.

175. Pingree 2002: 31. Burnett 2007 offers evidence that Ṯābit ibn Qurra used talismans like those described in the Pseudo-Aristotelian Hermetica.

religion of the ancient philosophers.[176] The Arabic Hermetica explaining talismans and astral magic claim to have been related from ancient books by Aristotle himself, the most prestigious of them. For now, the sources of classical Arabic scholars' understanding of paganism, and paganism's partial rehabilitation by certain Muslims as Ṣābianism, must await a separate, fuller, investigation. In any case, we are left without a securely identified, definitely Ḥarrānian tradition of Hermetic books. Therefore there are no known grounds for considering the Ḥarrānians to be "Hermetists."

3.5 The End of the Ḥarrānian Ṣābians

Any attempt to locate an alleged Hermetic literature of Ḥarrānian origin must take the chronology of their religious community into account. While historians have generally focused on the survival and continuity of the Ḥarrānian pagans *into* the ninth and tenth centuries, surprisingly little discussion, since Chwolsohn, has been raised about their history *in* this period and their eventual demise.[177] Yet any supposed Ḥarrānian Ṣābian literature in Arabic, Hermetic or not, must of course have been composed or translated while the Ṣābians of Ḥarrān still existed. To establish a chronological framework for future discussion—including section 5.3, where the topic is resumed—it is important to determine when, and under what circumstances, the Ḥarrānian Ṣābian religion flourished and disappeared.

There must have been strong social pressures against the Ḥarrānian Ṣābians to assimilate, both among Christians and Muslims, and this would likely have been felt most acutely when Ṣābian individuals in Syria ventured outside of their communities or had to deal with non-Ṣābians in legal disputes. One must suppose that their numbers slowly dwindled in size relative to the overall population, as did other non-Muslim communities of the region from the eighth century onward, Jewish, Christian, and Magian. The smaller the community, the more vulnerable they must have been, and the Ḥarrānian Ṣābians were surely the smallest. Living in important urban centers like Ḥarrān and Baġdād, they were probably more conspicuous in the eyes of Muslims and Christians than the pagans persisting in the countryside.

After al-Ma'mūn, whose threat reportedly induced them to claim to be the Qur'ānically sanctioned Ṣābians, the earliest indication that their status was a problem in dealing with Muslims is from a century later in a report on the life of Ṯābit ibn Qurra's son, Sinān.[178] Sinān was a physician for three consecutive caliphs: al-Muqtadir

176. A parallel can be found in the magicians of Renaissance Europe, described by Yates 1964, who thought that they were recovering the primordial religion in the form of the Hermetica and Kabbala in an era when ancient texts were being published and studied anew in great numbers. On the prestige of the Greek philosophers in this period, see 1.3 above and Gutas 1998.

177. Chwolsohn 1856: 1.657–676; Tardieu 1986: 11–12.

178. The report was written within about a half century of Sinān's death: *Fihrist* 359.28–360.1 = Ibn al-Qifṭī 190.13–16.

(r. 908–932), in whose reign he became Chief of the Physicians of the caliph, al-Qāhir (r. 932–934), and ar-Rāḍī (r. 934–940). The story has it that al-Qāhir, during his brief and notorious reign, urged Sinān to convert to Islam. Sinān declined repeatedly (*fa-mtana'a mtinā'an kaṯīran*). Then the caliph intimidated him, so that Sinān became afraid and actually did convert. Fearing more threats, however, he fled to Ḫurāsān, eventually returning to Baġdād, presumably only after al-Qāhir was deposed. Al-Qāhir was remembered for his brutality and vindictive behavior and Sinān was not his only victim. Even his vizier, Ibn Muqla, felt compelled to flee from the court, eventually inducing one of the Turkic regiments to overthrow him. Al-Qāhir was imprisoned and blinded within a short time.[179]

Muslim jurists relate a report amplifying the context of this incident with Sinān.[180] In 933 the caliph al-Qāhir went to the Šāfiʿī jurisconsult Abū Saʿīd al-Iṣṭaḫrī (d. 940), the *walīy al-ḥisba* in Baġdād at the time, and asked him for a *fatwā*, or legal ruling based on an interpretation of the sources of Islamic authority, on the status of the Ṣābians.[181] Al-Iṣṭaḫrī ruled that the Ṣābians should be put to death because they clearly differed from the Jews and Christians and worshipped the planets. But al-Qāhir soon dropped the matter upon receiving "a lot of money" (*māl kaṯīr*) from the Ṣābians. It is safe to assume that the caliph's pressure on Sinān ibn Ṯābit to convert, and Sinān's flight soon after, was connected to this event. Al-Qāhir's threat, however, was evidently short-lived, because the Ṣābians persisted in Baġdād for generations thereafter, many of them successful in the court. Therefore this did not amount to a new general policy of persecution of the Ḥarrānian Ṣābians. It was rather a money-making scheme of al-Qāhir, perhaps also designed also to win the support of a certain religious faction, all at the expense of a small and vulnerable group.

Yet al-Iṣṭaḫrī's *fatwā* against the Ṣābians remained in the books of jurists for centuries, long after the Ṣābians of Ḥarrān had ceased to exist. Probably some Muslims continued to feel justified in mistreating the Ṣābians in the meanwhile. There is an indication of an ambience of general harassment implicit in an edict decades later from the caliph aṭ-Ṭā'iʿ (r. 974–991), guaranteeing the rights and protections of *ḏimmī*s (licit non-Muslims who pay the poll tax) to the Ṣābians of Ḥarrān, ar-Raqqa, and Diyār Muḍar (the latter two being a town near Ḥarrān and

179. D. Sourdel, "al-Ḳāhir," *EI*² 4.423b–424a.

180. an-Nawawī (d. 1277) 1977, 2.239.3–5; cf. Chwolsohn 1856: 2.541. According to Massignon (1950: 651–652), the event is described in as-Siǧzī's astrological compendium *al-Ǧāmiʿ aš-Šāhī*; but he cites no manuscript, section title, or folio number; I have not been able to verify it. Chwolsohn (2.561) further gives a similar report from Abū ʿAbdallāh Muḥammad ibn Aḥmad al-Anṣārī al-Ḫazraǧī al-Andalusī, saying that al-Iṣṭaḫrī declared their *kufr* because they believed in the influence of the stars and that the stars had agency (i.e., were *faʿʿāla*). Both al-Maḥallī and Zaynaddīn Abū Yaḥyā Zakarīyā' ibn Muḥammad al-Anṣārī mention the *fatwā* (Chwolsohn 1956: 2.579 and 2.582).

181. The *walīy al-hisba*, or *muḥtasib*, is an official in charge of regulating the markets and public life in general according to the morality of Islamic law.

the district around them). The edict is directed at government officials who are reminded that the Ṣābians should remain free from injury of any kind, retain their inheritances, and keep the freedom both to attend their places of worship and to fulfill the duties required by their religious customs. The existence of the document implies that the Ṣābians had suffered some abuse at the hands of officials and others in the mid-tenth century, presumably the kind of mistreatment that can be expected against a people having a marginal status as a minority of disputed legality.[182] If the specific issues addressed by the edict are an indication, the Ṣābians in and around Ḥarrān were, at least sometimes, deprived of their inheritances, interfered with and perhaps even attacked while traveling, and prevented from using their temples and holy sites. An official edict was needed to ensure their security. However, the very fact that the caliph himself issued this edict shows that all Muslims were expected to tolerate the Ṣābians; indeed it is directly stated in the text.[183]

The edict survives in the collected epistles of Abū Isḥāq Ibrāhīm ibn Hilāl aṣ-Ṣābi' (d. 994), Sinān ibn Ṯābit's grandson, and also the chief secretary in Baġdād at different times of his life; he himself wrote the edict in the caliph's name. There are a number of anecdotes hinting at the largely amicable experience of Ḥarrānian Ṣābians living and working among the Muslims of the caliphs' court. This is probably because the Ṣābians there conformed to the norms around them and made their differences with others inconspicuous. For example, about this secretary Ibrāhīm ibn Hilāl aṣ-Ṣābi' we read, "He used to be good company for the Muslims and would provide the highest service for the magnates. He used to join them in fasting in the month of Ramaḍān and would memorize the Qur'ān so that it came readily to the tip of his tongue and the bit of his pen."[184] It is likely that the success of the Ṣābians in Baġdād and, in the case of the edict just discussed, the influential position of Ibrāhīm the Ṣābian chancellor were instrumental in ensuring the rights of the Ṣābians in Syria, in and around Ḥarrān.

For the most part, later Arabic authors found the Ṣābians of Baġdād noteworthy for their literary or scientific accomplishments; usually in discussions of their lives and works, their peculiar religion receives only brief and passing attention, no more than in the case of Jewish or Christian scholars working in Arabic. The exception is the heresiographers and theologians, for whom the Ṣābian doctrines were a matter of difficulty and who present us with partisan accounts of their religion.[185] Even

182. I use the term *minority* here with reservations, not wishing to import all of its modern senses. I refer primarily to their smaller numbers and their conspicuous differences with those around them.

183. Reproduced by Chwolsohn 1856: 2.537–538. De Goeje and Houtsma 1888–1907: 1.186–188 show that this edict occurs the section of Ibrāhīm's letters entitled *Mā nufiḏa ilā l-ʿummāli wa-l-mutaṣarrifīna wa-n-nawāḥī*, "Communications to the Governors, Administrators, and Districts." For a fuller account of the letters of Abū Isḥāq Ibrāhīm ibn Hilāl see Hachmeier 2002 (where the letter under discussion is listed as no. 319—pp. 248 and 269—but not edited).

184. aṯ-Ṯaʿālibī 2.241.20–242–1.

185. Hjärpe 1972 remains the most developed analysis of these discussions. See section 3.1 in this volume.

here, they do not address individual Ṣābians by name; they treat only the theological problems of Ṣābianism as they understood it. Nevertheless, the Ṣābian kind of paganism, in this late permutation and understood in the terms explained above, seems to have been simply an accepted religion in the court during the tenth century, even if, as seen in the *fatwā* of al-Iṣṭaḫrī and al-Qāhir's short-lived persecution, there were some predatory people who saw opportunities for personal advantage in making trouble for them in the name of religion.

One of the surviving poems of Abū Isḥāq Ibrāhīm ibn Hilāl indirectly discusses the status of the Ṣābians.[186] The poem is a *ġazal*, or love song, addressed to a young woman. The following translation makes no claim to capture any hint of its musicality:

> Every person, Muslim and confederate,[187] has in you the justest witness to his religion.
>
> When the Muslims see you they are certain of the *ḥūrī*s of the gardens in everlasting ease.
>
> When the Christians consider you to be a gazelle fawn stretching its full moon, bending on a branch,[188]
>
> they praise their trinity and testify about you, since you united three in one.[189]
>
> When the Jews see your shining forehead, they will say to the repudiator and denier of their religion:
>
> "This is the splendor of the All-Merciful, [as] when he showed it to his interlocutor Moses, the God-worshipping prophet."
>
> The Magians see your face shining white under black curls like a pool of darkness,
>
> so between the darkness there and the light here stand proofs that they make ready for every opponent.
>
> You become their morning sun; how many of them bend in prayer to you and prostrate at the darkness!
>
> The Ṣābians, holding that you are unique in beauty, confess a glorious single one (*fard*).
>
> You are for them like shining Venus when she is fortunate with Jupiter and Mercury.

186. aṯ-Ṯaʿālibī, *Yatīmat ad-dahr*, 2.259.7–19. Chwolsohn 1856: 2.540 gives the three verses on the Ṣābians.

187. *muʿāhad*, synonymous with *ḏimmī* (cf. Lane 1863–1893: 5.2184a).

188. This line is a case of multiple *istiʿāra* (roughly, "metaphor;" see S. A. Bonebakker, "Istiʿāra," *EI*², 4.248b–252b). The gazelle fawn frequently stands for the beloved in ancient Arabic poetry. Here its face is likened to the full moon, a normal description for the beloved's face; the branch is the fawn's slender body.

189. This also can refer cleverly to the three images combined metaphorically in the previous line: the fawn, the moon, and the branch.

So, by your power, all of them understand religion, whether they stray from the path or follow it rightly.

You were good to them but tempted me, and abandoned me going about among them with a corrupt religion (*asʿā bi-dīnin fāsidī*).[190]

In the song, Abū Isḥāq Ibrāhīm addresses a beautiful beloved playfully as if she were God, admired for her different characteristics by the different sects, and in each case rightly: the Magians adore the striking contrast between her jet-black hair and her white skin, like the two Zoroastrian principles of original light and darkness; the Jews see the light of God in her shining forehead, like the flame Moses saw on Mount Sinai; the Muslims see in her the promised maidens of Paradise; for the Ṣābians, she is like the bright planet Venus at her most auspicious. As in a typical *ġazal*, the poet complains at the end that the beloved has mistreated him; in this case, she has abandoned him to go about with a "corrupt religion," meaning that he worships *her* in his longing. Leaving the genre-specific significance aside, we find implicit in these lyrics several ideas about religion and the status of the Ṣābians: that the divinity is truly one and loved by all, but appears differently, and is admired differently, depending on how one looks at her; that the Ḥarrānian Ṣābian religion is just another legitimate point of view, comparable to and on a par with the other four major religions; the song assumes, and argues for, the legitimacy of the Ṣābians by asserting their confederate (*ḏimmī*) status in the first line. Moreover, the Ṣābians appear as unquestionably monotheistic: the Christians have God as a trinity, the Magians have two principles, but the Ṣābians see God as one. At the same time something of the astral character of the Ṣābian religion in Baġdād is confirmed, by this tenth-century Ṣābian's own words, although his poetic voice here is that of a lover overcome by longing, not that of a Ṣābian as such. It shows that the descriptions making the Ṣābian Ḥarrānian religion to be an astral cult cannot be entirely incorrect and purely polemical, at least as far as the Ṣābians in Baġdād are concerned.[191]

The last adherent of the Ḥarrānian Ṣābian religion known by name was Abū Isḥāq Ibrāhīm's grandson, Hilāl ibn al-Muḥassin (d. 1055), also an important figure in the court, but now in a time when the ʿAbbāsid caliphs had lost most of their former territory and actual power to generals, amīrs, sulṭāns, and the counter-caliphate of the Fāṭimids in Egypt. The sources say that Hilāl converted to Islam of his own choice. According to the most detailed account, told in the first person at three removes from Hilāl himself, he claimed to have some moving dreams, beginning in 1008/1009, in which the prophet Muḥammad entreated him to convert to Islam. After his father al-Muḥassin died, a Ṣābian until the end, in 1010, Hilāl followed the guidance of his dreams and finally converted in 1012. The king (probably the Būyid

190. My translation has benefited greatly from several pointers on its interpretation by Everett Rowson and Bilal Orfali.

191. Hjärpe 1972: 43–49 regards the astrolatry of the Ṣābians as a fiction developed in Christian antipagan polemic.

amīr Bahā'addawla), pleased and surprised, offered him a gift on the occasion, but Hilāl refused it, saying that he did not wish to mix worldly things with this event. He named his next son Muḥammad.[192] Their family continued to flourish in Baġdād thereafter, the rest of them presumably converting to Islam over the next generation. Thereafter we never hear of another Ṣābian among them. A few of his later, Muslim descendants are noted in the Arabic biographical compendia. The latest is Hilāl's great-great-great-grandson Muḥammad (d. 1212).[193] By his time, their "pagan" ancestry had become irrelevant. So the Baġdādī Ṣābians came to an end early in the eleventh century. Around the same time, the jurist Abū l-Ḥusayn Aḥmad al-Maḥāmilī (d. 1024) issued another *fatwā* against the Ṣābians, the nature, contents, and circumstances of which are unknown.[194] Presumably it renewed the opinion of al-Iṣṭaḫrī as the number of Ṣābians increasingly dwindled.

At Ḥarrān itself the Ṣābians persisted through very different political conditions. Syria in the later tenth and eleventh centuries was divided among many independent rulers in the midst of invasions of the Byzantines and the Fāṭimid caliphs of Egypt, whereas the Būyid amīrs ruling in the name of the 'Abbāsid caliphs lost direct control over the region. Pastoralist bedouins of Syria had been recruited by 'Abbāsid governors and the Ḥamdānid amīrs (themselves of bedouin origin) for military defense against these outside threats, but the bedouin leaders formed petty dynasties vying for autonomous control of relatively small pieces of territory. The Banū Numayr were one of five Syrian branches of the Qays; in 990, when their kin the Banū 'Uqayl drove the Ḥamdānids out of al-Ǧazīra, the Numayrid leader Waṯṯāb made himself an independent amīr with Ḥarrān as his seat of power. The line of Waṯṯāb was ultimately one of the least important of the Syrian dynasties of this period, but they retained control over Ḥarrān, and at times other neighboring towns such as Edessa, ar-Raqqa, and Sarūǧ, for almost a century.[195] The fate of the Ṣābians at Ḥarrān seems to be connected with this Numayrid dynasty.

The Ṣābians, along with the general population at Ḥarrān, were endangered in 1031, after Waṯṯāb's Numayrid successors gave up Edessa to the encroaching forces of the Byzantine general Georgios Maniakes. A very large alliance of Muslim rulers of Syrian cities responded with a joint siege of the Byzantines at Edessa, which proved inconclusive.[196] About the same time, and perhaps just while the Numayrids were with their allies besieging Edessa, a political crisis ensued at Ḥarrān.[197] As the

192. Ibn al-Ǧawzī, *Muntaẓam* 8.177–179. Yāqūt (*Iršād* 7.255.14–15) says slightly differently, "He was a Ṣābian and then converted to Islam at the end of his life, and his conversion was proper (*wa-ḥasuna islāmuhū*)."

193. F. de Blois, "Ṣābi'," *EI*² 8.672a–675a.

194. Chwolsohn 1856: 1.224–225 and 2.582.

195. On the Numayrids, see Rice 1952: 74–84 and Bosworth 1996: 93.

196. The Armenian account of Matthew of Edessa is fairly detailed (Dostourian 1993: 53–54).

197. The events in question are quite complicated. For an overview, see Rice 1952: 77, and W. A. Graham, "al-Ruhā," *EI*² 8.589a–593b (590a–b).

chronicle of the contemporary Christian Yaḥyā ibn Saʿīd of Antioch explains, "Some of the *ašrāf*[198] wrested control of Ḥarrān. They sought the support of its local militia[199] and through them got power over the others. They harmed the population of the city, robbed them, and made their conditions horrible. Most of them left it in flight. They also seized a *maǧmaʿ* of the Ṣābians; it was the temple in the name of the moon; they no longer had another temple in the whole world.[200] They made it into a stronghold. Many of the Ṣābians at Ḥarrān, having been a numerous community, converted to Islam out of fear of them."[201] This source does not make it seem as if the Ṣābians were singled out as victims during this upheaval. All Ḥarrānians not part of the usurpers' group were at risk. Nevertheless, conversion to Islam apparently was felt by many Ṣābians to bring an added degree of security. The testimony also indicates that the Ṣābians of Ḥarrān were no longer numerous after this time, but the wording implies that they *did* continue to exist as a very small community. It is not clear how long these *ašrāf* held power, though the Numayrid Šabīb ibn Waṯṯāb must have reestablished control of Ḥarrān sometime probably soon thereafter; his family retained control of the city until 1081. It is also unclear whether the remaining Ṣābians ever regained the same temple.

Our final source of information looks back on the eleventh century from two hundred years later. Ibn Šaddād (1217–1285) wrote a description of the buildings at Ḥarrān, including brief descriptions of formerly Ṣābian property. Inadvertently he indicates the persistence of Ḥarrānian Ṣābians at least until 1081. First he mentions a building called al-Mudawwar ("the rotunda") that had been of old (*qadīman*) a temple of the Ṣābians, but which had been rebuilt as a fortress (*qalʿa*) by the Ayyūbid al-Malik al-ʿĀdil (d. 1218).[202] He also says that when ʿIyāḍ ibn Ġanam first captured the city for the Arabs (in the seventh century), the conqueror had taken a great temple of the Ṣābians and made it into the *ǧāmiʿ* mosque of the city, but "gave them a place in exchange that they built up in Ḥarrān; it was in their control until Yaḥyā ibn aš-Šāṭir, governor of Ḥarrān, destroyed it on behalf of Šarafaddawla Muslim ibn

198. *Ašrāf* (sing. *šarīf*) are, strictly speaking, people enjoying noble rank due to their descent from the Prophet. See C. van Arendonk and W. A. Graham, "Sharīf," *EI*² 9.329b–337b. The term might refer here simply to "old families" of the town or it might refer to an ʿAlid movement.

199. On the *aḥdāṯ*, see Cl. Cahen, "Aḥdāth," *EI*² 1.256a–b. Another possible interpretation is that the *aḥdāṯ* here are people of families that came to importance more recently.

200. The word *maskūna* is equivalent to the Greek οἰκουμένη, "the inhabited world" (Ullmann 2002–2007: S1.754).

201. Kratchkovsky 1997: 150.15–19: *wa-taġallaba ʿalā Ḥarrāna baʿḍu l-ašrāfi fa-staʿānū bi-aḥdāṯihā wa-taqawwaw bihim ʿalā ġayrihim wa-staḍāmū ahla l-madīnati wa-nahabūhum wa-afsadū aḥwālahum wa-ḫaraǧa akṯaruhum ʿanhā hāribīna wa-aḫaḏū ayḍan maǧmaʿan li-ṣ-Ṣābiʾati wa-huwa l-haykalu llaḏī ʿalā smi l-qamari wa-lam yakun baqiya lahum fī l-maskūnati haykalun siwāhu wa-ǧaʿalūhu maʿqilan wa-aslama kaṯīrūna mimman fī Ḥarrāna mina ṣ-Ṣābiʾati wa-kānū ǧamāʿatan wāfirata l-ʿadadi maḫāfatan minhum.*

202. Lloyd and Brice 1955 present numerous photographic plates of the ruins of the *qalʿa* and a plan of the town.

Qurayš."[203] This Yaḥyā ibn aš-Šāṭir had ended the Numayrid reign over Ḥarrān in 1081, working for the ʿUqaylid Šarafaddawla Muslim ibn Qurayš, who in turn had the backing of the Salǧūqs at the time. The destruction of this Ṣābian site must therefore have been in the 1080s, perhaps during the repression of the anti-ʿUqaylid Ḥarrānian rebellion incited by a Numayrid in 1083.[204] Ibn Šaddād says that the Ṣābians retained control of this place until then, which must mean that some Ṣābians still existed at that time.

There are a several lingering questions about these reports. Was the building they had kept until the 1080s in fact the Mudawwar? Was the building seized by the *ašrāf* in 1031, presumably returned to them sometime soon after, the same building? Or did they move the locus of their worship to another site after 1031, only to see it destroyed in the 1080s? There is no reason to suppose either of the two sources to be incorrect; Yaḥyā of Antioch was a contemporary of the first events; Ibn Šaddād had carried out an official inspection of Ḥarrān for al-Malik an-Nāṣir, the Ayyūbid ruler of Aleppo, in 1242 and must have learned the details he presents during that trip.[205] Yet the information is not complete enough to identify the sites. What is clear is that the Ṣābians of Ḥarrān were a vulnerable group in this century of frequent conflict in Syria, and their vulnerability became critical at times of political upheaval. It was not a strictly intellectual, doctrinal persecution induced by a supposed general "Muslim hostility to freedom of thought" that ended the Ṣābian community, as some have imagined. It can be inferred rather that the Ṣābian leaders in Ḥarrān supported the Numayrids, who protected them in turn, because, as far as these few reports indicate, they were put at risk only when the Numayrids lost control of the town. The events of 1031 appear to have been a turning point; Ṣābians converted in large numbers out of fear to help safeguard their lives and property in the midst of a violent political crisis. It is a matter of speculation how long the Ṣābians of Ḥarrān survived as a sect without any place of worship after the 1080s, but because they are never mentioned again as a living group at a later time, it seems safe to say that ultimately the wars ending in the arrival of the Salǧūqs in Syria were instrumental in the definitive end of the Ṣābian community at Ḥarrān. The survival or demise of the

203. Ibn Šaddād 40–42: *wa-ǧaddada ǧāmiʿahā Nūru d-dīni Muḥammadu bnu Zankīyi wa-zāda fīhi wa-kāna haykalan li-ṣ-Ṣābiʾati ʿaẓīman aḫaḏahū minhum ʿIyāḍu bnu Ġanamin lammā fataḥa Ḥarrāna wāʿawwaḍahum ʿanhu mawḍiʿan āḫara ʿamarūhu bi-Ḥarrāna wa-kāna bi-aydīhim ilā an ḫarabahū Yaḥyā bnu š-Šāṭiri mutawallī Ḥarrāna min qibali Šarafi d-dawlati Muslimi bnu Qurayš.* Rice 1952: 37–52 gives a translation of the whole passage and commentary.

204. Rice 1952: 82; G. Fehervari, "Ḥarrān," *EI²*, 3.227b–230b.

205. Rice 1952: 36. Much later, ad-Dimašqī reports, "Ḥarrān was the city of the Ṣābians. Of their ancient buildings they retained al-Mudawwar, that is the fort, which was the temple of the moon. The Ṣābians persisted there until the year 424 (i.e., 1032–1033); at that time the Egyptians conquered this temple, and the Ṣābians had no temple left" (Chwolsohn 1856: 2.412–413). Ad-Dimašqī's report is clearly derivative of Yaḥyā of Antioch's and confused, but he might be correct to suppose that it was the temple called al-Mudawwar that was confiscated in 1032. He also knows that it was eventually destroyed by the Mongols (Chwolsohn 1856: 2.397).

Ṣābians was simply not important enough to attract more historical notice. By the 1080s their community was probably vanishingly small.

Ibn Šaddād mentions that Maḥmūd ibn Zangī, who took control of Ḥarrān in 1127, rebuilt and expanded the *ǧāmiʿ* mosque of the town, which had been converted from a great Ṣābian temple at the time of the Arab conquest.[206] In 1184, when the Andalusian traveler Ibn Ǧubayr wrote an account of his earlier visit to Ḥarrān, he described this *ǧāmiʿ* mosque in some detail, marveling at the wide arches and high, thick marble columns around its dome and remarking that the dome was a "Roman construction" (*hāḏihī l-qubbatu min bunyāni r-Rūm*),[207] but he does not mention Ṣābians once in his description of the town. They had totally disappeared. As for the city itself, the end came in 1271. After Ḥarrān changed hands several times over many decades, the Mongols under Hülegü eventually conquered it. As Ibn Šaddād, a contemporary of these events, wrote, "When the Tatars realized that they could not protect Ḥarrān and those in it, they deported all those in it to Mārdīn and elsewhere. They destroyed the *ǧāmiʿ* mosque at Ḥarrān, walled up the city gates, and left it empty but for the birds in their nests."[208]

This historical survey provides *termini* for dating any Ḥarrānian Ṣābian writings, Hermetic or not, that can be discovered, as well as a picture of their historical context. In fact, the Ṣābians at Ḥarrān itself produced no well-known authors and have left very little literature that has been identified: just parts of two apocalypses of Bābā of Ḥarrān, in Syriac and Arabic, and some scraps of a garbled Arabic translation of a liturgy of sorts that defy scholarly interpretation.[209] None of this is Hermetic in any normal sense of the word. On the other hand, the Ṣābians of Ṯābit's family at Baġdād, the preeminent center of Arabic intellectual life throughout the tenth century, included famous scholars and scientists, renowned authors. Yet hardly any of the known and surviving works of the Ṣābians of Baġdād have anything in particular to do with their religion. Their own histories, poetry, works of belles lettres and scientific writings in Arabic bear no significant, clear sectarian sign yet evident, with the exception of the minimal evidence already discussed. But *if* any extant Arabic work of Hermes were proved to be of Ḥarrānian Ṣābian provenance, it would probably belong to the Baġdādī branch, and to the century and a half from the arrival of Ṯābit ibn Qurra, sometime before approximately 870, until the death of al-Muḥassin, Ṯābit's great-great-grandson and the last unconverted Ṣābian of importance, in 1010. During this period the Ṣābians of Ḥarrānian origin in Baġdād had their efflorescence and defined themselves as members of a doctrinally coherent, ancient, and indeed original religion, amenable to Muslims and recognized by the

206. Rice 1952: 38.
207. Ibn Ǧubayr 246.12–13.
208. Trans. Rice 1955: 47, slightly modified.
209. The prophecies of Bābā are presented by Rosenthal 1962. *Fihrist* 390.22–391.25 contains the liturgy.

highest authority of the land. Once the Ṣābians were gone, their religion had its echoes only in works on sectarian classifications (*firaq*, loosely "heresiography") as one of the many sects to be refuted anew with every generation, and in the books of history as an ancient curiosity, and in the mere idea that Hermes was an ancient prophet.

3.6 Conclusions

Around 600 CE, during the lifetime of Muḥammad, the Ḥarrānians were known to hold the pagan philosophers in high esteem, including Hermes Trismegistus. The author of the Syriac *Prophecies of the Pagan Philosophers* cited discourses from the Greek *Corpus Hermeticum* as texts that might persuade them of Christianity's truth. However, there is no certain evidence indicating that Hermes or Hermetic works held a *special* place in their doctrines at this time. Rather, Hermes appears as one of a large number of ancient philosophers, most of them Greek, and at least one of them, Bābā, a local Aramaean figure.

In the early ninth century, Theodore Abū Qurra provides the earliest clear Arabic testimony to a specifically astral religion in Ḥarrān, with which Hermes, now their prophet, is explicitly connected. Unfortunately, the development of the Ḥarrānian religion for the previous two centuries is totally unknown, but one may guess that much had changed in the meanwhile. Around the time that Ṭābit of Ḥarrān came to Baġdād, the Arab philosopher al-Kindī (d. ca 870) knew an Arabic Hermetic work related to or resembling in its form the Greek *Corpus Hermeticum*, and he stated specifically that the Ḥarrānians, now known as Ṣābians, regarded its doctrines as authoritative. Three and a half centuries later, Ibn al-Qifṭī had heard of dialogues between Hermes and his son Tat, and Ibn al-ʿIbrī claimed to know Syriac Hermetica that fit the same description. These may well have been the Hermetica present in the Greek *Corpus Hermeticum*, and possibly even the text from which al-Kindī's Arabic version derived. But as far as evidence shows, the Arabic work of Hermes known to al-Kindī did not survive to be read by other authors or at least to merit mention elsewhere.

The Ḥarrānian Ṣābians in Baġdād seem consciously to have adopted Islamic norms of the caliphal court in their religious practices. Sinān ibn Ṭābit ibn Qurra (d. 942/943) was responsible for an Arabic work, *The Laws of Hermes*, describing this religion. It is possible that he translated it from Syriac. Mubaššir ibn Fātik's *Kitāb Muḫtār al-ḥikam* might incorporate portions of this, by way of unknown intermediaries, in its description of the religious laws of Hermes (a problem taken up again in section 5.3). In any case, the idea that Hermes was a prophet whom the Ḥarrānians acknowledged was widely known to Arabic scholars from the ninth century onward, but they never describe any scriptures of Hermes in enough detail for us to identify them. It appears that very few authors had such texts, if they existed, at their disposal.

Several Arabic Pseudo-Aristotelian Hermetica survive, including the *Kitāb al-Isṭamāḫīs*, containing instructions for rituals to procure the powers of the planets. Some of these were cited by the *Ġāyat al-ḥakīm*, which associates them with descriptions of Ḥarrānian practices, thus possibly tying the *Kitāb al-Isṭamāḫīs* and its siblings to Ḥarrānian Ṣābianism. Critical editions and thorough study of these Pseudo-Aristotelian Hermetica are therefore required because they may be the first definitely Ḥarrānian Ṣābian Hermetic works to become known to us. On the other hand, there is the real likelihood that these works have nothing properly to do with Ḥarrān but rather concern "Ṣābianism," meaning here not the religion of Ḥarrān as such, but rather "paganism," as imagined and reconstructed in the ninth and tenth centuries by intellectuals with such interests writing in Arabic.

In the end, there is exceedingly little evidence of Hermetic literature as such among the Ḥarrānian Ṣābians. What we find instead is just the repeated statement that they considered Hermes as an authority and as a prophet. A few brief notices indicate the existence of discourses of Hermes to his son and between Hermes and Tat, which perhaps points to some of the texts known from the Greek *Corpus Hermeticum*, or texts related to it, but these were evidently not widely known. No use seems to have been made of these works in Arabic, and they apparently left no trace there insofar as anyone has yet found. Even philosophers like al-Kindī, who wrote about magic and claims to have examined Hermetic chapters, were really more interested in Euclid, Galen, and Plotinus. By the end of the tenth century, the late antique systematization of Aristotle had come to dominate Arabic philosophy, leaving little room for anything very different that might have been found in an Arabic translation of the Greek *Corpus Hermeticum*; even if it was more widely known than appearances suggest, it could hardly have offered anything as systematic and comprehensive as the system of philosophy that eventually came to prevail. Rather it was the Arabic Hermetica concerned with talismans and astrological lore that became well known in the domain of Islam. These texts may well not be the product of Ḥarrānian authorship, as has sometimes been assumed, though only further study will be able to answer this question. It was these talismanic and astrological texts, along with Hermetic alchemical texts, and not the *Corpus Hermeticum* more familiar to scholars today, that gave Hermes his fame in Arabic and, in Latin translation from the twelfth century until Ficino's time, in Europe. The difficulty in finding any trace of the Greek *Corpus Hermeticum* in Arabic demonstrates the difference between the Hermetica known in Arabic and those that became famous in early modern Europe.

Appendix to Chapter 3

The Ḥarrānians and India

In his studies touching on the Ḥarrānians, D. Pingree argued repeatedly that there was some connection between the pagans of Ḥarrān and India.[1] He tended to designate this as an "influence" without ever specifying how the supposed influence actually happened. Nevertheless he concluded that these Syrian pagans had adopted Indian planetary rituals including even some Sanskrit terminology.[2] As far-fetched as such a connection might seem, the hypothesis of this influence seemed to explain a phenomenon that Pingree could not otherwise explain: the appearance of genuinely Indian elements in early Arabic talismanic magic.[3] In important contributions he clearly demonstrated that Indian iconography and even Sanskrit words appear as a part of certain rituals in the eleventh-century magic book *Ġāyat al-ḥakīm*, which in turn has among its sources the Talismanic Pseudo-Aristotelian Hermetica.[4] The *Ġāyat al-ḥakīm* describes these rituals as Ṣābian.

Pingree's misstep, in this case, was motivated by two assumptions. First, he assumed that the term *Ṣābian*, unqualified, regularly meant specifically "Ḥarrānian," so that these "Ṣābian" Indian elements must have come from Ḥarrān. The second assumption was not Pingree's alone, but has been pervasive in modern scholarship for a number of decades: that the Ḥarrānians were "Hermeticists," so that any Hermetic books, such as the works of Hermes cited in the *Ġāyat al-ḥakīm*, must have come from these Ḥarranians. Indeed, a close reading of Pingree's works shows that he regularly assumed almost any reference to Ṣābians to indicate the

1. 1976: 178–189; 1980: 12; 1989b: 9–11; 2002: 22. His two important articles of 1963, dealing with the transmission of Indian astral sciences, do not yet invoke the Ḥarrānians to explain the Indian elements found in Arabic talismans or astral iconography.

2. Pingree 1980: 12: "One is not in a position to decide whether or not Greek planetary worship influenced the Indian *grahapūja*, or to what extent that *grahapūja* influenced the planetary worship of the Ḥarrānians. One can only conclude that Ḥarrānian worship was undoubtedly influenced by Indian practices, at least to the extent of incorporating the Sanskrit names of the planets into the prayers."

3. Nevertheless Pingree 1976 demonstrates a fairly large number of mathematical traditions transmitted from Indian astronomy into Arabic *without* Ḥarrānian intermediaries.

4. Pingree 1976 and 1980.

Ḥarrānians. As seen, however, the term *Ṣābian* had acquired a generic sense by the tenth century, and certainly by the time of al-Mas'ūdī (one of Pingree's major sources on the problem). *Ṣābian* came to mean just "pagan" and little more. The connection between the Indian characteristics of these talismans and Ṣābianism lies not in Ḥarrān, but rather in the Arabic use of that particular term: it is a neutral, if not sympathetic, designation for poorly understood religions characterized by the veneration of statues and images, such as those of India as well as that of Ḥarrān.[5] Moreover, as also seen, there is little evidence to connect the Ḥarrānians with the Talismanic Pseudo-Aristotelian Hermetica, notwithstanding the Ḥarrānian Ṣābians' esteem of Hermes as their prophet. In fact, the more Indian elements we find in such works, the weaker the likelihood of any connection with the Ḥarrānians becomes. We are left needing an explanation of just how elements of Indian planetary worship were incorporated into talismanic magic in early Arabic texts. The starting point for solving the problem will lie in the study of the many early works of talismanic magic in Arabic and sorting out their chronology.

Pingree's connection of Ḥarrān with India also led him, in a circuitous way, to suppose the influence of the Ḥarrānians on the astrologer Abū Ma'šar. In his view, this made Abū Ma'šar a major conduit for Ḥarrānian theories.[6] He first formulated a connection between the Ḥarrānians and Abū Ma'šar on the basis of his analysis of a short passage in al-Hāšimī's *Kitāb 'Ilal az-zīǧāt* (*The Reasons for the Astronomical Tables*, written ca 890), in which India plays a very minor role. He seems to have made the connection in two steps, though it is not explained explicitly.[7]

First step: Al-Hāšimī (fol. 107v) attributed to a certain "group of Persians" the idea that the concept of a world-year composed of 360,000 solar years was brought to India from Babylon (in the Persian Empire) by Būḏāsaf.[8] In fact the world-year of 360,000 years, as it occurs in Arabic, does appear to derive from India. Būḏāsaf, the alleged source of this world-year according to al-Hāšimī's unnamed "group of Persians," is an Arabic rendering (through the intermediary of an Iranian language) of the Sanskrit Buddhist term *Bodhisattva*. Now, Abū Ma'šar's system is based on a world-year of 360,000 solar years, and is called "the world-year of the Persians."[9] Pingree supposed that al-Hāšimī's reference to the Persians therefore indicated Abū Ma'šar, so that this story about Būḏāsaf must come from Abū Ma'šar himself. This would mean that Abū Ma'šar claimed that his system derived from Būḏāsaf: an Indian source.[10]

Second step: In order to show that Abū Ma'šar's source of information about this was Ḥarrānian, Pingree also cited several Arabic authorities stating that Būḏāsaf was

5. This was recognized already by, for example, Walbridge 1998: 397–399.

6. Pingree 2002: 26–29.

7. Pingree 1968: 405.

8. This passage is cited in Pingree 1968: 4.

9. Pingree 1968: 30–31.

10. It seems more likely that a group of Persians would have claimed that Indian science derived from Iran. See chapter 2 on the Iranocentric history of science.

a prophet of astrology in India according to the Ṣābians.[11] Assuming that these Arabic authorities were correct, *and* that they were referring specifically to the *Ṣābians of Ḥarrān*, Pingree therefore concluded that Abū Maʿšar must have taken this legend about Būḏāsaf, along with the world-year of 360,000 years, from the Ḥarrānians.[12] This would strengthen his theory that the Ḥarrānians had adopted Indian practices. He repeated this general conclusion in slightly varying forms in different works.[13]

The first step described above is a *possible* conclusion but not proved. Pingree's second step in this argument is more problematic. The Arabic sources do not support it because, as already indicated repeatedly, references to Ṣābians in Arabic works are not necessarily also references to Ḥarrānians. Rather, the sources for this step in his argument are just references to "pagans" as imagined by Arabic scholars in the tenth century and later. Pingree cited five passages from three authors to make the argument that the Ḥarrānians followed the Bodhisattva. Here are summaries and brief analyses of their contents to make this clear.

1. Al-Masʿūdī (*Tanbīh* 90.15–91.1) states that Būḏāsaf began the Ṣābian religion, but he does not mention Ḥarrān here. This is thus not a reference to the Ḥarrānian religion but ultimately to Buddhism as understood in an Arabic milieu in the tenth century.
2. Al-Masʿūdī (*Murūǧ* Pellat §535 = 1.263.4–14) says that Būḏāsaf was the first to give rise to the views of the Ṣābians of Ḥarrān and of the Ṣābians of the Marshes (i.e., Mandaeans). It is clear that the Ḥarrānians are given here merely as one example of a Ṣābian sect, of which there were many. We have no reason to regard Ḥarrānians as the source of the report.
3. Al-Masʿūdī, (*Murūǧ* Pellat §1371 = 2.380.13) states that Būḏāsaf was the first to give rise to the doctrines or sects (*maḏāhib*) of the Ṣābians, but again, not the Ḥarrānian sect in particular.
4. Ṣāʿid al-Andalusī (17.2–3) says that Būḏāsaf the Easterner brought the religion of the *ḥunafāʾ*, who are the Ṣābians. Again, there is no reference to Ḥarrān. As seen above and in section 5.2 in this volume, the term *ḥunafāʾ* refers in such a context only to pagans.
5. Al-Bīrūnī (*al-Āṯār* 204.18ff., trans. 186.15ff.) credits Būḏāsaf with bringing the Ṣābian religion and says that the Ḥarrānians are the surviving group of this sect. However, even though his list of Ḥarrānian prophets does not include Būḏāsaf, al-Bīrūnī later does go on to say that Hermes might be identified with Idrīs, but that some of them (perhaps the Ḥarrānians, perhaps the "pagans") have claimed that Būḏāsaf is Hermes (206.8–9, inaccurate trans. by Sachau 188.24–25).

11. Pingree 1968: 5n3.
12. On Abū Maʿšar as a transmitter of this ultimately Indian doctrine of the world-year, see Pingree 1963a: 243.
13. Most recently in Pingree 2002: 25–26.

To these sources I add a brief statement from the student's handbook of Muḥammad ibn Aḥmad al-Ḫwārizmī (late tenth cent.), that "the remnants of the Ṣābians are at Ḥarrān and ʿIrāq [again, the latter group is the Mandaeans], and they claim that their prophet is Būḏāsaf who went off to India, while some of them say Hermes."[14] This occurs under an entry on the Sumanīya (Šamanīya), that is, the Buddhists.

No source unambiguously says that Būḏāsaf was a special figure for the Ḥarrānians in particular, as opposed to the Ṣābians in general. Even if it did, that would have to be regarded as part of the later explanation of Ṣābianism frequent in Arabic sources, which states that it was a primordial religion of idolatry common to all ancient people. Instead, all of the accounts make the Bodhisattva the founder of idolatry, the religion of the Ṣābians (= pagans) in general, imagined as a primordial religion. In this case, the real, or at least the original, reference is to Buddhists, as understood or misunderstood by medieval Arabic scholars, and not to Ḥarrānians. Similarly, one can see how the word is used later for another idolatrous group, this time the ancient Egyptians, by Ibn al-Qifṭī (348.5–7): "The Egyptians long ago were Ṣābians (= pagans) worshipping idols and cultivating temples; then they converted to Christianity when the Christian religion appeared." The term *Ṣābian* was similarly used to refer to the ancient Greeks[15] as well and to all ancient idolaters. A more refined picture might be produced by a detailed study of the development of the term Ṣābian to indicate a member of the primordial religion of idolatry in all its sects, showing its historical development and taking into account the Christian, Muslim, and Zoroastrian etiologies of idolatry. As I currently understand the issue, it seems to me that this idea of Ṣābianism is a construct in Arabic literature dating to the ninth century that became widespread in the tenth. It is conceivable that the Ḥarrānian Ṣābians in Baġdād were the ones who first promoted the idea.[16] But in any case, it is not safe to conclude on the basis of the few vague and conflicting reports cited above about Ṣābianism, a subject of confusion for more than a thousand years, that the Ḥarrānians transmitted to Abū Maʿšar a system of astronomy or astrology attributed to Būḏāsaf, or that the Ḥarrānians had adopted Indian religious practices.

14. *Mafātīḥ al-ʿulūm, bāb* 2, *faṣl* 5.
15. See section 3.2 in this volume.
16. On this possibility see Pingree 2002: 35.

PART II

History of the Arabic Hermes

4

The Three Hermeses

FROM ANDALUSIA TO INDIA, WHEN A PRE-MODERN ARABIC scholar read a treatise attributed to Hermes, he understood its contents to be a part of an extremely ancient wisdom transmitted from many centuries ago, long before the revelation of Islam. To his knowledge, the Hermes who wrote the text could be one of three different individuals, each of them a great sage of the remote past. The first of them founded the sciences before the great Flood; the second, a Babylonian, lived after the Flood; and the third lived later in Egypt. Medieval Arabic histories gave an especially prominent place to the first Hermes, the antediluvian prophet of science, reporting that he received revelations from God and ensured the preservation of knowledge through the cataclysmic Flood through his Egyptian inscriptions. It is a peculiar historical accident that the messenger god of Homeric Olympus became, after long centuries, in Arabic accounts, not only the first philosopher, physician, and astrologer, but also a true prophet, identified with the Qur'ānic figure Idrīs.

These legends about Hermes, or rather the Hermeses,[1] were passed on from the Arabophone countries of the Mediterranean and Southwest Asia to Latin Europe, where they formed in turn part of the background for the reception of Hermetic texts in Latin.[2] Knowledge of a version of this legend of the ancient Hermes may have helped to induce Marsilio Ficino in the fifteenth century to set aside his translation of Plato's works and give precedence to the newly-discovered *Corpus Hermeticum*, for Hermes was considered a bearer of a primordial philosophy more ancient and closer to God than that of Plato.[3]

But why these *three* Hermeses? The task of this chapter is to uncover the origins of the triple biography. Some modern scholars have thought that these tales resulted

1. Although the plural "Hermeses" sounds a bit strange, it is a regular form more recognizable in English than Greek Hermai (Latin Hermae) or Arabic Harāmisa.

2. On Latin versions of this Arabic legend see Burnett 1976 and Delp 2006: 5–9.

3. Ficino finished this translation in 1471 (Faivre 1995: 184).

merely from an imaginative interpretation of the Greek epithet *trismegistos*, most commonly rendered in Arabic as *al-muṯallaṯ bi-l-ḥikma* ("triplicate in wisdom").[4] One has even proposed the idea of three Hermeses in light of theories that postulate a connection between a theoretical mythic archetype of Hermes and the number three.[5] I will show, however, that the Arabic biographies of Hermes are really continuations of specific and identifiable traditions already developed in pre-Islamic late antiquity, synthesized together in Arabic in ways reflecting a genuine effort to understand the pre-Islamic past. At the same time I will prove that late antique Greek chronography was an important component of the Arabic reception of ancient texts.

4.1 The Earliest Form of the Arabic Legend of Three Hermeses

The earliest extant version of the Hermes legend in Arabic is found in the history of philosophy and medicine written by the Andalusian physician Ibn Ǧulǧul of Cordova (d. after 994). In his *Kitāb Ṭabaqāt al-aṭibbā' wa-l-ḥukamā'* (*The Generations of the Physicians and Philosophers*), composed in 987, Ibn Ǧulǧul stated that he took his account of the Hermeses from the *Kitāb al-Ulūf* (*The Book of the Thousands*) of Abū Maʿšar of Balḫ (d. 886), the most influential astrologer of the Middle Ages. This book, the *Thousands*, is the main source of the Hermes legend in Arabic, cited by numerous later historians and biographers. Unfortunately, the *Thousands* survives today only in excerpts and very much abbreviated summaries, although David Pingree was able to reconstruct a good deal of its mathematical contents in his fundamental work *The Thousands of Abū Maʿshar* (1968).

The *Thousands* of Abū Maʿšar was a work of historical astrology. The investigation of the Arabic legends of Hermes must therefore begin with this book and its genre. Historical astrology was developed on the basis of Greek and Indian astrological systems by Iranian astrologers of the Sasanian period (third–seventh cent.) who sought to interpret historical events of the past and to predict events of the future in terms of major astrological conjunctions, particularly of Jupiter and Saturn.[6] Such explanations were made to fit into a chronological scheme that

4. Pingree 1968: 14 suggested that this was an invention of Abū Maʿšar. "Abū Maʿshar, presumably thinking of the epithet *trismegistos*, developed the myth further by hypothesizing three Hermes..." Interpretation of the *trismegistos* epithet may have been a factor in the development of the story but does not explain the origin of the specific elements of the three individual linked Hermes stories.

5. Sladek 1988.

6. See Pingree 1963a, especially 245, and Pingree, "Astrology and Astronomy III. Astrology in Islamic Times," in *EIr*, 2.868b–871b, esp. 870. Another short and clear explanation of the theory of conjunctions in historical astrology is Pingree's "Ḳirān," *EI²*, 5.130b–131a. For a detailed study of Abū Maʿšar's historical astrology in another work using similar concepts, see Burnett and Yamamoto 2000: 1.573–613. For the more complicated historical astrology of the *Thousands* in particular, see Pingree 1968: 57–68.

sometimes claimed to give the duration of the world, from Creation to the end of time (or of the present cycle of time). In this way the endeavor of historical astrology brought astrology and chronography together. Its main goal was to show that the major changes of history corresponded to cycles of movement of the planets and stars, thereby explaining the past and predicting future changes in dynasty, the appearance of new religions, and major disasters such as universal floods. The chronological place of the Hermeses was especially important in the historical astrology of Abū Maʿšar both because the first two Hermeses circumscribe the date of the Flood, the most important point in Abū Maʿšar's timeline of history and the era from which he dated events, and also because the original, antediluvian Hermes was the founder of astrology, and hence effectively the ancient source of Abū Maʿšar's knowledge, giving his own practice the authority of antediluvian and even revealed wisdom.[7]

Before further discussion, it is necessary to take a close look at the text of the legends of Hermes from Abū Maʿšar's *Thousands*. Ibn Ǧulǧul's account of the Hermeses is the earliest extant in Arabic, but it was also widely repeated in slightly altered forms by later authors, including almost every historian of science in Arabic after him, either directly from the *Thousands* or through an intermediary. A few later works cited Ibn Ǧulǧul by name for their version. Although the task of tracking the three Hermes stories in Arabic is greatly simplified by the identification of the *Thousands* of Abū Maʿšar as the Arabic source from which these legends derive, one should keep in mind that there were also at least two abridgements of the *Thousands*, one of which, made by Abū Maʿšar's own student, is lost.[8] It is possible that those who cited the surviving accounts of Hermes sometimes knew them by way of the lost abridgement. Nevertheless, all of the surviving versions of the Arabic legend of three Hermeses are to a great extent similar to each other, giving essentially the same information with slightly different wording. Because many of the same words and phrases occur in each account, the relationship between all the surviving versions is clearly apparent. The account follows below in Arabic and in my English translation. Ibn Ǧulǧul is taken as the base, with variants from two other sources indicated in notes to the English translation by the following sigla:

Ǧ = Ibn Ǧulǧul (d. after 994), *Kitāb Ṭabaqāt al-aṭibbāʾ wa-l-ḥukamāʾ* (The Generations of the Physicians and Philosophers), 5–10.

7. See Pingree 1968: 13–18, and the chart on 19.

8. The lost one is by "Ibn al-Māzyār," discussed below; this was described by al-Masʿūdī as a *muntaḫab*, or selections of important passages (al-Masʿūdī, *Murūǧ*, Pellat §1419 = 2.408.11–12). The other extant summary, the *Muntaḫab Kitāb al-Ulūf* by as-Siǧzī, is outlined by Pingree 1968: 23–25. I have inspected a manuscript copy of this latter *Muntaḫab* in British Library Or. 1346, 80b–91b; it is a concise outline of the astrological theory and method of the *Kitāb al-Ulūf* without any of the chronographic data that the original work contained or the story of Hermes. A section of it is translated into English by Kennedy and van der Waerden 1963.

Z = Ibn al-Qifṭī (d. 1248), *Kitāb Iḫbār al-ʿulamāʾ bi-aḫbār al-ḥukamāʾ* (*Apprising the Scholars of Reports about the Philosophers*), first Hermes 6–7, second Hermes 346–7, third Hermes 347–50. This work is known only in the epitome of az-Zawzanī, *Kitāb al-Muntaḫabāt al-multaqaṭāt min Kitāb Iḫbār*... (*Selections Garnered from "Reports for Scholars about the Philosophers"*) made in 1249. It cites Ibn Ǧulǧul explicitly as the source of its reports on the first Hermes .[9] The excerpts on the second and third Hermeses relate the text given by Ibn Ǧulǧul almost exactly, with the addition of some notes on the books of these latter Hermeses.

IAU = **Ibn Abī Uṣaybiʿa** (d. 1270), *Kitāb Ṭabaqāt al-aṭibbāʾ* (*The Generations of the Physicians*), 1.16.24–17.16. IAU seems to have used Ibn Ǧulǧul's history of the Hermeses, although he must have had access to sections of the *Thousands* in some other form, or in excerpts from other works, since he cites an otherwise unknown passage of the *Thousands* just a page before, IAU15.7ff.

Unless indicated, Ǧ, Z, and IAU agree (except in a few unnoted insignificant variants).

1. **قال أبو معشر البلخي المنجّم في كتاب الألوف** الهرامسة ثلاثة أوّلهم هرمس الذي كان قبل الطوفان ومعنى هرمس لقب كأن يقال قيصر وكسرى وتسمّيه الفرس في سيرها ابنجهن وهو الذي تدّعي الحرّانية حكمته وتذكرالفرس[10] أنّ جدّه جيومرث وهو آدم ويذكر العبرانيون أنّه خنوخ وهو بالعربية إدريس

2. **قال أبو معشر** هو أوّل من تكلّم في الأشياء العلوية من الحركات النجومية وأنّ جدّه جيومرث علّمه ساعات الليل والنهار وهو الأوّل من بنى الهياكل ومجّد الله فيها وأوّل من نظر في الطبّ وتكلّم فيه وأنّه ألّف لأهل زمانه قصائدَ موزونةً وأشعاراً معلومةً في الأشياء الأرضية والعلوية وهو أوّل من أنذر بالطوفان ورای أنّ آفة سماوية تلحق بالأرض من الماء أو النار وكان مسكنه صعيد مصر تخيّر ذلك فبنى هنالك الأهرام ومدائن التراب وخاف ذهاب العلم بالطوفان فبنى البرابي وهو الجبل المعروف بالبربا بأخميم نحته وصوّر فيه جميع الصناعات وصنائعها نقشاً وصوّر جميع آلات الصناع وأشار إلى صفات العلوم برسوم حرصاً منه على تخليد العلوم لمن بعده وخيفةً أنْ يذهب رسم ذلك من العالم

3. وثبت في الأثر المروي عن السلف أنّ إدريس أوّل من درس الكتب ونظر في العلوم وأنزل الله عليه ثلاثين صحيفة وهو أوّل من خاط الثياب ولبسها ورفعه الله مكاناً علياً

4. **وحكى عنه أبو معشر** حكايات شنيعة أثبت بأحقّها وأقربها وبالله تعالى التوفيق.

5. هرمس الثاني من أهل بابل سكن مدينة الكلدانيين وهي بابل بعد الطوفان في زمن نبريزباني الذي هو أوّل من بنى مدينة بابل بعد نمروذ بن كوش وكان بارعاً في علم الطبّ والفلسفة وعارفاً بطبائع

9. The account of the First Hermes begins, "Sulaymān ibn Ḥassān, known as Ibn Ǧulǧul, said..." and closes with, "This ends Ibn Ǧulǧul's statement" (*inqaḍā kalām Ibn Ǧulǧul*).

10. الفرس supplied by IAU16.26.

الأعداد وكان تلميذه[11] فيثاغورس الأرتماطقي وهرمس هذا جدّد من علم الطبّ والفلسفة وعلم العدد ما كان قد درس بالطوفان ببابل **ذكر ذلك أبو معشر**

6. ومدينة الكلدانيين هذه هي مدينة الفلاسفة من اهل المشرق وفلاسفتهم أوّل من حدّد الحدود ورتّب القوانين وهم فلاسفة حذاق الفرس

7. هرمس الثالث سكن في مدينة مصر كان بعد الطوفان وهو صاحب كتاب حيوان ذوات السموم وكان فيلسوفاً طبيباً عالماً بطبائع الأدوية القتّالة والحيوانات المعدية وكان جوّالاً في البلاد طوّافاً بها عالماً بنصبة المدائن وطبائعها وطبائع اهلها وله كلام في صناعة الكيمياء نفيس يتعلّق منه إلى صناعات كالزجاج والخرز والغضائر وما أشبه ذلك

8. وكان له تلميذ يُعرف اسمه اسقلابيوس له أخبار شنيعة وقصص كثيرة نستجلب ما صحّ منها مما وقع في الكتب المعروفة إن شاء الله

Ibn Ǧulǧul[12]

1. Abū Maʿšar al-Balḫī the astrologer said in the *Book of the Thousands*: "The Hermeses are three. The first of them[13] is Hermes who was before the Flood. The significance of "Hermes" is a title, like saying "Caesar" and "Ḫusraw" [which are titles]. The Persians named him Wīwanghān, meaning "the Just,"[14] in their biographies of the kings (*fī siyarihā*). He is the one to whose philosophy the Ḥarrānians adhere.[15] The Persians[16] state that his grandfather was Ǧayūmarṯ, that is Adam. The Hebrews state that he is Enoch, which, in Arabic, is Idrīs."
2. Abū Maʿšar said, "He was the first to discuss[17] the celestial phenomena of the movements of the stars, and his grandfather Ǧayūmarṯ taught him the hours of the nychthemeron. He is the first who built temples and glorified God in them. He was the first to investigate medicine and to discuss it. He composed for his contemporaries odes in poetic meter[18] and well-known verses[19]

11. Suffix pronoun -*hū* supplied by IAU17.9.

12. Other English translations of this passage, along with different analyses, are found in Plessner 1954a: 50–52; Pingree 1968:14–15 (First Hermes only); A. Fodor 1970: 336–337 (gives the translation of the version in IAU). There is an Italian translation in Pappacena 2004, 266–267, together with a general summary of the Hermes legends in Arabic.

13. IAU16.24–25 adds here "he who is the Triplicate in Blessings" (*al-muṯallaṯ bi-n-niʿam*). It is not clear whether this is IAU's insertion or original in the *Thousands*.

14. The text of Ǧ gives *ʾbnǧhḏ* (Z6.13, *ʾyhḥl*; IAU16.26 *ʾllhǧd wa-tafsīruhū ḏū-ʿadl* "*ʾllhǧd*, which means the Just"). Pingree recognized that "the name is clearly identical with Awanjhān = Wīwanjhān, the father of Jamshīd according to Ibn Nawbakht, which is a mistake for Ūshhan = Hūshank," Pingree 1968: 14–15. On the name Wīwanjhān, see Justi 1895: 144a–b and 274a. The *Muntaḫab Ṣiwān al-ḥikma* contains the same words (ed. Badawī 184.5, ed. Dunlop 1262). *Pēšdād* was a title of Hoshang, though eventually all the earliest kings of Iran were called Pēšdādians. The word was interpreted in Iran by folk etymology to mean "just," although actually meaning "first created," cf. Avestan **paraδāta*). On Hoshang, the discoverer of fire in medieval Iranian legend, and the interpretation of his epithet see CHI 3(1), 420.

15. IAU16.26 "he is the one whose prophethood the Ḥarrānians declare."

16. "The Persians" supplied here by IAU 16.26; Ǧ lacks this word.

17. Z6.15 "to prognosticate from" (*taḥakkama fī*)

18. IAU16.30 "many books in poetic meter."

19. IAU16.31 "and well-known rhymes in the language of his contemporaries."

on terrestrial and celestial things. He was the first to give advance warning of the Flood, and he thought that a celestial catastrophe of fire or water would overwhelm the earth. His home was Upper Egypt; he chose that [place] and built the pyramids and cities of clay there. He feared that knowledge would pass away in the Flood, so he built the monumental temples (*al-barābī*);[20] it is a mountain known as *birbā* in Aḫmīm,[21] which he chiseled out, portraying in it in carvings all the arts and their uses, and pictures of all the instruments of the artisans, indicating the features of the sciences by illustrations,[22] out of desire thereby to preserve the sciences forever for those after him, fearing that all trace of it would perish from the world."

3. It is reported by the account transmitted from the early Muslims (*as-salaf*) that Idrīs was the first to study books, to investigate the sciences. God sent thirty scrolls (*ṣaḥīfa*) down to him. He was the first to sew clothing and to wear it. God raised him to a high place.[23]
4. Abū Maʿšar told some absurd reports about him, of which I related the most true[24] and most likely. Success is in God the Exalted.
5. "The Second Hermes, of the people of Babylon: He lived in the city of the Chaldaeans, Babylon, after the Flood in the time of Naburīzbānī, who was the first to build the city of Babylon after Nimrod son of Kush.[25] He was skilled in the knowledge of medicine and philosophy, knew the natures of numbers, and his student was[26] Pythagoras the Arithmetician. This Hermes renewed the knowledge of medicine, philosophy, mathematics that was lost during the Flood at Babylon." That is what Abū Maʿšar stated.
6. This city of the Chaldaeans was the city of the philosophers of the people of the East. Their philosophers were the first to make definitions and to organize laws; they were the philosophers of the Persian sages.

20. This was an Egyptian loan word for *temple* in Arabic and used to refer to all the great stone monuments of Egypt (Coptic ⲡⲣ̄ⲡⲉ *p-ərpe*). See G. Wiet, "Barbā," *EI²*, 1.1038b–1039a.

21. Akhmim, the Panopolis of Roman Imperial times. The famous Egyptian temple there was finally dismantled for materials to build a *madrasa* in the fourteenth century (G. Wiet, "Akhmīm," *EI²*, 1.330a).

22. IAU17.3 adds "for those after him."

23. Paraphrase of Qur'ān 19:57; (56) *wa-ḏkur fī l-kitābi Idrīsa innahū kāna ṣiddīqan nabīyan* (57) *wa-rafaʿnāhu makānan ʿalīyan*, "And mention Idrīs in the Book. He was true, a prophet. We raised him to a high place."

24. Reading *aḥaqqahā* (Z) rather than *aḫaffahā* ("the least significant of them").

25. Nimrod son of Kush is the founder of Babylon according to the Hebrew Bible (Genesis 10:8–10). Evidently a tradition with its roots in Christian Greek chronography was also available to Abū Maʿšar. On his chronographic sources, see section 4.5. It is probable that the king intended here, whose name is distorted in the text, is Nabukhodonosorus (Nabuchadnezzer II, r. 604–562 BCE, famous from the book of Daniel), whose building activities at Babylon, including the Hanging Gardens, were related by Berossus in a report known from a citation appearing both in Josephus' *Antiquitates Judaicae* (10.220–228) and from George Syncellus (416–418) (for English translation and notes, see Verbrugghe and Wickersham 1996: 57–59). The name Naburīzbānī is omitted in some later citations of this story, and the second, Babylonian Hermes himself is made into the one who built Babylon after Nimrod: Ibn al-Qifṭī, 346.17–18; Ibn al-ʿIbrī, *Ta'rīḫ*, 11.14–15.

26. Following IAU17.9; the omission of a letter makes him the student of Pythagoras in Ǧ.

7. The Third Hermes: He lived in the city of Egypt. He was after the Flood. He is the author of *Poisonous Animals* (*K. al-Ḥayawān ḏawāt as-sumūm*). He was a philosopher and a physician, knowledgeable in the natures of lethal drugs and infectious[27] animals. He traveled around in different countries, wandering in them, knowing the foundations of cities, their natures, and the natures of their peoples. He is the author of a valuable discourse on the art of alchemy; part of it is related to crafts like glass, stringing precious stones, implements of clay, and such things.
8. He had a student who is known, whose name was Asclepius.[28] There are absurd reports and many stories about him; we will seek to procure the ones that are true from what is available in well-known books.

In the English translation above, I have enumerated paragraphs to indicate what I think are its separate parts, but these divisions do not occur in the original. Paragraphs 1 and 2 are both explicitly quotations from Abū Maʿšar. Paragraph 3 may or may not be by him, but gives a report known to Muslims since the early eighth century (due to Wahb ibn Munabbih, discussed below). Paragraph 4 clearly marks the omission of material originally in Abū Maʿšar's *Thousands*, evidently regarded as fabulous, and paragraph 5 returns us to the quotation of the original text. This includes material definitely derived from a late antique Greek work of chronology, because it contains information transmitted from Berossus. The end of 5 is marked as the end of the quotation from Abū Maʿšar. Paragraph 6 seems to be an inserted explanation of the reference to the city of the Chaldaeans in paragraph 5. Paragraph 7, on the third Hermes, is not explicitly named as derived from Abū Maʿšar, and this fact will require discussion below.

A number of derivative sources also present different abridged or differently phrased versions of the same triple Hermes tale. A couple of sentences about Hermes in the *Muntaḫab Ṣiwān al-ḥikma*, an abridgement of a gnomological history of Greek and Arabic philosophers written ca 1000, are derived from Abū Maʿšar's Hermes story.[29] The erudite aṯ-Ṯaʿālibī of Nīšāpūr (d. 1038) relates Abū Maʿšar's story of the first Hermes almost totally reworded but with hardly anything new added.[30] The Christian scholar Abū l-Farağ Ġrīğūriyūs Ibn al-ʿIbrī (Bar ʿEbrāyā) (d. 1286) introduced his brief reference to the three Hermeses with no more specific

27. IAU17.13 *muḍiya*, "harmful;" Ǧ *muʿdiya* (infectious?).

28. IAU17.16 adds "His residence was in Syria (*aš-Šām*)."

29. *Muntaḫab Ṣiwān al-ḥikma* ed. Dunlop, section 90, lines 1261–1268; ed. Badawī 184.4–13.

30. "The first person who expounded the science of the stars, pointed out the patterns of the heavenly spheres, plotted the paths of the stars, revealed their various influences on human affairs and brought forward information on the wonders of their construction, was Idris, peace be upon him. He was also the first to use writing and to sew garments, whereas previously people had only worn skins and hides; and he was the first to make use of weapons, make war and capture people as slaves." Translation by Bosworth 1968: 39. Cf. aṯ-Ṯaʿālibī 1960: 6.5–10.

attribution than "it is said."[31] Although he is able to cite the name Trismegistus fairly intact (*Ṭrīsmīǧīsṭīs*), and to give a Christian exegesis of the name—that Trismegistus was so called because he described the creator's three essential characteristics: existence, wisdom, and life[32]—his account is derived mostly from Abū Maʿšar's. The Išrāqī (Illuminationist) philosopher aš-Šahrazūrī (fl. late thirteenth century) compiled many reports and sayings of Hermes in his history of philosophy, at the head of which he gave a slightly edited version of the same legend of three Hermeses.[33] The Egyptian polygraph as-Suyūṭī (1445–1505) referred to the story in his treatise on the pyramids.[34] In short, the stories of Hermes originating with Abū Maʿšar were widely known over many centuries and indeed were part of common scholarly knowledge.

Only a few versions of the story of the first Hermes appear to contain information that might have belonged to the original of Abū Maʿšar not related by Ibn Ǧulǧul. The *Muntaḫab Ṣiwān al-ḥikma* (discussed at greater length in section 5.3) cites Abū Maʿšar by name as its source on the identity of Hermes and relates a slightly variant version of paragraph 2 of the Hermes stories given above. Without any break, the *Muntaḫab* also adds a little more information; it is possible, though not certain, that this also derives from Abū Maʿšar.[35]

وأنّه علم أنّ آفة سماوية تصيب بعد وفاته سكان الأرض من الغرق بالمياه والاحتراق بالنيران والحرارات فبنى هو وأهل زمانه في الناحية التي يسكنها من المغرب في الأرض المعروفة بيونان فيما بين صعيد مصر المتصل ببلاد السودان الى الاسكندرية وأسفل اهراماً كثيرة من حجارة على رؤوس الجبال والمواضع المرتفعة ارتفاع كل هرم منها بين الثلاثين ذراعا الى الخمسين ذراعا عريضة الرؤوس وجعلوا من بنائها هرمين أرفع سكما

[Abū-Maʿšar also stated that] he [i.e., Hermes] knew that a heavenly ruin would befall the inhabitants of the earth after his passing from an inundation of waters and burning by fires and heat, so he and his contemporaries in the region of the west that he inhabited—in the land called Greece (*Yūnān*) between upper Egypt adjacent to the Sūdān and Alexandria and lower [Egypt]—built pyramids of stone on the mountain peaks and elevated places, each pyramid being between thirty and fifty cubits, broad at the top. In their construction they made two pyramids among them with higher tops.

31. Ibn al-ʿIbrī, *Taʾrīḫ*, 11.8ff. It is likely that Ibn al-ʿIbrī was using Ṣāʿid al-Andalusī as his source on the Hermeses. He cites Ṣāʿid by name as an authority elsewhere in the work and gives at least one important variant reading following Ṣāʿid (*al-ǧawāhir* instead of *al-ašyāʾ*; see note 36 below).

32. Cf. similar Christian explanations of the title *trismegistos*, seeing in it trinitarian and other meanings: Malalas 26 (Scott 4.233), *Chronicon Paschale* 85, John of Antioch 6, Cedrenus (Scott 4.235), *Suda* (under Ἑρμῆς ὁ Τρισμέγιστος; also Scott 4.235).

33. aš-Šahrazūrī, *Kitāb Nuzhat al-arwāḥ*, 131.3–134.2.

34. Nemoy 1939: 27.

35. *Muntaḫab* Badawī 184.7–13, Dunlop 1264–1268.

There are essentially only two new pieces of information supplied here. The first locates Hermes generally in Egypt, mistakenly named Greece or considered a part of Greece. The second adds more details about the pyramids, and it includes other contemporaries of Hermes in their construction. It is difficult to know whether these are later additions or part of Abū Maʿšar's original account.

Most of the other versions of the Hermes legends in Arabic are just reworkings of the version presented by Ibn Ǧulǧul, known at one or more removes. Below I present as a representative of these later versions just one of the more interesting uses of the account by Ṣāʿid al-Andalusī (1029–1070), *qāḍī* of Toledo, who wrote an influential history of science organized by nation. According to his arrangement of the data, science passed over the centuries from the Eastern nations toward the Western nations, culminating, not surprisingly, in the Andalusia of his day. Abū Maʿšar is the only authority he names specifically in his account of Hermes.

Ṣāʿid al-Andalusī, 18.18–19.6 (ed. Cheikho).
[Section on the Sciences among the Chaldaean Nation]
According to us, the most famous and most respected of their [the Chaldaeans'] scholars is Hermes the Babylonian. He was in the time of the Greek philosopher Socrates. Abū Maʿšar Ǧaʿfar ibn Muḥammad ibn ʿUmar al-Balḫī stated in the *Book of the Thousands* that he was the one who restored many of the books of the ancients on the astral sciences and other kinds of philosophy that had perished, and that he composed many books on various sciences. Abū Maʿšar said, "The Hermeses (*al-Harāmis*) are a group of different individuals. Among them is the Hermes who was before the Flood, who the Hebrews claim is the prophet Enoch, who is Idrīs, peace be upon him. After the Flood were a number of them [*scil.* the Hermeses], knowledgeable and discerning. The preeminent of these were two, the first of whom was the Babylonian we mentioned, and the second was the student of the philosopher (*al-ḥakīm*) Pythagoras, and an inhabitant of Egypt."

Ṣāʿid said, "Information has reached us about the doctrine (*maḏhab*) of the Babylonian Hermes indicating his preeminence in science. This includes his doctrine on projection of the rays of the stars and his doctrine on the uniformity of the Houses of the celestial sphere. It also includes his books on astrology such as the *Book of Latitude* and the *Book of Longitude* and *The Rod of Gold*."

39.7–16. [Section on the Sciences among the Egyptian Nation. In this section *italics* are used to illustrate the correspondence between this source and the account as presented by Ibn Ǧulǧul.]

A group of scholars reported that it was from the First Hermes, who lived in highest Upper Egypt, that all of the sciences that appeared before the Flood came. He is the one whom the Hebrews name Enoch, son of Jared, son of Mahala'il, son of Cainan, son of Enosh, son of Seth, son of Adam, upon him be peace. He is the prophet Idrīs, peace be upon him. They said that *he was the first to discuss celestial* substances[36] *and the movements of stars,*[37] *and he is the first who built temples and glorified God* the Exalted *in them. He was the first to investigate medicine.*[38] *He composed for his contemporaries odes in poetic meter on terrestrial and celestial things.* And they say that *he was the first to give advance warning of the Flood, and he thought that ruin would overtake the earth from water and fire.*[39] *He feared that knowledge would pass away* and that the arts would perish *in the Flood, so he built* the pyramids and *the monumental temples* in highest Upper Egypt. *He portrayed* in them all the arts and the instruments, *indicating the features of the sciences by illustrations,*[40] *out of desire thereby to preserve the sciences forever for those after him, fearing that all trace of it would perish from the world.*

40.5–7.
Among the ancient scholars in Egypt is the Second Hermes. He was a philosopher, traveling in the lands, wandering the cities, knowing the foundations of the [cities?][41] and the natures of the peoples. He wrote a great book on the art of alchemy and a book on poisonous animals.

It can be seen from the other relations of the legend that Ṣāʿid al-Andalusī has presented a heavily edited version of the passages on the Hermeses from the *Thousands* of Abū Maʿšar, arranged to fit his program of giving the history of science in order by nation. Therefore the first Hermes appears in the section on Egypt, the second, Babylonian, Hermes is in the section on the Chaldaean nation, and the third (whom he calls the second) is again in the section on Egypt. In all, it does not add much of importance to the present discussion besides showing one example of how easily this triple biography of Hermes could be altered and adapted in later works. However, there are a few details deserving emphasis. First, according to this account, "a group of scholars" related the tale of the first Hermes. This indicates that by the time of Ṣāʿid, the eleventh century, the report of Abū Maʿšar was in fairly wide circulation and had been cited by a number of others. Second, he states that the Babylonian Hermes recovered the knowledge that was lost in the Flood, emphasizing the relationship between the first and second Hermeses in Abū Maʿšar's account.

36. *ǧawāhir* جواهر, substances; a misreading for *ḥawādiṯ* حوادث, events.

37. The passage referring to Iranian national history has been omitted ("and that his grandfather Ǧayūmarṯ taught him the hours of the nychthemeron").

38. Omitted: "and to discuss it."

39. The following was omitted because Ṣāʿid had already mentioned it: "His home was Upper Egypt; he selected that [place] and built the pyramids and cities of clay there."

40. Many words were edited out in this passage.

41. The edition reads, "knowing the foundations of their people," clearly an error. Cf. the distantly parallel text of Ǧ above: "He was skilled in the knowledge of medicine and philosophy, knew the natures of numbers…"

We turn now to examine the contents of the tale more critically. One of the great puzzles about Hermes in Arabic is the appearance of three different figures of the same name, all recognized as important sages. In our base account, the earliest extant one, presented by Ibn Ǧulǧul, the first two Hermeses are connected together in their narratives in that the second, Babylonian, Hermes recovered the knowledge, lost in the Flood, that the first, antediluvian, Upper Egyptian Hermes tried to preserve in writings and carvings. The third Hermes is said to be an Egyptian from later times, and he appears to be unrelated to the first two. Moreover, only the first two Hermes stories are connected explicitly with the name Abū Maʿšar. These are the initial clues. How, then, did the idea of three Hermeses originate?

In his 1968 work on Abū Maʿšar, David Pingree proposed an answer to this question, one that has been widely accepted since then. He argued, in what must be described as a convoluted manner,[42] that this legend of Hermes was developed by Abū Maʿšar himself under the influence of his predecessor Ibn Nawbaḫt. If I have understood it, his case is as follows. The astrological book called the *Pentateuch* of Dorotheus of Sidon (first cent.), which survives only in the Arabic translation of a Middle Persian recension translated from the original Greek, is addressed, in the extant version, to Hermes, the son of Dorotheus, who is king of Egypt (discussed earlier in section 2.2). But, wrote Pingree, this figure of Hermes was "perverted by Ibn Nawbakht, under Ḥarrānian influences, into that of the Babylonian sage who becomes king of Egypt and brings learning to that land," in Ibn Nawbaḫt's *Kitāb an-Nhmṭ'n*.[43] He went on to write that "Abū Maʿshar, presumably thinking of the epithet *trismegistos*, developed the myth further by hypothesizing three Hermeses, of whom the last two correspond to Ibn Nawbakht's Babylonian-Egyptian sage."[44] Pingree characterized Abū Maʿšar as a charlatan attempting to foist a total fabrication, a "brazen imposture," on his contemporaries.[45]

I believe that this is mostly incorrect. The verdict on Ibn Nawbaḫt was certainly false because, as Pingree discovered for himself when he edited it (at the latest), a reference to another Hermes Trismegistus "King of Egypt" is found in the very book of Dorotheus already mentioned.[46] Certainly Ibn Nawbaḫt, who had the *Pentateuch* of Dorotheus at his disposal, should also be credited with awareness of this reference.[47] The only thing that Ibn Nawbaḫt could have added is that Hermes was Babylonian in origin. It seems to be this Babylonian epithet that is supposed to

42. As Toomer wrote in his review of the book (1970: 156), "Anyone not already familiar with the subject matter would probably find most of it unintelligible. Even for the initiate the author's allusiveness is baffling...The general organization of the material seems to be more for the author's convenience than for the reader's."

43. Pingree 1968: 14. This passage by Ibn Nawbaḫt was studied in chapter 2.

44. Ibidem.

45. Idem, vii–viii.

46. Dorotheus Sidonius, *Carmen Astrologicum*, ed. D. Pingree, 2.20.1. *ka-mā qāla l-mušarrafu l-maḥmūdu bi-ṯalāṯi ṭabāʾiʿa Hirmisu maliku Miṣra.*

47. Pingree 1968: 11.

have constituted "Ḥarrānian influence—direct or indirect,"[48] although one is at a loss to demonstrate how that could be, or why one should blame the Ḥarrānians for it. Pingree's tendency to exaggerate the importance of the Ḥarrānians in ninth-century intellectual life has already been encountered in chapter 3. Moreover, as shown in chapter 2, the reference to Babylon points to the idea that this Hermes was from the capital of the Iranian realm, where Dahāk gathered the world's scholars before they were dispersed to different lands, including Egypt. There is no trace of the idea of multiple Hermeses in Ibn Nawbaḫt's account. Instead, there is the view that Hermes was really originally from Babylon, the capital of the Persian Empire. This Iranocentric view of the history of science is attested in other sources, too, as already seen. Ibn Nawbaḫt can therefore hardly have been responsible for "perverting" the legend of Hermes "under Ḥarrānian influence." It seems rather that he was presenting an Iranian tradition that made the astrologer Hermes a Babylonian so that the Iranians could claim him for their own.[49] This point of view must have come naturally to Ibn Nawbaḫt, since he was himself an Iranian scholar and a translator of Middle Persian books into Arabic at a time when Arabic-speaking Muslims were still a small ruling minority. He is indeed one of the main sources about this Iranocentric history of science.[50] But it is true that the Babylonian epithet of his Hermes was later to be applied to the second Hermes of Abū Maʿšar.[51]

As for the charge against Abū Maʿšar, that he fabricated the legend of the three Hermeses, I will show in the next sections that he was actually just combining a couple of late antique chronographic traditions with the Iranian legend. He is almost certainly responsible for the wording of the Arabic version of the biographies of the first two Hermeses that we possess in his name, but not for that of the third Hermes. The next step is to look for ancient sources suggesting more than one Hermes.

4.2 Ancient Greek Antecedents to the Arabic Legend

The preservation of the ancient writings in Egypt through catastrophes is a motif at least as old as Plato's *Timaeus* (23a). The Egyptian Theuth (Thoth), later identified with Hermes, appears briefly in Plato's *Philebus* (274c–d) as the discoverer of mathematics and astronomy and the revealer of writing to Ammon. But the role of

48. Idem, 10.

49. Pingree seems to have thought that the alleged Ḥarrānian reason for making the second Hermes a Babylonian would be for "thereby including Mesopotamia in the religion of Ṣābianism" (Pingree 2002: 25).

50. *Fihrist* 333.12–16 (ed. Flügel 274.8–12); Ibn al-Qifṭī, 255.1–11 (and at 409.3–13 the amusing story is related from Ibn Nawbaḫt's own account of how his cumbersome Iranian name—given as Ḫwarrašāḏmāh Ṭīmāḏāh Mābāzār Dubād Ḫusrawānahšāh—was replaced by the caliph al-Manṣūr with the polite Arabic *kunya* Abū Sahl).

51. On the references to Hermes in Dorotheus, see section 2.3.

Hermes as the *preserver* of this Egyptian knowledge is not found in Plato. It is implied, rather, in the Greek Hermetica, where Isis teaches Horus that Hermes saw and understood everything, and wrote it down in *charaktēres* (Egyptian hieroglyphs) that he then hid. Hermes had successors, she says, in Tat, Asclepius-Imouthes, and Ptah.[52] In the ancient Greek Hermetica themselves there is a notion of multiple Hermeses, though it is not possible to demonstrate that this was transmitted directly in translation to Arabic authors. One of the most widely cited Hermetic books of antiquity, *The Perfect Discourse*, better known today in its Latin version as *Asclepius*, mentions an "ancestor" Hermes. Originally written in Greek, *The Perfect Discourse* was used by early Christians (e.g., it was known in some form by Lactantius ca 310 CE)[53] and was translated into Latin and Coptic. The Greek text survives only in citations by other authors, whereas the Coptic survives only in a partial translation; only the Latin version preserves the passage now relevant.[54] One finds Hermes Trismegistus there addressing his three disciples, Asclepius, Ammon, and Tat. The discourse is particularly directed toward Asclepius and continues in dialogue with him. Hermes here refers to Asclepius' ancestor (*avus*) who first discovered medicine (*medicinae primus inventor*) and furthermore to his own namesake, the ancient Hermes, whose ancestral name he himself has (*Hermes cuius avitum mihi nomen est*).[55] Here is evidence in one of the early Hermetica for the recognition of two Hermeses, and also, perhaps, of multiple Asclepii (Asclepius being Hermes' disciple also in Arabic sources), written at the latest in the third century CE.[56] But because the work itself left no trace in Arabic literature, and Abū Maʿšar shows no sign of having encountered any part of it, we can set it aside and continue the search. In any case, it is important to keep in mind that multiple Hermeses were conceived to exist long before Abū Maʿšar.

There are a few other references to multiple Hermeses in Greek and Latin from the Roman Imperial period.[57] Only one of these, however, brings us closer to the Arabic biographies. In fact, it sheds light on a major source of Abū Maʿšar's legend, as will be seen. The report is found in the Byzantine Greek world chronicle of George Syncellus, who worked in Constantinople around 808–810,[58] extensively citing earlier Christian chronicles that are now lost; his work therefore is of great importance

52. SH 23.5. On Egyptian *charaktēres* see Dieleman 2005: 96–101.

53. Scott 1.77.

54. Nagʿ Ḥammādī VI.8 (including sections 21–29).

55. *Asclepius* 37 = CH 2.347.20–348.6.

56. The appearance of two Hermeses here and in related works is discussed by Fowden 1993a: 29–31. Fowden takes *avus* to mean "grandfather" rather than "ancestor," which is a possibility. See also the Fowden's discussion of the devotion of the Hermopolitans to Hermes, the divine founder of their city, 174–176.

57. Cicero, *De natura deorum* 3.56, cited by Lactantius, *Divinae Institutiones* 1.6, refers to different Hermeses as known to several local cults. Cf. Fowden 1993a: 24.

58. See Syncellus trans. by Adler and Tuffin, xxix.

for the modern understanding of late antique historiography and chronography. One of the sources he cites (indirectly, through another source), mentioned in no other surviving ancient work, is a *Book of Sothis* ascribed to the Egyptian priest and chronicler of the Ptolemaic period, Manetho. The ascription to Manetho is apparently false, although George Syncellus seems to have believed it. He quotes several lines from the beginning of the *Book of Sothis*, introducing it as follows:

> Now it remains to treat small passages from the writings of Manetho of Sebennytos concerning the dynasty of the Egyptians. In the time of Ptolemy Philadelphos, he was serving as a high priest of the temples of idols in Egypt, and, on the basis of monuments lying in the Seriadic land that were inscribed, as he says, in a sacred language and priestly characters by Thoth, the first Hermes, and translated after the Flood from the sacred language into Greek (εἰς Ἑλληνίδα φωνὴν) with hieroglyphic characters, and committed to books, by the second Hermes, Agathodaimon's son, father of Tat,[59] in the shrines of the holy places of Egypt—[on the basis of this] Manetho addressed the above Philadelphos, Ptolemy II, in the *Book of Sothis*, writing in precisely the following words:
>
> "To the great king Ptolemy Philadelphos Augustus, Manetho the high priest and scribe of the holy shrines in Egypt, from Sebennytos and dwelling at Heliopolis, greetings to my lord Ptolemy. It is our obligation, greatest king, to consider all those matters about which you wanted us to inquire. So, then, as you are doing research about the future of the universe, in response to your request, I will produce for you those sacred books composed by our forefather Hermes Thrice-Greatest of which I was aware. I bid you farewell, my lord King."
>
> [End of quotation; Syncellus' commentary resumes:] This is what he says about the translation of the books by the second Hermes. Now after this, he also narrates about the five Egyptian classes [of kings] in thirty dynasties, called by them gods, demigods, and mortal men . . .[60]

The affinity between this passage and the story given by Abū Maʿšar is readily apparent. Here are the first two Hermeses of his tale, one before the Flood, one after the Flood, presented to us in an ancient Greek source. The details of the preservation of

59. It is important to note here the ambiguity of the Greek. The string of genitives, ὑπὸ τοῦ Ἀγαθοδαίμονος υἱοῦ τοῦ δευτέρου Ἑρμοῦ may have been intended to mean "by Agathodaemon, son of the second Hermes," which is how Adler and Tuffin have translated it. My modification to their translation makes the second Hermes the one who recovered the antediluvian knowledge because it is evident that the version of the story transmitted into Arabic must have been understood to mean that it was the second Hermes, and not Agathodaemon, who recovered the ancient knowledge.

60. Syncellus 72–73 (ed. Mosshammer 40.31–41.22); the English translation given here is that of Adler and Tuffin, 54–55, with some modifications. Adler and Tuffin's translation gives notes regarding the textual difficulties presented by this corrupt passage.

knowledge through the Flood, put into writing by the first Hermes and recovered by the second Hermes, cannot have been coincidentally invented twice, so we can conclude that the Arabic account and the Greek account must be related. The next task then is to determine what this relationship is.

Martin Plessner was among the scholars to draw attention to the similarity of the story in this Greek passage to the Arabic Hermes stories. In his perspicacious article of 1954 on the Arabic Hermetica and the state of research on the subject, he concluded that the stories about Hermes related by Abū Maʿšar had to be of Babylonian origin because they contained the story of the Deluge. "It is self-evident," he wrote, "that the missing link has to be found in a Greek text dealing with Aegyptiaca." Although one might not agree that it is self-evident, he was certainly on the right track, and he cited the beginning of the *Book of Sothis* as reported by Syncellus as his missing link.[61] As long ago noted in modern scholarship, it is clear from the opening text of the *Book of Sothis* that the work is from Roman Imperial times, because Ptolemy is addressed by the anachronistic title Augustus (σεβαστῷ). It may well have been written by a Roman Egyptian trying to illustrate the superiority of his ancient country in competition with Berossus' chronography of ancient Mesopotamia.[62]

William Adler suggests that Plessner exaggerated the importance of the *Book of Sothis* as the "missing link" here.[63] However, the citation of the *Book of Sothis* is in fact the only relevant evidence of a closely parallel account of multiple Hermeses before Abū Maʿšar. It is not the proximate link, but it is nevertheless a link between us and the lost account preceding Abū Maʿšar, his source. The account from the *Book of Sothis*, in its specific details, clearly represents a tradition available somehow to Abū Maʿšar. But how did the account of the ancient Hermeses reach him? This question is very important for the Arabic legends of Hermes, but it has not been investigated before.

4.3 A Trail of Chronicles

The chronicle of Syncellus, in which the story of the antediluvian Hermes and his postdiluvian successor appears, is a Christian world chronicle, telling the history of the world from its beginnings down to the Roman period. To explain the relationship between Syncellus' text and Abū Maʿšar's, it is necessary to introduce some basic information about late antique Christian world chronicles. This requires a little excursus, after which Abū Maʿšar's story of Hermes will begin to make more sense.

61. Plessner 1954a: 55–57.

62. See Adler 1989 (chapter 2) on the contest for the chronological priority of nations among late antique chronographers.

63. Adler 1989: 62n73.

Christian world chronicles were a successful genre of Greek historiography that sought, in its formative centuries, to prove the truth and superiority of the Christian nation by showing the relative antiquity of its scriptures and the persons and events described therein. This meant establishing the early dates of the patriarchs and the prophets, especially Abraham, with whom God founded his covenant, and Moses, the supposed author of the Pentateuch. Simultaneously, it attempted to show that other nations, especially the Hellenic, were younger and derivative. In this effort, some of the most intractable debates concerned the precise dates of the most important ancient events, such as the creation of the world, the creation of Adam, Adam's Fall from the Garden, and the Flood. Christian chroniclers were especially divided about the possibility of knowing any history of the period before the Flood beyond the very little told in the book of Genesis. The Bible itself provided the basis for their chronology of ancient times with its many indications of the number of years passing at each stage, particularly the life spans of the patriarchs and reigns of kings. Sometimes a millennial scheme was imposed on the timeline, so that Jesus was born in the middle of the sixth and last millennium of the world. However, inconsistencies and ambiguities within the text of the Bible, and differences in the numbers between its Hebrew, Greek, and Aramaic versions, made the establishment of a universally accepted timeline impossible. Attempts to revise the numbers of years to a finally correct form never ceased, and Syncellus himself was a late participant in this effort. Later chronographers, including Syncellus, often recorded the variant opinions of their predecessors in an attempt to be comprehensive or to refute opposing views.

Another problem facing Christian chroniclers was the dilemma of whether, and how much, to rely on unorthodox sources for very ancient times. These included the Babylonian history of Berossus and the Egyptian history of Manetho, which claimed to give information about kings who, according to those chronologies, should have long antedated the Flood, as well as noncanonical Jewish and Christian texts, such as *1 Enoch*, which elaborated at length on antediluvian events described only briefly in the canonical Bible. It is not necessary to dwell much on the factors involved in these debates here, though they are interesting.[64] It is important to note, however, that the periods of Babylonian and Egyptian kingship provided by Berossus and Manetho, respectively, were extremely ancient, beyond tens of thousands of years, and were regarded as incredible by both Hellenic and Christian Greek authors. Besides these remarks, suffice it to say that Christian world chronicles included excerpts of numerous passages drawn from ancient authorities, some biblical and some not, levied to prove their arguments about the dates and significance of this or that ancient event. In this way they become storehouses of excerpts derived from texts otherwise lost to us today. One stylistic feature of these chronicles is that their authors regularly cite sources from citations in earlier chronicles *without ever seeing*

64. Adler 1989 provides a helpful and detailed treatment of all these issues.

the entire work originally cited, and *without mentioning the intermediary chronicler* from whom they took these citations. They give the appearance of knowing many ancient sources not actually at their disposal directly.

Eusebius' world chronicle, completed in 325, was the most influential of all of them.[65] It was arranged in two parts. The first part, surviving entire today only in Armenian, was an essay discussing the validity of different ancient sources and the best method to approach chronography. Among other things, Eusebius argued against using the ancient Egyptian and Babylonian dates of Manetho and Berossus, but in the course of this argument, he cited these works at length. He began his dates not with the creation of the world but with Abraham, whom he considered as early a historical figure as could be reliably discussed. The second, and more influential, part of his chronicle was the application of the conclusions reached in the first part in "chronological canons," a long timeline arranged in columns, with a different column for each nation, and with the nations synchronized according to Eusebius' understanding. The years of the era of Abraham formed the backbone of the timelines of kings and nations, running from beginning to end. All of this was annotated in the margins and spaces between the columns with notes about the important events that took place and the famous historical persons who lived in a given year. Subsequent world chronicles tended to be arranged in a similar tabular form. They also took over large amounts of Eusebius' collected information, often without attribution.

Around 400 CE, an Egyptian chronicler named Panodorus was writing in Alexandria, using and criticizing Eusebius' chronicle in different ways. His own work is lost, and what is known about it is based solely on the citations that Syncellus made of it. The story of the two Hermeses came to Syncellus from this Panodorus. He makes this clear in his discussion following the passage about the two Hermeses translated above. In fact, he cites it mainly in order to refute Panodorus' novel use of this *Book of Sothis* as a source of ancient dates.[66] In other words, the story of the two Hermeses from the *Book of Sothis* was cited by Panodorus, who was interested in the ancient Egyptian dates it provided. Syncellus knew it not from the *Book of Sothis* directly, but from his source, Panodorus (or rather from an intermediary that cited Panodorus in turn).[67] The only reason the passage survives is that Syncellus wanted to attack it.

Because Panodorus, and not Syncellus, was the source of the story of two Hermeses, naturally the next step is to see whether Abū Maʿšar had any of Panodorus'

65. There are several excellent introductions to the work in English alone: Mosshammer 1979, Adler 1992, Croke 1982, Witakowski 1999–2000 (this last one is particularly relevant to the Eastern reception of the Eusebian tradition).

66. Panodorus' treatment of Egyptian and Babylonian antediluvian chronology is explained by Adler 1989: 75–80.

67. Adler 1989: 132–158 treats the kinds of sources at Syncellus' disposal, including whether or not he saw his sources directly.

writings at his disposal. As mentioned, however, the chronicle of Panodorus is lost, and Syncellus is the only witness to its existence. But Panodorus had a contemporary named Annianus, another Alexandrian monk who wrote another chronicle now lost, apparently only a few years later and relying on Panodorus' work. Syncellus also includes some discussion of Annianus' chronicle along with that of Panodorus. Syncellus is one of the most important sources of information also about Annianus. He regards Annianus' chronicle as more sound in respect to its overall tally of years from creation, but he makes it clear that the chronicles of Panodorus and Annianus shared a great deal of material, some of it blameworthy. In particular, they had common recourse to sources on antediluvian history that Syncellus (as Eusebius before them) criticized as false: both accepted the existence of an antediluvian Babylonian kingdom (according to the work of Berossus as cited by the earlier, non-Christian historian of the first century BCE, Alexander Polyhistor), and they also both accepted the views of Manetho, along with those of Abydenus and Apollodorus, three historians "who," Syncellus says, "told the same kinds of stories (as that of Berossus' antediluvian history) about a dynasty of Egyptian kings before the Flood."[68] In short, this Annianus relied on the same sources as his fellow Alexandrian for antediluvian Egyptian dates, including the *Book of Sothis* attributed to Manetho.[69] Whether Annianus did so directly from the *Book of Sothis* or indirectly, taking his information from Panodorus, is irrelevant. There is every reason to assume that Annianus mentioned the ultimate origin of his ancient Egyptian dates, the two Hermeses, in his chronicle, even if they appeared only as annotations on his chronological tables, on the column of Egypt, before and after the Flood. In fact, it is possible that Syncellus knew the theories of Panodorus, cited above, only through the intermediary of Annianus.

The question then is whether Abū Maʿšar had this information from Annianus at his disposal. The short answer is yes. Abū Maʿšar had the story of the antediluvian Hermes and his postdiluvian successor through the transmission of Annianus, although indirectly.[70] As I will show below, Abū Maʿšar actually cites a date from Annianus, by name, in one of his works, and he proves to have been well acquainted with the Christian chronographic tradition in Arabic translation. Before discussing Abū Maʿšar's use of ancient chronographic sources, however, some words about the Eastern reception of Annianus are necessary.

68. Syncellus 35. Trans. adapted from Adler and Tuffin, 46. See also Gelzer 1885–1888: 2.191. Syncellus 63.19 Panodorus was first: Sync 74.8ff. Annianus finished in 412. Panodorus flourished ca 395–408.

69. Scholars (e.g., Verbrugghe and Wickersham 1996: 101–102) doubt that Manetho was really the author of this book, but the accuracy of the attribution is beside the point here.

70. In an earlier publication (van Bladel 2003) with close constraints of space, I presented a much more concise version of the following discussion, from which a reader might infer that Abū Maʿšar had direct access to a complete text of Annianus. As will be evident here, I do not think it likely that he actually had the full text of Annianus before him.

4.4 Annianus in Syriac and Arabic

The reception of Annianus' chronicle was more favorable than that of Panodorus, but it also was subject to the demands of its genre.[71] Once another chronicler mined it for information, and incorporated whatever he wanted out of it into his own chronicle, the original could be discarded and the more up-to-date one preserved. In other cases, authors of chronicles regarded the discussion of the earlier dates as a finished subject, referring the reader to those works; their concern was to update the older chronicles with more recent events. The result of this method of working is that there are quite a few traces of Annianus' chronicle surviving in numerous later works, but it is difficult to tell how much of Annianus' original writing was actually available to any given later author. Sometimes these traces are nothing more than the citation of a length of time between two events as given by Annianus: for example, the number of years he believed to have elapsed between the Flood and the beginning of the Seleucid era. In other cases, but more rarely, whole paragraphs are cited on Annianus' authority. When a chronicler names Annianus as one of his authorities, it is likely that a great deal of other information related by that chronicler is owed ultimately to Annianus without citation. Annianus himself had evidently used his sources, including Eusebius, Alexander Polyhistor, and others, in a similar fashion. Such is the character of the late antique chronographic tradition.

One innovative feature of Annianus' chronicle was that in it he formulated a cycle of 532 years, according to which the dates of the year and the days of the week on which Easter falls are repeated.[72] These Paschal tables were admired and became standard.[73] Annianus used the cycle to prove not only that the birth of Jesus was exactly in the middle of the sixth millennium of Creation, but also that the creation of the world and the resurrection of Jesus occurred on the same date of the calendar year, the first day of the first month of the religious calendar of the Jews (1 Nisan), being also the twenty-fifth of March in both instances.[74] Another important characteristic of Annianus' chronicle that appears to have made it an important supplement to Eusebius' chronicle in the eyes of later historians was its willingness to relate a history of antediluvian times. Eusebius had declined to include the antediluvian period because he considered such early history unreliable. But when citations from Annianus turn up in Syriac and Arabic works, most of them are concerned with the primordial times before and after the Flood, which Eusebius left out. Annianus' treatment of antediluvian times is just what Syncellus says his chronicle had in common with that of Panodorus. All of the extant citations from Annianus concern the earliest history of the world.

71. It is not clear whether Panodorus' chronicle really circulated at all; it may have been only Annianus who related its contents. Witakowski 1999–2000: 434–435 discusses the Eastern reception of Annianus in the context of Eusebian chronographic tradition.

72. Syncellus 35 and 382.

73. Neugebauer 1989: 27–30 explains briefly the mathematics behind this cycle and its influence.

74. Syncellus, 1.

The resulting picture may appear complicated, but fortunately the Syriac chroniclers, who based their work directly on the Christian Greek antecedents, were often scrupulous in listing the sources that they used, even if they used them only as cited by other, unnamed authors. One of the most scrupulous was the eleventh-century chronicler Elias of Nisibis, a cleric of the Church of the East, who composed a bilingual Syriac and Arabic chronicle that survives in a single lacunose manuscript. Though whole sections are missing where pages have fallen out, the remaining text has much to say about its sources. In the Syriac passage where he discusses his predecessors' treatments of Easter datings, he gives a good idea of how Annianus came to be used in Syriac tradition.[75] The beginning of the passage, presumably discussing Annianus himself and his Easter tables, is lost in the manuscript. What does survive is the following. In the days of Justinian (r. 527–565), a chronicler named Andronicus used Annianus for his Paschal tables, but disagreed with the exact dates given by Annianus and Eusebius for the biblical patriarchs.[76] Then, in 691–692, Jacob of Edessa (d. 708), a renowned Syrian scholar who read Greek, wrote his own chronicle in Syriac, agreeing with both Annianus and Andronicus on the Paschal tables, and also criticizing Eusebius.[77] Moreover, Elias says that in the days of the Persian Emperor Ḫusrō II (r. 590–628), a chronicler named Šemʿōn Barqāyā made a translation—apparently of Andronicus' chronicle—that the Easterners ("Nestorians") were still using in the time of Elias, though he says it was mistakenly thought to be by Eusebius.[78] Almost all of these intermediary chronicles, including Andronicus', are lost, but they clearly used Annianus and Eusebius a great deal, and must have cited them extensively. Jacob of Edessa's chronicle survives in what appear to be excerpts or reworked portions, including some of his chronographic discussions and his table of dates from Constantine onward.[79] Again, these chroniclers cited their earlier sources through an unknown number of removed stages. Although Elias mentions them in a context where he discusses the cycles of Easter dates, it is clear that all of these authors

75. Elias text pt. 2, 99.12–26, Latin trans. pt. 2, 111.

76. Witakowski 1999–2000: 435–436 suggests that this Andronicus wrote in Syriac, not Greek, because no references to this Andronicus survive in Greek. However, Elias makes it appear that Šemʿōn Barqāyā translated the text into Syriac, which implies that Andronicus did write it in Greek. He says that Šemʿōn "made a translation of (less likely: a commentary on) this chronicle which the Easterners presently use" *ʿbad Šemʿôn Barqāyâ pûššāqâ da-krônîqôn hādê d-bāh metḥaššbîn Madnḥāyê b-hānâ zabnâ*. "This chronicle" is apparently the chronicle just mentioned, that of Andronicus.

77. The surviving portions of Jacob's chronicle are edited and translated (Jacob of Edessa 1903). On Jacob of Edessa's chronicle in general see Witakowski 1999–2000: 429–431 and 2008. Jacob cites "Annianus, the monk of Alexandria" (*Aniyânôs dayrāyâ haw d-Aleksandriyâ*) and Andronicus among other chronographic authorities in one of his letters to John the Stylite (Nau 1900: 584.35–37, trans. 590).

78. Elias text pt. 2, 99.18–22, Latin trans. pt. 2, 111.

79. Palmer 1993: 36–42 presents an English translation of one of the fragments of Jacob's work.

wrote chronicles continuing the earlier ones at their disposal and correcting the earlier ones where they deemed necessary.

Jacob of Edessa's and Šemʿōn Barqāyā's chronicles were evidently important sources for Elias of Nisibis, from which he probably learned most of the foregoing information. Elias is able to cite historical dates and period lengths from the works of all of his main predecessors, including Eusebius, Annianus, Andronicus, and Jacob of Edessa, indicating that his proximate sources must have related the views of his predecessors on a number of points as well as their own views.

Another important Syriac chronicler, whose work must be pieced together from the extensive citations made from it by later chroniclers, was Dionysius of Tell Maḥrē (d. 845).[80] The introduction to his chronicle, listing its sources, was cited by two later chroniclers, Michael the Elder (d. 1199) and the anonymous author of the *Chronicle to 1234*.[81] Among these sources are Andronicus, Annianus, Jacob of Edessa, and quite a few others. From Dionysius, later chroniclers had access to information from many earlier works. It is probably from Dionysius that both Michael the Elder and Bar ʿEbrāyā (Abū l-Farağ Ġrīġūriyūs Ibn al-ʿIbrī) (d. 1286) cite Annianus, Andronicus, and the other commonly cited earlier chronographers, just as Elias did. Often they relate the chronologies of Annianus and Andronicus side by side as two possible views on things, both of which needed to be taken into account. They do this usually without crediting their immediate source at every point. The real source for any report may have been obscured by an intervening chronicler in the case of any given report.

Unfortunately, neither Elias nor Michael have anything to say about Hermes that might tell us about the coverage of the antediluvian sage in Annianus. The text of the former is lost where one would expect any mention of Hermes, if it had occurred there. Michael's huge chronicle has plenty of room for such a notice but he does not give it, although he does mention that Enoch introduced writing into the world.[82] Ibn al-ʿIbrī (Bar ʿEbrāyā), the learned Syrian Orthodox maphrian who used both Syriac and Arabic sources, presents a different case requiring special attention. This thirteenth-century scholar wrote two chronicles, one in Syriac and one in Arabic, which contain much of the same information but significant differences as well. They include somewhat different treatments of Hermes in Syriac and in Arabic. The Hermes of Ibn al-ʿIbrī's chronicle is a modified version of that of Abū Maʿšar, supplemented with the account of al-Mubaššir ibn Fātik (studied in section 5.2 in this volume).[83] Here there are three different individuals named Hermes. Whatever information Ibn al-ʿIbrī had at his disposal from earlier chronicles, he preferred that

80. Palmer 1993: 85–221 reconstructs a good deal of Dionysius' chronicle.

81. Palmer 1993: 91–92.

82. Michael the Elder 1.5: text 4, trans. 9.

83. Ibn al-ʿIbrī, *Kitāb Taʾrīḫ ad-duwal*, 11.6–13.6.

of Abū Maʿšar instead, which was by then the standard in Arabic literature. In his Syriac chronicle, however, written for a Christian audience, Ibn al-ʿIbrī mentions only one Hermes. Hermes here is identified with the patriarch Enoch, and the report comes in connection with the genealogy of Enoch in the midst of the patriarchs. In both the Syriac and the Arabic text, Hermes-Enoch is said to be the teacher of Asclepius. Ibn al-ʿIbrī did not translate the Arabic of Abū Maʿšar into Syriac, although he paraphrased some parts of it.[84]

Information deriving from this Syriac Christian chronographic tradition is easy to find in Arabic, and Muslims scholars derived it from learned Christians. Dates and king lists that had passed through Eusebius' chronicle, or his later critics, are common in Arabic world histories.[85] The historian and traveler al-Masʿūdī (d. 956) reported that he himself saw a copy of Annianus' chronicle.[86] He mentions it in a list of the Christian chronicles that he has seen as "the book of Annianus, the Alexandrian monk, in which he ordered the kings of Rome and those of other nations, their lives and histories, from Adam to Constantine son of Helen." Al-Bīrūnī (d. after 1050) had chronicles at his disposal from which he could cite some of the same king-lists as given by Eusebius. For this information he names his authorities as "a certain book" (*baʿḍ al-kutub*),[87] "the people of the west" (*ahl al-maġrib*),[88] and "a certain historian" (*baʿḍ ahl al-aḫbār*).[89] But in his *al-Qānūn al-masʿūdī*, he cites both Annianus and Andronicus by name as authorities on ancient dates, indicating that his source was part of this Christian chronographic tradition.[90]

It should not come as a surprise, therefore, to find that Abū Maʿšar himself cites Annianus by name in one of his surviving works on historical astrology, *Kitāb al-Qirānāt* (*Book of Conjunctions*), also known as *Kitāb al-Milal wa-d-duwal* (*Religious Nations and Dynasties*). There has been some confusion about the identity of the author of this work since the Middle Ages. According to an-Nadīm, Abū Maʿšar wrote this book for the little-known astrologer Ibn al-Bāzyār.[91] One recension of manuscripts attributes the work to Ibn al-Bāzyār himself,[92] and al-Bīrūnī (d. ca 1050) followed the latter attribution. Al-Bīrūnī also stated specifically that the *Book of Conjunctions* of Ibn al-Bāzyār cited Annianus, and he related this citation in

84. For the accounts of Hermes-Enoch, see Bar ʿEbrāyā, Syriac chronicle trans. Budge 5, Bodleian MS Hunt. 52 facsimile 2v; Ibn al-ʿIbrī, *Taʾrīḫ* 11.6–12.12.

85. For examples, see Stern 1972.

86. al-Masʿūdī, *Tanbīh*, 154.16–155.2. The correct reading of the name there was recognized by E. Honigmann 1952, who also supplied a very useful bibliography on Annianus. The work is mentioned in a passage in which al-Masʿūdī is listing books of the histories of the Christians that he has seen. Among them are the histories of Ḥabīb al-Manbiğī and Saʿīd ibn al-Biṭrīq.

87. al-Bīrūnī, *Āṯār* 84.14.

88. al-Bīrūnī, *Āṯār* 87.1.

89. al-Bīrūnī, *Āṯār* 87.6.

90. al-Bīrūnī, *al-Qānūn al-masʿūdī*, I.171.1. The name is given by the editors as *ʾbnʾnws*. As I will show below, al-Bīrūnī's source here may have been Abū Maʿšar's *Thousands*.

91. *Fihrist* 336.2 (ed. Flügel 277.12).

92. Pingree, "Abū Maʿshar," in *DSB* 1.36; cf. GAS 7.146, 154.

his own book.[93] Al-Masʿūdī, on the other hand, wrote of a certain Ibn al-Māzyār, a student of Abū Maʿšar, who compiled a summary of his teacher's *Thousands*. This is probably an error for Ibn al-Bāzyār.[94] It is easy to see how confusion over authorship arose given the relationship between the two astrologers, student and teacher. However, the Western tradition of the manuscripts, and modern scholarship, credits Abū Maʿšar, and not his student, with the *Conjunctions*.[95]

To our good fortune, this *Book of Conjunctions* has recently been edited in both the original Arabic and the Latin version.[96] It shows that al-Bīrūnī was exactly right in his citation of Annianus through it (except for the omission of two words, lost by the manuscript tradition).[97] However, the editors were unable to identify the correct reading of the name Annianus.[98] Pingree noted the trouble with this name already in 1963, when he drew attention to the "name, corruptly preserved in Arabic" cited in the *Book of Conjunctions*.[99] Following the suggestion of A. J. Sachs, Pingree suggested that the obscure name should be read "Abydenus,"[100] a known early Greek chronographer, although such an interpretation led to other difficulties. The name should, however, be read as Annianus, as Sachau's translation of the *Chronology* of al-Bīrūnī rendered it long ago.[101]

More problematic is that Pingree misunderstood the text to say that Annianus had proposed a date for the Flood of -3101 (i.e., 3102 BCE).[102] The text does not actually say that Annianus proposed a date for the Flood of -3101 at all; it says rather that

93. Al-Bīrūnī, *Āṯār* 21.19, says that Ibn Bāzyār reported this statement from "Annianus, one of the historians," *ʾnynws aḥad aṣḥāb al-aḫbār* in his *K. al-Qirānāt* (Book of Conjunctions). Sachau read *ʾṯnyws* in the 1878 Arabic edition of the *Āṯār*, but rendered the reading "Anianus" according to the correct pointing (with a single "n," a common alternative spelling) in the index to the Arabic edition and in his English translation (25.27) of 1879. The difficulty in reading this name in Arabic script becomes especially important in the discussion immediately below. Rowson 1988: 196–197 also noticed and discussed the reading of this name.

94. *Murūğ* (Pellat §1419 = 2.408.11) gives the voweled reading "Ibn al-Māziyyār." Pingree 1968: 23n4 and *DSB* vol. 1, 36 questioned the reading of Ibn al-Māzyār, suggesting Ibn al-Bāzyār. As far as I can determine, only one Māzyār (or Māziyār) is attested in Arabic literature: the rebel leader Māzyār ibn Qārin, who seized power from the Bāwandids of Ṭabaristān, 825–839 CE (Bosworth 1996: 164–165). The name at least is contemporary to Abū Maʿšar, and a son of this man could (in theory) have been his student. The little that is known about Ibn al-Bāzyār, student of the better-known mathematician Ḥabaš (fl. mid-ninth cent.), is summarized in GAS 7.154.

95. Burnett and Yamamoto 2000: 1.xxiii–xxv.

96. Burnett and Yamamoto 2000.

97. The words omitted by al-Bīrūnī are *wa-šahr*. As shown below, this reading is the correct one, and al-Bīrūnī's *Āṯār* should be emended at 21.21 with this addition after *sana*.

98. Their text reads *ʾbṯnws* (22.209), which is correct except for the pointing. The facing translation includes a note saying, "It is difficult to identify this authority, whose name appears in several different forms in the Arabic and Latin manuscripts" (23n26).

99. Pingree 1963a: 243.

100. Idem: 243n114.

101. The Arabic letters *ʾbṯnws* and *ʾnynws*, deprived of their diacritical dots, are identical in Arabic script. (Compare ابثنوس and انينوس.) The reading of al-Masʿūdī, al-Bīrūnī, and Ibn al-ʿIbrī is slightly different: *ʾnyʾnws* انيانوس (see Honigmann 1952)

102. He says that "this information is from a translation by E. S. Kennedy."

"Annianus and others have stated" (*wa-qad ḏakara Anniyanūs wa-ġayruhū*)[103] that the period between the creation of Adam and the Flood was two thousand, two hundred and twenty six years, one month, twenty-three days, and four hours. Abū Maʿšar's date of -3101 for the Flood comes from the Indian tradition of the *Sindhind*, not from Annianus.[104] In fact, the context of the citation makes it clear that Annianus has no readily apparent bearing on Abū Maʿšar's discussion whatever, because in this passage he was concerned not with the period between the creation of Adam to the Flood, but that between the Flood and the advent of Islam. What is important here is that Abū Maʿšar cites Annianus by name in a parenthetical remark. It shows without any doubt that he had access, in some form, to Christian chronographic traditions, and it is only the first of several indications to be discussed.

Decisive proof that Annianus is the correct reading of the name in the *Conjunctions* is provided by Ibn al-ʿIbrī in his already mentioned Arabic chronicle, *Kitāb Taʾrīḫ ad-duwal*. Ibn al-ʿIbrī cited Annianus several times by name as an authority on specific dates. Among these citations he says, "Annianus the Alexandrian Monk said that the period between the beginning of the creation of Adam and the Friday night on which was the Flood was two thousand, two hundred and twenty six years, one month, twenty-three days, and four hours."[105] First, the reading of the name Annianus is just as al-Masʿūdī reported it, with the epithet "the Alexandrian monk," confirming the identity. Second, Abū Maʿšar's *Book of Conjunctions* gives exactly the same period, down to the hour, between Adam's creation and the Flood. This establishes the identity of the name in the *Conjunctions* beyond any doubt.[106]

It is interesting to note that the eastern Iranian philosopher al-ʿĀmirī (wr. 985–986, d. 992) cites the same information in a discussion of eras. He refers to the Era of the Deluge used by Abū Maʿšar in his *zīǧ*, and adds, "It is reported by one of the historians that between the beginning of the creation of Adam and the Thursday night on which the Flood began were 2226 years, 1 month, 2[3] days, and 4 hours—but this man wrote his compositions arbitrarily, without proving what he talked about."[107] It is not clear whether "this man" refers to the historian, clearly Annianus, or to Abū Maʿšar, from whose work this information came to al-ʿĀmirī.

103. Burnett and Yamamoto 2000: 1.22.209.

104. Pingree himself makes this clear in his discussion of the passage (1963a: 243).

105. Ibn al-ʿIbrī 9.25–10.2. See Annianus in the index of the edition for the citations.

106. The only mystery remaining is that Syncellus (33) and Michael the Elder (1.4, trans. 8) report that that Annianus' period between Adam and the Flood is 2,242 years. The antediluvian tally of 2,242 years was also given by Eusebius (Adler 1989: 47 and 119–120). There is therefore a difference between the two periods reported for Annianus—that given by Abū Maʿšar and Ibn al-ʿIbrī on the one hand and by Syncellus and Michael on the other—of sixteen years, less one month, twenty-three days, and four hours. What is the source of this difference? How could Annianus have followed both tallies? The discrepancy led Rowson 1988: 196 to doubt the identity of the two figures (Annianus and "Athenaeus"). The difference must come from chronographic considerations that remain, for now, lost or undiscovered.

107. Translation adapted slightly from Rowson 1988: 67 (*Kitāb al-Amad ʿalā l-abad* 2.16).

The Persian history written by al-Ǧuzǧānī in India, *Ṭabaqāt-i Nāṣirī* (wr. 1259–1260),[108] contains more evidence for Abū Maʿšar's detailed familiarity with material from the chronicle of Annianus.[109] William Adler discovered that this Persian work incorporated into its ancient history an influential account derived from Annianus about the origin of kingship before the Flood, "Annianus' euhemerizing interpretation of Berossus, *1 Enoch*, and Genesis 6." Adler was not in a position to explain how the account reached a Persian writer in northern India in the thirteenth century.[110] Al-Ǧuzǧānī does indeed tell the story of the angelic Watchers, their fall, their mating with human women, and their troublesome giant offspring, a story derived ultimately from the first part of *1 Enoch*, the *Book of the Watchers*.[111] The first king elected by the people to resist the giants appears there as *Aylūrūs*. This is Aloros, the first king from Berossus' Greek version of the Sumerian king list, which was incorporated into the chronicle of Annianus and many later chronicles.[112] But al-Ǧuzǧānī explicitly states that this story in his account derives *from Abū Maʿšar*, beginning the account with the words, "and in the transmitted account of the *Qānūn-i masʿūdī*, that Abū Maʿšar the Astrologer made..."[113] Sven Hartman interpreted this phrase to mean that al-Ǧuzǧānī strangely cited Abū Maʿšar as the author of the *Qānūn-i masʿūdī*; this last work is in fact the *maximum opus* rather of al-Bīrūnī, who lived two centuries after Abū Maʿšar.[114] Yet the Persian is ambiguous here (as reflected in the English translation above), and the intended meaning is probably rather that it was the transmitted account (Arabic *riwāya*, Persian *rivāyat*), provided by al-Bīrūnī, that was Abū Maʿšar's. In other words, what Abū Maʿšar made was not, of course, al-Bīrūnī's book, but the transmitted account, or tradition, contained therein. Again, Adler showed that this tradition definitely derived from Annianus.

The trail leads then to the source that al-Ǧuzǧānī cited, al-Bīrūnī's *al-Qānūn al-masʿūdī*. Like all *zīǧes* (astronomical handbooks), this work contains a section on calendars and the conversion of dates, one of the specialties of astronomers.[115] Al-Bīrūnī had already written his major work on chronology (*al-Āṯār al-bāqiya*)

108. The dates of composition given here are from A. S. Bazmee Ansari, "al-Djuzdjānī," *EI*², 2.609a–b.

109. al-Ǧuzǧānī 132–133 (near the beginning of the fifth *ṭabaqa*). The relevant passages of this text are given in abbreviated form by Hartman 1953: XXVII–XXX (Persian text) and 151–155 (French translation with notes).

110. Adler 1989: 127–128.

111. "The Watchers," in Greek ἐγρήγοροι, is rendered in Arabic quite accurately as اليقظى *al-Yaqẓā*, "the Vigilant," but the name is corrupted in the manuscript tradition of the *Ṭabaqāt-i Nāṣirī* to النبطى *an-Nabaṭī*, "the Nabataean" (in Arabic, *Nabaṭī* means "ʿIrāqī Aramaean" and does not refer to the ancient kingdom of Petra). On this misreading cf. Hartman 1953: 153n1.

112. al-Ǧuzǧānī 132.20.

113. idem, 131.25–132–1: *va ba rivāyat-i Qānūn-i masʿūdī ki Abū Maʿšar-i Munaǧǧim karda ast....*

114. Hartman 1953: 153n2. Hartman translated the passage differently as "Et d'aprés le récit de Qānūn-i Masʿūdī, que l'astronome Abū Ma'šar *a écrit...*" (idem: 153, emphasis added).

115. King and Samsó 2001: 19–20.

when he composed his *zīǧ*, the *Qānūn*. Predictably, it contains well-informed chronological tables. However, some of the information here is not found in his *Āṯār*, at least as it is known today. We find in the *Qānūn* an abbreviated account of the Watchers and their fall at the head of its chronological charts, themselves distant descendants of Eusebius' canons. This is followed directly by a remarkable table of all the ancient Chaldaean kings, as listed originally by Berossus, beginning with Aylūrūs/Aloros.[116] But problems remain. The story told at length and in detail by al-Ǧuzǧānī appears in al-Bīrūnī's *Qānūn* only in the form of a schematic, chronological canon. Al-Bīrūnī gives neither the full story of the Watchers related by al-Ǧuzǧānī, nor does he cite Abū Maʿšar as an authority for this section of the work. But al-Bīrūnī *does* cite Annianus and Andronicus as authorities on ancient dates in the commentary following his chronological charts![117]

The simplest solution to this puzzle might be to assume that al-Ǧuzǧānī had at his disposal a fuller version of al-Bīrūnī's *al-Qānūn al-mas'ūdī* than that presented in the modern edition. Perhaps a longer commentary or narrative, now lost, once accompanied al-Bīrūnī's chronological charts. Since it has already been shown that al-Bīrūnī cited Annianus through Abū Maʿšar's *Book of Conjunctions*, one may assume that something similar happened here, where al-Bīrūnī used another work of Abū Maʿšar's. Whatever the case, al-Ǧuzǧānī clearly makes Abū Maʿšar responsible for the story proved by Adler to derive from Annianus. It would be hard to explain otherwise the appearance of Abū Maʿšar's name in his work at this point.[118]

All this is a clear indication that Abū Maʿšar took into his chronology more than a few details about ancient history derived from Annianus. Abū Maʿšar was the source, for both al-Bīrūnī and, indirectly, for al-Ǧuzǧānī, of the story of the angelic Watchers, their fall, and the origin of kingship with Berossus' first Babylonian, antediluvian king, Aloros. In other words, Abū Maʿšar not only had dates cited from Annianus, but he also had some of the narrative commentary derived from Annianus' chronological canon, specifically for antediluvian times. I think that knowledge of two Hermeses, one before the Flood and one after—Annianus' authorities for dates on antediluvian *Egypt*—also reached Abū Maʿšar indirectly from Annianus in just the same way. This explains his knowledge of the two Hermeses of the *Book of Sothis*.

For those who are skeptical of this conclusion on the grounds that one cannot in any case be certain about a lost work known from scattered citations, there is still a general point to consider. Because both of the earliest reports on the two Hermeses

116. al-Bīrūnī, *al-Qānūn al-mas'ūdī*, I.149. Aloros الوروس is mispointed in the edition as ايلولزوس Aylūzūs.

117. Annianus appears in the edition as ابنانوس Abnānūs (idem, I.171.1).

118. It may even be suggested that the text of al-Ǧuzǧānī requires emendation to the following effect: "and in the transmitted account of the *Qānūn-i mas'ūdī and the transmitted account* which Abū Maʿšar the Astrologer made…"

are found in works of authors concerned with chronology, Abū Maʿšar in Arabic and Pseudo-Manetho as related by Syncellus in Greek, it becomes extremely likely that the story of two Hermeses was also conveyed by a work on chronology. There are no other known sources for the coincidence between the two stories of Hermes. Moreover, the surviving fragments of Abū Maʿšar's *Book of the Thousands* amply show that Abū Maʿšar was well informed about Christian world chronography, and that he treated the subject in detail in that very work.

4.5 Abū Maʿšar as a Chronographer

Pingree's important study of the *Thousands* of Abū Maʿšar does not recognize that this work was indebted to Christian chronographic tradition,[119] but focuses on two tasks: a study of the astronomical parameters and astrological doctrines that Abū Maʿšar employed, and the attempt to prove that Abū Maʿšar had deliberately concocted a false history of science that linked his astrological practice to the antediluvian Hermes. His analysis is based correspondingly on two main sources of information. One is several surviving unpublished mathematical abridgments of the *Thousands*, which Pingree consulted in manuscript. These were the sources for his study of Abū Maʿšar's astronomical parameters.[120] The other is a number of extant fragments of the *Book of the Thousands* showing that Abū Maʿšar discussed a variety of chronological problems and ancient eras at fuller length.[121] Apparently none of the chronological discussions is present in any of the aforementioned abridgments in the manuscripts, because Pingree did not mention any.[122]

There is no intention here of discussing the mathematical or astrological aspects of Abū Maʿšar's *Thousands*. What is of concern is rather Abū Maʿšar as a historian, or rather chronicler, because it is in his historical discussion that the story of Hermes was related. The second of these two types of witnesses, the fragments of the historical part of the *Book of the Thousands*, provided the grounds for Pingree's allegation that Abū Maʿšar had concocted a fake history of science. In this section, the reëxamination of these fragments leads to a different conclusion.

Most of the citations of the *Thousands* made by other Arabic historians make it readily apparent that Abū Maʿšar's lost work presented a fairly detailed discussion of ancient chronology. Not only did he have at his disposal at least one Christian world chronicle in Arabic translation, but he also read somewhere an explicit discussion of

119. In fact, Pingree disregarded Plessner's study, discussed above, which pointed to the chronographic tradition as Abū Maʿšar's source. Apparently Pingree thought that Plessner did not understand the sources at Abū Maʿšar's disposal: "[Plessner] does not deal with the more contemporary theories of the transmission of science that Abū Maʿshar actually had at hand" (Pingree 1968: 2–3n4).

120. Pingree 1968: 21–69.

121. Pingree 1968: 1–21.

122. Such chronological information is entirely omitted at least in the *Muntaḫab* of the *Kitāb al-Ulūf* by as-Sigzī, which I examined in British Library Arabic manuscript Or. 1346 80b–91a.

the problems at stake in Christian world chronography. One of the most important witnesses to this is the Iranian scholar Ḥamza al-Iṣfahānī, who lived about a hundred years after Abū Maʿšar. Ḥamza includes several passages of Abū Maʿšar in his own excellent chronographic work (wr. 961) and specifically states in one of these that he is excerpting from the *Book of the Thousands*. The other passages, although not citing the *Thousands* by name, nevertheless clearly come from the same work. Ḥamza uses one long passage as an introduction to the chronography of ancient Iran:

> Abū Maʿšar the astrologer said: "Most chronicles (*tawārīḫ*) are confused and corrupt. The reason that the corruption takes place is that they present the years of one of the nations from times gone by (*min al-azmina*) and deal with long periods. When they copy it from book to book, or from language to language, errors occur by addition to it or subtraction from it. This is like what happened to the people of the Jewish community (*milla*) in the years that are between Adam and Noah and between other prophets and nations that they tell about in chronography (*at-ta'rīḫ*), for they are in disagreement about them. Many of the people of various regions of the earth disagree with them about this [antediluvian period] also. Likewise, much confusion and corruption is evident in the years of the kings of Persia and their chronography, as well as in the continuity of the days of their rule from the beginning of time until their rule disappeared. At the same time, they claim that the earth remained for many years, in successive periods, without having a king from among them or from anyone else."[123]

In the rest of the citation, Abū Maʿšar goes on to discuss the periods during which the ancient Iranians had no king for an unknown length of time. "Because of this," he says, "they hold the years of the chronicles of their ancient kings differently" between the different periods. "They differ also on the lifespans of their kings." Ḥamza quotes further:

> Abū Maʿšar said: "Similarly, the years of the Greeks contain discrepancies like those in the years of the Persians. That is because their years and the ancient chronicles (*tawārīḫ*) of those years were translated from Hebrew, and the Hebrews differ on it because that which the Samaritans hold differs with what the generality of Jews hold, and the translation into Greek also contains discrepancies because the translation of the Seventy differs with the translation of others."
>
> He said: "The discrepancy in the number of years from the beginning of procreation to the year of the Hiğra remains an issue (*qā'im*). For the Jews follow the account of the Torah to 4042 years and three months. The

123. This and the following excerpts from Ḥamza al-Iṣfahānī 1921/2: 10–12 (*faṣl* 1, *bāb* 1).

> Christians follow the account of the Torah, too, to 5909 years and three months. The Persians follow the account of the book that Zoroaster (*Zaradušt*) brought, called Avesta (*Abistā*), it being the book of their religion, that from the age of Gayōmarṯ, father of mankind, to the year of King Yazdiğird are 4182 years, ten months, and nineteen days."

The long citation from Abū Maʿšar ends with his statement that the astrologers use a quantity of years far surpassing those of the chronographers. Here he is referring to the *kalpa* system of the Indians, consisting of 4,320,000,000 years. This and other very long periods used in calculating planetary longitudes were current in Arabic astrology in his time. No doubt Abū Maʿšar mentioned them to prepare the way for his own proposal of a world-year of 360,000 years, which is described by al-Bīrūnī. It is to these 360 millennia, and the millennial system of his historical astrology, that the name of his book, *The Thousands*, refers.

The passages above shed light on several important aspects of Abū Maʿšar's work. First, they show that in his conception, the work of historical astrology was closely related to that of chronography. The rationale might have been that if historical astrology was to interpret history in terms of conjunctions of planets over many centuries, the practitioner required accurate tables of ancient dates and important events that could fit into the astrological scheme to make it meaningful. Abū Maʿšar was also promoting his system in another way. His date of the Flood was in the very middle of his world-year of 360,000 years. As he says, dates before the Flood were uncertain, and this might have helped to justify using the Flood as the central event of history, the most important certainly known event in the past, and the date of the era from which he calculated dates. He knew his Flood-date from the Indian tradition of the *Sindhind*. Or perhaps it was the reverse, that his experience with chronography, and the uncertainty of antediluvian events, induced him to make the Flood-date his point of reference. One also wonders whether he felt that astrological dating was superior to tradition. It may be for this reason that he emphasizes from the outset that ancient dates of the chronicles are poorly known and corrupt, the implication being then that astrologers know better and more exactly than the transmitters of traditions.

Another important aspect of the passage is that Abū Maʿšar, himself of eastern Iranian origin, was particularly interested in establishing the relative chronology of ancient Iran. In fact, Abū Maʿšar is an early witness to the attempt by Iranian scholars to reconcile the chronography of ancient Iran with Christian world chronicles.[124] I have left out the details of the passage wherein Abū Maʿšar discussed the periods of Iranian history of unknown length. Most striking is the explicit comparison of the periods of ancient Iran with those undetermined periods of the Jews, "between

124. The Sixteenth Book of the Right Ginza of the Mandaeans, a seventh-century composition, is strikingly similar in several respects, having a 480,000 year world-period and revealing an attempt to identify biblical figures with ancient Iranian kings.

Adam and Noah," that is, the antediluvian period. In other words, Abū Maʿšar was fully aware that the dates of events before the Flood, such as the one he cited from "Annianus and others" in his *Book of Conjunctions*, were *controversial* in Christian world chronicles. Moreover, he is using Christian chronography as a model for understanding the chronography of ancient Iran. That is the very reason why Ḥamza preserves the passage, for Ḥamza, too, was dealing with the same subject. Indeed, in the following sections of his work, Ḥamza goes to great lengths to compare numerous witnesses in an attempt to determine the years of ancient Iran, but not before including Abū Maʿšar's warning at the outset.

Ḥamza also cites Abū Maʿšar's *Thousands* as one of several sources of information at his disposal on the chronology of ancient Rome: "Abū Maʿšar states in his *Book of the Thousands* the years of two of the Roman kings: Diocletian and Augustus, and that between Alexander and Augustus there were 280 years, and between him and Diocletian 596 years."[125] Ḥamza cites this in the midst of much fuller reports on the dates and events of ancient Greece and Rome from other, more detailed sources. The excerpt shows merely that Abū Maʿšar, like every astronomer, studied the conversions of calendars, including the eras of Augustus and Diocletian.[126] But a few sections later, another citation shows that Abū Maʿšar had substantial information about this particular subject. Ḥamza writes:

> What I relate in this place is from what Abū Maʿšar the Astrologer has in his *Book of the Thousands* on the discussion of the eras of the Greeks, together with the Romans.
>
> He claims that Philip (*Fylfs*) was the last of the kings of the Greeks. He lived in the city of Macedonia, of the Roman land. The Greeks made the first year of the years of his reign an era, going forward through the years. All the Romans divide the years of Philip into three parts. From the first year of them to the end of 294 years they name the years Greek, because the Greeks were their kings and were governing them. There were twelve kings, the first of whom was Philip, and the second Alexander. After Alexander were nine of their kings, each one of them having the epithet Ptolemy. It is a name derived from "war." Each one of them had a name differing from the previous one. After nine of them the kingship came to a woman queen named Cleopatra (*Qlwfṭr'*). Then there was after those twelve a group both from those of rank and from the common people of the sages (*ʿawāmm al-ʿulamā'*), each one of them bearing this epithet. One of them was Ptolemy, author of the *Almagest*. Then, after that, were 313 more years—a total of 607—that the Romans call

125. Ḥamza 1921/1922: 48 (*bāb* 2, *faṣl* 2).

126. Even the elementary *Ǧawāmiʿ ʿilm an-nuǧūm* of al-Farġānī, a contemporary of Abū Maʿšar, begins with explanations of the calendars and definitions of five eras: the Hiǧrī, the era of Yazdǧird, the era of Buḫtnaṣṣar used by Ptolemy, the era of Philip, and the era of Alexander (*Ǧawāmiʿ*, 1–7 Arabic, 1–7 Golius' Latin trans.).

> years of Augustus, because he was the first of their kings. Then, after that, to the present time, they name them the years of Diocletian, because the reign transferred to him and stayed among his progeny.[127]

Although Abū Maʿšar evidently did not list every Greek and Roman king, he knew their numbers and many of their names. This information would have been available to him partly through Ptolemy's *Table of Kings*, which included a king list of the Ptolemies and others. It was standard information for an astrologer to know. It had also been normal for the Christian chronographers to take Ptolemy's king lists into account ever since Panodorus, whom Syncellus blames for misusing "the astronomical canon" (i.e., of Ptolemy), whereas he himself cites Ptolemy's king list for comparison with other sources.[128]

The thirteenth-century historian of medicine, Ibn Abī Uṣaybiʿa, cites another passage from the *Thousands* of Abū Maʿšar because of his own interest in the etymology of the name of ancient Greece. For the present discussion, it definitely proves Abū Maʿšar's knowledge of Christian world chronography:

> Abū Maʿšar said in the second section (*maqāla*) of his *Book of the Thousands* that a country of the west was called Arġos in ancient times, and its people were called Argives (*Arġayū*). Afterwards this city was called Ionia (*'ywny'*) and they named its people Yūnānīyīn [meaning, in Arabic, Hellenes] after the name of their land. Their king was one of the petty kings.[129] It is said that the first one for whom the kingdom of Ionia was united, among the kings of the Greeks, was named Julius (*Iyūliyūs*). His epithet was Dictator (*Diqṭāṭur*). His dominion was eighteen years. He established for the Greeks many laws in use among them.[130]

The reference to Argos and the Argives is decisive. The kings of Argos had figured in Christian chronography in Eusebius' canons as one of the nations having its own column. (Eusebius had his information on Argos from Castor of Rhodes, fl. first cent. BCE.) The kings of Argos are also listed by Syncellus in Greek, Elias of Nisibis in Syriac, and others indirectly dependent on Eusebius through Annianus.[131] Abū

127. Ḥamza 1921/1922: 45 (*bāb* 2, *faṣl* 5).

128. Syncellus 378 blaming Panodorus, 244ff, comparing Ptolemy's list with others. Toomer 1998: 11 conveniently reconstructs this table.

129. The term "petty kings," *mulūk aṭ-ṭawā'if*, refers to minor kings who did not have world rule or empire, but were one of many kings ruling simultaneously across the earth. This shows again dependence on a Christian canon of years arranged in columns, which indicated graphically many nations ruling simultaneously.

130. Ibn Abī Uṣaybiʿa, 1.15.7–12.

131. Syncellus' reports on the Argives are scattered through 144–183, not in one table. Elias of Nisibis 30–31 (trans. 17–18) names Eusebius and Andronicus as authorities on the Argives, "the kings of the Greeks known as Argives" (*malkê d-Yônāyê d-metîdʿîn b-Argāyê*); he names their city as *Argôs*. Jacob of Edessa (262.7, trans. 199) knew of the Argives and probably included them in his full canon, no longer extant.

Maʿšar's discussion of Argos proves that he knew at least one such source. There is no other medium from which Abū Maʿšar might have learned this information, and no other reason that he might have found it relevant. Moreover, it appears in the context of his discussion of the length of rule of Julius Caesar, showing that his source, direct or indirect, was a chronicle.

Pingree knew and discussed all of the same passages cited here, but he assumed that all of them were part of an "astrological history" connecting Abū Maʿšar directly to Hermes though a bizarre chain of transmission. As he put it, "Abū Maʿshar's astrological history must have been hilariously inaccurate."[132] What was hilarious, for him, was that (so he considered) Abū Maʿšar invented a wild and totally fictional historical scheme according to which Abū Maʿšar's own astrology came into being and was transmitted through the ages. For example, he thought that the references to Argos and Julius Caesar meant that Argos and Julius Caesar must have figured as links in the transmission of astrology to Abū Maʿšar.[133] This would perhaps be in keeping with Abū Maʿšar's mathematical inconsistency and relative incompetence, a subject on which Pingree, following the criticisms of al-Bīrūnī, was indeed an expert. It seems rather that the attribution of such motives to Abū Maʿšar results from a misreading of the passages in the contexts of their citation. The part of the *Book of the Thousands* under discussion here was not devoted to claiming the *transmission* of knowledge. Abū Maʿšar appears instead to have been trying to establish a historical basis for testing the theory of conjunctions. Specific information of the kind Abū Maʿšar demonstrates, about ancient kings and kingdoms, was precisely the kind of historical data that astronomers dealt with and had at their disposal, and many astronomical handbooks (*zīǧes*) contain canons of kings and rules for converting calendars. In short, a different reconstruction of the cultural meaning of the *Thousands* of Abū Maʿšar and its historical narrative is required. Although Abū Maʿšar obviously was interested in the origins of astrology, the art he practiced, and thought that it had been revealed to Hermes-Enoch, whose prophetic character validated astrology, he did not construct a history of its *transmission* from Hermes, through Argos, Caesar, and the rest, down to his own time. That would indeed have been nonsense, and any astrologer or historian would have been able to recognize it as such.

Another indication of the chronographic context of Abū Maʿšar's legend of Hermes comes from the *Muntaḫab Ṣiwān al-ḥikma*, an abridgement of a tenth-century history of philosophy. As seen above, this work gives somewhat different information about Abū Maʿšar's Hermes legend than we have from Ibn Ǧulǧul. It introduces the story as follows: "Abū Maʿšar said, in the history of the bygone nations of the westerners, that Hermes..." (*fī aḫbāri l-umami s-sālifati mina*

132. Pingree 1968: 18n3.
133. Pingree 1968: 18.

l-Maġribīyīna . . .).[134] This shows that the story of Hermes probably occurred in the same section of *Book of the Thousands* as did the history of Argos and the kings of the Roman Empire: the second *maqāla* (chapter) of the book. For as already seen in the citation from Ibn Abī Uṣaybiʿa, the second *maqāla* dealt with the "Western nation" of Argos. If the second *maqāla* of the *Thousands* dealt with "Western nations" in general, and Hermes was mentioned among those nations, then Hermes was mentioned in the second *maqāla*. As one can therefore see, the three Hermeses came up in a *chronographic context*, dealing not with a history of science as such, but with the main events of different eras.[135]

Another passage of the *Thousands* surviving in a later citation discusses the calendar of the pre-Islamic Arabs and their system of intercalation.[136] Here again we see the chronographic interest of Abū Maʿšar. Two aspects of the passage are especially noteworthy. First, Abū Maʿšar elaborates supposed etymologies of the Arabic names of the lunar months in order to show that they originally corresponded to specific seasons. Because they were part of a lunar calendar, however, they fell out of sequence with the seasons of the solar year as long as they lacked intercalated months. The meanings of the names ceased to correspond to the seasons to which they referred. Abū Maʿšar's interest in such interlocking natural cycles is ordinary in any astronomer, but the means of his explanation, etymology and the exegesis of ancient names, is one of his favorites. As already seen, Abū Maʿšar gave the etymology of Ptolemy, connecting it correctly to war (Epic Greek πτόλεμος), and tried to explain the etymology of Yūnān, the Arabic word for Greece, relating it, again correctly, to Ionia. In the Hermes legend, Abū Maʿšar explains the ancient Iranian name Wīwanghān, interpreting the epithet of ancient Iranian kings, *Pēšdād*, by a common folk etymology to mean "the Just." Furthermore, the Persian poet Niẓāmī (d. 1209) records that the *Book of the Thousands* offered an explanation of the origin of the Qurʾānic epithet of Alexander the Great, "the two-horned one" or *D̲ū l-Qarnayn*, a subject of persistent debate among exegetes.[137] Names and their meanings seem to have been one of Abū Maʿšar's abiding interests.

134. *Muntaḫab Ṣiwān al-ḥikma*, ed. Dunlop section 90, line 1261; ed. Badawī 184.4.

135. Pingree recognized, but did not explain clearly, that the Hermes story must have come in the second *maqāla*, but he thought it was a history of science: Pingree 1968: 14: "(The *Kitāb al-Ulūf*), apparently in the second of its eight *maqālāt*, contained a complete history of the transmission of the sciences."

136. Published in Arabic with a French translation by Mahmoud Effendi 1858: 168–178 from a manuscript of *Kitāb Muntahā l-idrāk fī taqāsīm al-aflāk* by Bahā'addīn Abū Muḥammad ʿAbdalǧabbār al-Ḫaraqī aṯ-Ṯābitī, composed near Marw in 1132 (the details of the authorship supplied by Pingree 1968: 131–132).

137. Niẓāmī says that Abū Maʿšar claimed the name *D̲ū l-qarnayn* was a misunderstanding of Arabs who saw illustrations of Alexander depicting him with two angels of glory situated by his head. The Arabs mistook these designs for horns (Niẓāmī 2001: 73–74, section 70(9)). Most early explanations for this Arabic epithet related it to Alexander's deeds (his conquering of the eastern and western "horns" of the earth, for example), but Abū Maʿšar is keenly aware of the distortions involved in translating information from one language and culture to another, and he regards this kind of confusion to be the source of the name.

Second, this passage on the pre-Islamic Arabic calendar proves to be taken over almost verbatim, but without attribution, by al-Bīrūnī in his masterful *al-Āṯār al-bāqiya.*[138] Al-Bīrūnī often mentions his sources, but elsewhere in the same work, he singles out Abū Maʿšar and his followers, and their system of 360,000 year cycles, for special criticism.[139] Perhaps al-Bīrūnī decided not to mention his source in this case in order to salvage Abū Maʿšar's useful discussion of the pre-Islamic Arabic calendar, without discrediting his own work by associating it with the astrologer he disliked so much. This leads to another question: how much of the rest of al-Bīrūnī's chronology actually derives from Abū Maʿšar's *Thousands*? To be sure, a large amount of al-Bīrūnī's scholarship is clearly original. Because the whole scope of the *Thousands* remains unknown, it is impossible to say, but it stands to reason that more passages of al-Bīrūnī's work, as already seen in the case of al-Bīrūnī's *Qānūn*, rely on Abū Maʿšar as their unnamed source. If so, Abū Maʿšar's *Thousands* may have been one of the most important chronographic texts in early Arabic literature. Indeed, Ibn al-Qifṭī recorded in the thirteenth century that Abū Maʿšar was "the most knowledgeable person in the tales of the Persian kings (*siyar al-Furs*) and the histories of all the nations."[140]

After all the foregoing, it is evident that Abū Maʿšar's *Thousands* owed a great deal to the world chronography of Christian scholars. He even cites Annianus by name, the chronicler criticized in Greek several decades earlier by Syncellus for his antediluvian history of Egypt. We can therefore be sure that story of the antediluvian, Egyptian Hermes and his postdiluvian successor, from the *Book of Sothis*, could easily have been available to Abū Maʿšar through a few intermediaries. Until more evidence comes to light, we must conclude that Annianus is almost certainly the mediating source for the tale given in the *Book of Sothis*, a work of Roman Egyptian chronography dating at the latest from about 400 CE, but probably written considerably earlier. When dealing with lost ancient and medieval works, this near certainty is unfortunately often the best one can obtain.[141]

138. al-Bīrūnī, *Āṯār* 11–12; cf. also 60 and 325.

139. al-Bīrūnī, *Āṯār* 25–26, a section dealing with the Era of the Flood, Abū Maʿšar's base era.

140. Ibn al-Qifṭī 152.19–20: *wa-kāna aʿlama n-nāsi bi-siyari l-Fursi wa-aḫbāri sāʾiri l-umam.*

141. Because the texts in question are lost, there is still a remote possibility that Abū Maʿšar may have known the story of two Hermeses presented by Annianus through other intermediaries, or even through notes in the texts of Hermes themselves. Abū Maʿšar cites the doctrines of Hermes (sometimes Hermes relating the doctrine of Agathodaemon) in his own works: his *Great Introduction to Astrology* (e.g., in *maqāla* 2, *faṣl* 7 and *maqāla* 4, *faṣl*s 4 and 7), his *De revolutionibus nativitatum* (Greek translation made around 1000 CE), ed. D. Pingree 1968: 189.7ff. and his *Abbreviation of the Introduction to Astrology*, ed. and trans. Charles Burnett, Keiji Yamamoto, Michio Yano, 1994: 74 (chapter VI, paragraph 30). A list of works, given on the authority of Abū Maʿšar by his student Šāḏān, includes a *Kitāb al-Asrār* by Hermes (Rosenthal 1963: 456). Abū Maʿšar cited the Hermetic text *Kitāb Asrār an-nuǧūm* (= *Liber de stellis beibeniis*), an Arabic translation from Middle Persian, in his *Kitāb Aḥkām al-mawālīd* (Pingree 1989a: 233a–b). None of the astrological works of Hermes that he had at his disposal, as far as are known, contain any reference to multiple Hermeses. It makes perfect sense for an astrologer already familiar with works attributed to Hermes to be especially interested in the Hermes legend of the chronographers.

4.6 Other Aspects of the Hermes Legend

If one does accept Annianus as the source, at however many removes, for Abū Maʿšar's legend of the first two Hermeses, then the solution to another problem becomes possible. As Adler showed, part of the project of Panodorus and Annianus was to reconcile biblical and Jewish records of ancient history with other historical traditions, including the Egyptian of Pseudo-Manetho's *Book of Sothis* and the Babylonian of Berossus. Moreover, as already discussed, it is clear from the surviving traces of their works that among the other important sources for Panodorus and Annianus were the apocryphal works known generally as the Books of Enoch.[142] Because Hermes and Enoch, both antediluvian sages, figured prominently in the sources of Panodorus and Annianus, whose principal effort was to reconcile different traditions of ancient history, one can hypothesize that it was in these lost intermediary chronicles, by the Alexandrian Christians, or in a work of their successors, that Hermes and Enoch were first identified one with the other. (In Arabic, we know that Hermes and Idrīs had been identified before the 840s.)[143] Therefore, Abū Maʿšar might also have found this identification of the two ancient teachers in his chronographic source.[144]

The identification of Hermes and Enoch gave the Egyptian Hermes the biblical genealogy that appears in other versions of the Arabic Hermes legend and accommodated him to the biblical tradition of ancient chronography.[145] On the other hand, the subsequent equation of Hermes-Enoch with Hoshang (or Wīwanghān), who often appears in Iranian accounts as the grandson of Gayūmarṯ, is probably Abū Maʿšar's innovation or the innovation of one of his recent predecessors. Similarly, in the Persian account credited to Abū Maʿšar by al-Ǧuzǧānī, the first king, Aloros, was equated with Gayūmarṯ, the first man or first king in Iranian mythology.[146] (One wonders, therefore, if Abū Maʿšar thought that Hermes was the grandson of the antediluvian first king.) In synchronizing Iranian and Biblical

142. Adler 1989: 80–97 and 117–122.

143. The *risāla* of al-Ǧāḥiẓ entitled *Kitāb at-Tarbīʿ wa-t-tadwīr*, composed between 842 and 845, mocks its recipient with unanswerable questions. Among them: "Inform me about Hermes: is he Idrīs?" (al-Ǧāḥiẓ 1955: 26 §40). The identity of Hermes with Idrīs was evidently a subject of discussion before this time. For more on this, see section 5.1 in this volume.

144. The similarity between the books of Enoch and the Arabic story of the antediluvian Hermes was noted and discussed by Plessner 1954a: 53–54, without definite conclusions.

145. Hermes' biblical genealogy is expressed most fully by al-Mubaššir ibn Fātik, whose biography of Hermes is studied in the next chapter.

146. al-Ǧuzǧānī 132.19–21: *va avval kas az pādšāhān-i ʿādil-i muṣliḥ ki barāyi dafʿ-i ẓālimān naṣb kardand ba zabān-i Yūnānīyān ū-rā Aylūrūs nām būd va dalālat ān mīkunad ki ahl-i ʿAǧam-u Fārs ū-rā Kayūmarṯ mīgūyand va laqab-i ū Gilšāh būd.* "The first of the just, reformer kings whom they appointed to drive back the oppressors (= the giant offspring of the Sons of God) had the name Aylūrūs in Greek; that means the one whom the Iranians and Persians call Gayūmart. His title was *Gilšāh*." On the legend of Gayūmart, see Hartman 1953 (French translation of this passage, idem, 154), Shaked 1987b, and CHI 3(1).370. The title Gilšāh is also related by al-Masʿūdī, *Tanbīh* 85.16, Ḥamza al-Iṣfahānī 1920/21: 12.4 (in *bāb* 1, *faṣl* 1), and al-Bīrūnī (*Āṯār* 99). Abū Maʿšar's *Thousands* is probably their common source, as it was used by all three of these historians. See also *Fihrist* 15.2

histories, Abū Maʿšar had antecedents in earlier Iranian Muslim astrologers, practitioners of Sasanian historical astrology, who had already added traditions about ancient Iran to the earlier mixture of Babylonian, Greek, and Judaeo-Christian chronographic traditions.[147] The effort to supply ancient Iran with an accurate chronology on a par with those of the other nations was evidently still incomplete in the time of Abū Maʿšar, because he was forced to confront the difficulties of the problem directly in discussion. Ḥamza was still dealing with the same problem a century later, and al-Bīrūnī still decades later.

There was also an Islamic component to the new synthesis of traditions about Hermes. Abū Maʿšar's account, as given by Ibn Ǧulǧul and others, states that Idrīs received thirty scrolls sent down by God. This information is said to be transmitted (orally?) from the early Muslims (*al-aṯar al-marwī ʿan as-salaf*). In this case, one of the early Muslims who transmitted this report can be identified: an important source of originally Jewish traditions (*isrāʾīlīyāt*) in Arabic, Wahb ibn Munabbih (d. ca 730), one of the *tābiʿūn* (i.e., a second-generation Muslim). He is credited by early sources with the statement that Idrīs is Enoch and that Idrīs received thirty scrolls from God.[148] Nearly the same statement appears in aṭ-Ṭabarī's *Taʾrīḫ* on the authority of one of the "people of the Torah" (*ahl at-Tawrāh*).[149] Because we know from an-Nadīm that Abū Maʿšar began his career as a *muḥaddiṯ*, a transmitter of Islamic traditions, it is not surprising to find that he was aware of this report about the prophet Idrīs.[150] In fact, it makes perfect sense that a specialist in the ancient histories in which the *isrāʾīlīyāt* were rooted would also take an interest in historical astrology. Both fields of inquiry offer rationalizations about formative events of the distant past and their ancient dates. Although we still do not know the precise

147. It is entirely possible that Judaeo-Christian and Iranian chronographic traditions had been interacting for a long time in pre-Islamic, Sasanian Iran, where pagans, Jews, Christians, Manichaeans and Mazdaeans had been neighbors for centuries. In the late seventh century, Jacob of Edessa had incorporated the Sasanian kings into his chronological canon; Jacob's Sasanian king list is related by Elias of Nisibis (text 42 and n3, trans. 26 and n1). There is still no sign in the extant pieces of Jacob's work of a coordination of ancient (pre-Alexander) Iranian legend with ancient Christian chronography. The Mandaean *Right Ginzā* 18, also from the late seventh century, contains a chronology for all of ancient Iran.

148. See Khoury 1972: 214–217, on the basis of *Kitāb at-Tīǧān* of Ibn Hišām (d. ca 830), the principal source for the work of Wahb ibn Munabbih. See also Ibn Rusta 191.12–14: *qāla Wahbun awwalu man ḫaṭṭa bi-l-qalami Idrīsu wa-smuhū Uḫnūḫu ʿalayhi s-salāmu wa-huwa awwalu man ḫāṭa ṯ-ṯiyāba wa-labisahā wa-kāna man qablahū yalbasūna l-ǧulūda* "Wahb said: the first who wrote with the pen was Idrīs, whose name was Enoch, peace be upon him. He was the first to sew clothing and to wear it. Before him people used to wear animal skins." Remarkably, the words *wa-huwa awwalu man ḫāṭa ṯ-ṯiyāba wa-labisahā* are repeated verbatim by Abū Maʿšar in part 3 of the Hermetic biography cited above, where he gives "the account transmitted from the early Muslims (*as-salaf*)." The verbatim citation probably indicates a written, not oral, transmission from Wahb, or at least from Ibn Hišām, to Abū Maʿšar.

149. aṭ-Ṭabarī, *Taʾrīḫ* 1.173.1–7.

150. *Fihrist* 335.21 (Flügel 277.2): *kāna awwalan min aṣḥāb al-ḥadīṯ*.

origins of the identification of Idrīs with Enoch, it can be dated back to Wahb in the early eighth century at least.

4.7 The Third Hermes

The foregoing discussion accounts for the first two Hermeses of Abū Maʿšar and several of the details of their stories—the stories derived from the *Book of Sothis*, via the lost chronicle of Annianus, mixed with other traditions—but the origin and reason for the third Hermes remains until this point a mystery. The first step to answering this puzzle is to remember that Ibn Ǧulǧul clearly ended the account of Abū Maʿšar after the first two Hermeses by saying, "that is what Abū Maʿšar said." It could be that this simply marked a break in the excerpt, where material from the *Book of the Thousands* is omitted. But Ibn Ǧulǧul did not continue with another "Abū Maʿšar said." Therefore the story of the third Hermes may come from another source.

Fortunately, the solution to the problem was preserved centuries later by Ibn Nubāta (1287–1366), a poet from Cairo who made his career in the courts and administration of Mamlūk Syria. His erudite *Kitāb Sarḥ al-ʿuyūn fī šarḥ Risālat Ibn Zaydūn* (*Commentary on the Epistle of Ibn Zaydūn*) explains in detail a famous Andalusian satire full of the names of important historical figures.[151] It consists essentially of biographies for each name, and because Hermes is among them, Ibn Nubāta had occasion to retell the legend of Hermes.[152] He seems to have been using the *Thousands* of Abū Maʿšar directly or through a source other than Ibn Ǧulǧul, because his account contains slightly different sections of that work. Most interesting is that he attributes the account that appears in the *Thousands* as that of the third Hermes unambiguously to al-Kindī, the renowned philosopher credited with introducing Abū Maʿšar to astrology.[153] Ibn Nubāta appears also to include information left out by Ibn Ǧulǧul and others. Because of these important details, I translate the text of Ibn Nubāta's discussion of Hermes below, divided by letter headings according to the divisions useful for the subsequent analysis. Notes to the following text, where it is the same as presented by Ibn Ǧulǧul, can be found above with the original Arabic.[154]

Ibn Nubāta

A. [after the lemma from the epistle of Ibn Zaydūn mentioning Hermes:] This Hermes is the one who some of the Ṣābians[155] claim was a prophet

151. On Ibn Zaydūn and his *ar-Risāla al-hazlīya*, see G. Lecomte, "Ibn Zaydūn," *EI*², 3.973b–974b.

152. Ibn Nubāta, 205ff.

153. *Fihrist* 335.22–23 (Flügel 277).

154. I have consulted a manuscript of Ibn Nubāta's *Sarḥ* housed in the Beinecke Rare Book Library, Yale University, MS Salisbury 64 (relevant section = 77b9–78a13). This presents a couple of significant variants to the published text of Ibn Nubāta, noted below.

155. Following the well-attested reading *qawmun mina ṣ-Ṣābiʾa* rather than merely *aṣ-Ṣābiʾa* (ed. 305n3).

sent [by God], and that he was Idrīs, peace upon him. They base on his authority their religious laws concerning worship of the seven planets and the twelve signs of the zodiac, their offering of sacrifices to them, and such doctrines of theirs as these.

B. Abū Maʿšar al-Balḫī said: He was the first to discuss celestial phenomena concerning movements of stars. His grandfather Ǧayūmarṯ, who is Adam, peace be upon him, taught him the hours of the nychthemeron, and he is the first who built temples and glorified God in them. He was the first to investigate medicine and to discuss it. He composed for his contemporaries many books in metric verses in their language on knowledge of celestial and terrestrial things. He was the first to give advance warning of the Flood, and he thought that heavenly ruin would afflict the earth from water and fire. His dwelling was Egypt; in that place (*ʿinda ḏālika*) he built the pyramids and the cities of dust. He feared that knowledge would pass away in the Flood, so he built the monumental temples (*al-barābī*) and the mountain known as the *birbā* of Aḫmīm, and portrayed in it, in carvings, the arts and their practitioners, indicating the features of the sciences for whoever was after him, out of desire thereby to preserve the sciences forever after him.

C. The Ṣābians [i.e., pagans] claim that Asclepius had the prophethood after him. His name was Balīnūs [Apollonius] but it was lengthened (*zīda fīhi*) as an honorific (*taʿẓīman*) for him [so that it became "Asclepius"].[156] Likewise this is said in the case of Arisṭūṭālīs, for his name was Arisṭū. Everyone who became an expert in his sciences had his name increased (*zīda fī smihī*). Balīnūs had taken the sciences and the secrets from this Hermes, Hermes of the Hermeses. Others claim that Hermes, the master of Balīnūs, was after the Flood, and was not this one [who was just described].

D. Al-Kindī[157] said: He is the author of *Poisonous Animals*. He was a physician and a philosopher, knowledgeable in the natures of medicines. He traveled around the earth, wandering in different countries, knowing the foundations of cities, their natures, and the natures of their peoples and their medicines. He is the master of the Andalusian talismans like the Sudanese [?], the brass, and others.[158]

E. This Balīnūs was his student. He traveled with him through the countries, and when they left India for Persia, he put him in charge of Babylon. He had taken from him all his sciences. Miraculous events appeared to him

156. MS Salisbury 64 adds here, "according to their custom," *ʿalā ʿādatihim*.

157. MS Salisbury 64 gives *qāla Yaʿqūb al-Kindī*, "Yaʿqūb al-Kindī said," making the identification of the philosopher al-Kindī explicitly clear.

158. Clearly there is a textual problem with the designations of the talismans in this passage.

> in medicine and curing the sick, to the point that people talked about it a lot, and they said, "He is a prophet," and [others] said "[He is] an angel." They claimed that his birth was spiritual (*rūḥānī*) and that God the Exalted raised him on a pillar of light.[159] Euclid traced his ancestry to him. [The reports about Asclepius continue.][160]

Though there is a little difficulty in deciding where citations of Abū Maʿšar begin and end, one may note immediately the solution to the problem at hand: the story of the third Hermes is attributed to al-Kindī. We will return to this point after some remarks comparing the account of Abū Maʿšar as related by Ibn Ǧulǧul and Ibn Nubāta.

Paragraph A is Ibn Nubāta's own introduction. Paragraph B is Abū Maʿšar's passage identifying the place of Hermes in various national histories, practically identical with what is also transmitted by Ibn Ǧulǧul. Paragraph C here is totally different from paragraph 3 in Ibn Ǧulǧul's rendering above. It appears to be what Ibn Ǧulǧul described (in paragraph 4 in his text above) as "absurd reports" about Hermes, a passage he left out, perhaps considering it too strange.[161] The bizarre attempt to provide an etymology for the names Aristotle and Asclepius-Apollonius is characteristic of Abū Maʿšar's already established interest in the etymology of foreign names.[162] The reference to Apollonius' taking "the sciences and the mysteries" (*al-ʿulūm wa-l-asrār*) from Hermes is of special importance, as this seems to constitute the earliest known reference to the *Kitāb Sirr al-ḫalīqa* of Pseudo-Apollonius.[163] It shows that already in the time of Abū Maʿšar, the *Kitāb Sirr al-ḫalīqa* was raising questions about the identity of the Hermes from whom Apollonius took the sciences of nature. Abū Maʿšar alludes to two views: one that the Hermes who was Apollonius' master lived before the Flood, and one that he lived after it. As we now know, he learned about these two Hermeses from a work of Christian chronography.

Ibn Ǧulǧul's paragraph 5, on the second Hermes, is entirely omitted by Ibn Nubāta. Instead, we have only the short reference to the possibility that Hermes

159. On the heavenly ascent of Asclepius in a pillar of light, see the beginnings of an investigation in Manzalaoui 1974: 189–191.

160. Ibn Nubāta, *Šarḥ*, 205.12–207.2.

161. This section was mentioned and translated into German by Weisser 1980: 25–26, with an inconclusive discussion.

162. Abū Maʿšar evidently saw special titles in many ancient proper names, including Hermes and Ptolemy as well as Caesar and Ḫusrō. Perhaps Abū Maʿšar had heard that ancient Greek names were mostly compounds of two words, and this led him to his mistaken notion of "augmented" names.

163. The *Sirr al-ḫalīqa* was composed around the time of al-Ma'mūn (813–833) (see chapter 5, note 23). Abū Maʿšar was writing ca 840–860. Weisser (1980: 25–26), the editor of the *Sirr al-ḫalīqa*, has noted this passage of Ibn Nubāta but seems less willing to attribute these sentences to Abū Maʿšar. I find no reason to doubt it, because the rest of the passage down to and including the citation of al-Kindī is evidently drawn from Abū Maʿšar as well. Moreover, Ibn Ǧulǧul, in his rendering, explicitly says that he has omitted some things, and there is also present in the immediately preceding lines of Ibn Nubāta the interest in etymology so characteristic of Abū Maʿšar.

lived after the Flood. Ibn Nubāta's paragraph D, corresponding with Ibn Ǧulǧul's 7, tells us that the source of the passage on the third Hermes is the philosopher al-Kindī (d. ca 870), a crucial piece of information. The final paragraphs of both accounts diverge, giving different accounts of the student of the third Hermes. Ibn Ǧulǧul's paragraph 8 tells us again that he has omitted something relevant to the story of these ancient sages.

Both Ibn Ǧulǧul and Ibn Nubāta give reports about Asclepius immediately after the stories of Hermes. One may surmise that this is also an original feature of Abū Maʿšar's account.[164] Ibn Nubāta's account notably removes any idea of multiple individuals named Hermes. Instead, he has three separate versions: the first Hermes of Abū Maʿšar, the reported belief that Hermes actually lived after the Flood, and al-Kindī's alternative Hermes. This is apparently a change made by Ibn Nubāta himself.

A note in the *Fihrist* of an-Nadīm relates that Abū Maʿšar had some kind of relationship with his contemporary al-Kindī (d. ca 870), at first antagonistic and then, later, less so, after al-Kindī induced him through an intermediary acquaintance to the study of mathematics.[165] Until recently, modern scholars thought that Abū Maʿšar was guilty of plagiarizing al-Kindī in his work on historical astrology, but this has been disproved by Yamamoto and Burnett in their new editions of both works, in which they conclude that the two authors shared a common source.[166] Nevertheless, one may see from Ibn Nubāta's excerpt that Abū Maʿšar took his report about the third Hermes directly from al-Kindī.

Abū Maʿšar's text must have originally cited al-Kindī, but al-Kindī's name disappeared early in a branch of the manuscript tradition of the *Thousands*. The tradition lacking the name of al-Kindī must have existed at least by the eleventh century, when Ṣāʿid al-Andalusī used it. The modern publication of Ibn Ǧulǧul's history is based on a unique manuscript of the late sixteenth century; perhaps other copies of Ibn Ǧulǧul's work contained the correct attribution. Because the excerpt from Abū Maʿšar begins, "The Hermeses are three," a copyist might have assumed that the story of the third Hermes must have come from the same source, not from al-Kindī, and therefore omitted it. Even a single manuscript could have preserved the original reading "al-Kindī said" for Ibn Nubāta. In any case, the text that became widespread omitted the reference to al-Kindī, for whatever reason, and this version became the most popular one, used by almost all of the later Arabic historians of science. The only alternative to accepting the name al-Kindī in the original text of Abū Maʿšar is to suppose that someone *added* it at a later date, but there is no

164. A comparison of the two accounts of Asclepius should form the basis of a further investigation.

165. *Fihrist* 335.22–23 (Flügel 277). Interesting but inconclusive evidence about their relationship is discussed by Dunlop 1971: 238.

166. Burnett and Yamamoto 2000: 1.525.

reason for anyone to have done such a thing. Clearly the attribution preserved by Ibn Nubāta is correct.

Because al-Kindī is the source of the passage about the third Hermes, then one must ask where al-Kindī found his own information about Hermes. In this case, one can make a good guess. As already seen (section 3.4.3 in this volume) al-Kindī was familiar with a philosophical Hermetic text, a work in which Hermes taught his son.[167] He was also the authority on which his disciple as-Saraḫsī wrote about the Ṣābians of Ḥarrān, their religious beliefs and cultic practices, and mentioned their affiliation with the name Hermes.[168] It is possible that a report by al-Kindī on the life of Hermes would have derived from a Ḥarrānian tradition, or at least a tradition about the Ḥarrānians. Then again, al-Kindī knew many translations from Greek and employed the services of knowledgeable scholars of the Greek sciences. Any one of them might have been able to tell him something about an ancient Egyptian scholar named Hermes. It may also be supposed that the information about the third Hermes' teaching Asclepius, connected with the stories of Hermes in the versions given by both Ibn Ǧulǧul and Ibn Nubāta, was derived from al-Kindī and his source, because the story of this last Hermes clearly derives from al-Kindī. Asclepius was portrayed as Hermes' disciple in the Greek Hermetica, addressed by Hermes as his son; al-Kindī may well have obtained this information from the work he read in which Hermes teaches his son. For now this must remain speculation.

Finally, the name *trismegistos* in its various Arabic calques (all beginning with *al-muṯallaṯ bi-* "triplicate in" wisdom, blessings, etc.) could have suggested three separate Hermeses, so that Abū Maʿšar was more inclined to combine two separate reports (one about two Hermeses, and another, different one about a single Hermes) rather than to assume that they contradicted one another, arriving at a triplicate Hermes. Yet this was no fabrication out of thin air. It was the work of a man trained as a transmitter of reports, who would rather combine all of them than omit any of them. Abū Maʿšar's craft is found in the combination and synthesis of ancient traditions, not so much in their invention.[169]

4.8 Conclusions

After so much sleuthing, the origin of the legends of the three Hermeses has become fairly clear. Abū Maʿšar took the story of the two earliest Hermeses from a lost Arabic chronographic work based on Christian sources. The Alexandrian chroniclers Panodorus and Annianus transmitted this tale of two Hermeses from the *Book*

167. *Fihrist* 385.1–2 (Flügel 320.7–9).

168. The three surviving passages (from two works composed ca 966 and in 986, respectively) drawing on as-Saraḫsī's "Description of the Religious Views of the Sabians" are compared and analyzed by F. Rosenthal 1943: 41–51.

169. Burnett and Yamamoto 2000: 1.609–611 draw a similar conclusion about Abū Maʿšar's astrological synthesis.

of Sothis. At the same time, they make much use of pseudepigraphic Enoch literature and combined Biblical and pagan (Greek as well as Hellenistic Egyptian and Babylonian) chronographic traditions. The chronicle of Annianus became an authority on ancient dates throughout Syriac chronographic literature. At some stage, perhaps as early as Annianus himself, it was easy to identify Hermes with Enoch. Abū Maʿšar did not have to invent anything in all this. He was apparently a faithful and often uncritical tradent in all his interests.

Abū Maʿšar also knew the Iranian astrological tradition represented by the Middle Persian recension of Dorotheus, and stated explicitly by Ibn Nawbaḫt, that all science came from Iran, and that Hermes was from Babylon (the capital of Iran) who became king of Egypt. The chronographic works at his disposal included dates of the ancient Iranian kings, and he, or one of his sources, thereby identified Hermes with Wīwanğhan (or Hōšang), the *Pēšdād*.

Al-Kindī related an account about a Hermes of more recent times who was the author of the more mundane Hermetica in circulation, such as the book on poisonous animals, that were perhaps unlike the great secret lore expected from the ancient Hermeses. This report, which did not match those received from the chronographic tradition, was included in the *Thousands* of Abū Maʿšar, perhaps in order to fill out an idea of "triplicate Hermes" (*Hirmis al-muṯallaṯ*). Whatever the motive, the result was three Hermeses: two Hermeses plus one more.

The survival of Abū Maʿšar's account to the present depends on later excerptors. Ibn Ǧulǧul indicated clearly that he was selective in his reports on the various Hermeses and omitted parts of the story he regarded as unlikely (including associations with Balīnas, or Apollonius of Tyana). He used the passage from the *Thousands* of Abū Maʿšar to form the basis for the reports on the Hermeses that circulated throughout Arabic biographical literature and histories of science. Al-Kindī's name dropped out of this version of the text at some stage either before or after Ibn Ǧulǧul completed his work. Ibn Nubāta's learned commentary made use of the reports of Abū Maʿšar and al-Kindī, but is unique in that it cited both authorities by name. It gives a slightly different version than what is found edited in Ibn Ǧulǧul, who cut out the references to Apollonius of Tyana. Nevertheless, Ibn Ǧulǧul's account, edited from that of Abū Maʿšar, became the standard account in later Arabic literature.

More briefly put, this analysis comes to the conclusion that these biographies of Hermes in Arabic are, at their basis, and excluding some details, composed from just three sources, at least the first two of them being pre-Islamic: (1) the *Book of Sothis* of pseudo-Manetho, mediated by the Egyptian chronographers Panodorus and Annianus, and Syriac chronographic tradition, (2) the Iranian astrologers' account of the Babylonian Hermes, and (3) al-Kindī, who had an independent knowledge of some Hermetica in Arabic that are lost today. Abū Maʿšar, or perhaps one of his sources, added another Iranian element by equating Hermes with a figure of ancient

Iranian legend, an obvious identification of two great inventors who lived among the first generations of humankind.[170]

The mystery appears to be solved. The legend of the three Hermeses in Arabic represents the intellectual atmosphere of early classical Baġdād, where scholars were actively translating and synthesizing all available and useful knowledge in Arabic, and where al-Kindī could write, in words famous today, "We ought not to be ashamed of appreciating the truth and of acquiring it wherever it comes from, even if it comes from races distant and nations different from us."[171] But in this case the traditions received were not historically true. They were all fabricated in late antiquity to support and explain the existence of a Greek literature originally from Egypt claiming to relate primordial wisdom.

170. In this way Pingree's description of Abū Maʿšar's astrological history as "a strange mingling of the traditions of Īrān and Ḥarrān" (Pingree 1968: 18) is correct, although the Ḥarrānian element is not what he thought it was, as explained above.

171. al-Kindi 1974: 58.4–7.

5

Hermes the Prophet

EARLY IN THE NINTH CENTURY, THE SYRIAN CHRISTIAN theologian Theodore Abū Qurra gives the first known reference in Arabic to Hermes explicitly as a prophet (*nabīy*) (see section 3.4.1 in this volume). Although Hermes became most famously the prophet of the pagans (*Ṣābi'a, ḥunafā'*), nevertheless Muslims, Christians, and others took an interest in the works, deeds, and sayings of the ancient sage. The accounts of Hermes claimed that he antedated the foundation of all the major religions. This made him potentially a forebear for anyone who claimed to follow him or to use the sciences he taught, regardless of religion. The Arabic myth of Hermes as an ancient teacher of universal appeal and a prophet of science resulted from the combination of several pre-Islamic and Islamic elements: the legends of Enoch, the role of Hermes as a teacher of astrological sciences in the ancient Greek Hermetica, the Ḥarrānian claim that Hermes was their prophet, the teaching of the early Ismā'īlī mission about Hermes, and the wisdom literature ascribed to Hermes.

5.1 THE HEAVENLY ASCENT OF HERMES

And mention Idrīs in the Book. He was true, a prophet. We raised him to a high place.

—Qur'ān 19:56–57

And Ismā'īl, Idrīs, and Ḏū l-Kifl were each among the patient. We brought them into Our mercy. They were righteous ones.

—Qur'ān 21:85–86

The name of the prophet Idrīs is first known from these two verses of the Qur'ān. As already seen (section 4.6), Idrīs was identified with the biblical Enoch at the latest by the time of Wahb ibn Munabbih (d. ca 730), in the first century after Muḥammad's death.

Enoch himself was a figure of much more ancient legend. The earliest extant references to him come from a few of the oldest canonical biblical texts of the Jews and Christians. In the Hebrew Bible and its translations he makes a few brief appearances in ancient genealogies.[1] The *Sēfer tôlədōṯ Āḏam*, or *The Book of the Generations of Adam*, incorporated in the fifth chapter of *Genesis*, briefly reports about Enoch the son of Jared (*Genesis* 5:18–24). Enoch begot Methusaleh. After 365 years of life on earth, he "walked with God and he was not, for God took him." The first-century Greek text *Epistle to the Hebrews*, which became part of the New Testament canon, listed Enoch as one in a series of several righteous and believing followers of God, including Abel, Noah, Abraham, and others: "By faith Enoch was removed from seeing death, and he was not found because God removed him. For before his removal it was testified that he had pleased God well" (*Hebrews* 11:5). Clearly this is an interpretation of the story in *Genesis*, and one can hardly suppose that it was new in the *Epistle to the Hebrews*.

From the third to the first century BCE, a literature destined *not* to become canonical in the Bible had already developed in Aramaic, greatly elaborating the legend of Enoch and how he was "taken by God." These *Books of Enoch*, subsequently translated into many languages, describe Enoch as a prophet who was raised to the heavenly heights and given visions of the shape of the cosmos, of the distant past, and of the future redemption and judgment before God. Although the collection of about five originally separate books known today as *1 Enoch*, or just *The Book of Enoch*, survives complete only in Ethiopic, fragments of the text have been discovered in Aramaic and Greek. It appears to have been read widely. Due to the influence of Enochic works on and their importance in the development of Jewish, Christian, and Manichaean literature, the modern bibliography of scholarship on Enoch and his books is immense.[2]

Chapter 14 of *The Book of Enoch* describes how Enoch was taken up to heaven to behold the house of God, surrounded by the angels.[3] Again in chapter 71, from what was originally a separate work, Enoch relates that he was brought up into the heavens to learn celestial secrets (*1 Enoch* 71:3–5): "And the angel Michael, one of the archangels, seized me by my right hand, and lifted me up and led me forth unto all the secret things, and he showed me all the secrets of mercy and he showed me all the secrets of righteousness. And he showed me all the secrets of the ends of the heavens, and all the storehouses of all the stars, and all the luminaries, whence they proceed in the presence of the holy ones. And he translated my spirit, and I, Enoch,

1. Hebrew Bible: *Genesis* 5:18–21 and *1 Chronicles* 1:3. Greek New Testament: *Luke* 3:37 and *Hebrews* 11:5. Another Enoch, son of Cain, is mentioned in *Genesis* 4:17–18.

2. For bibliographies and introductions see M. Black 1985 and Reed 2005.

3. The surviving Greek text has, Καὶ ἐμοὶ ἐφ' ὁράσει οὕτως ἐδείχθη· ἰδοὺ νεφέλαι ἐν τῇ ὁράσει ἐκάλουν καὶ ὁμίχλαι με ἐφώνουν, καὶ διαδρομαὶ τῶν ἀστέρων καὶ διαστραπαί με κατεσπούδαζον καὶ ἐθορύβαζόν με, καὶ ἄνεμοι ἐν τῇ ὁράσει μου ἐξεπέτασάν με καὶ ἐπῆράν με ἄνω καί εἰσήνεγκάν με ἐις τὸν οὐρανόν.... (M. Black 1970: 28). "Thus it was shown to me in a vision. Behold: Clouds and vapors in the vision were calling me and speaking to me, and shooting stars and lightning flashes were troubling and disturbing me, and winds in my vision lifted me and raised me upward and brought me into the sky...."

was in the heaven of heavens…"[4] In the next part of the *Book of Enoch* (chapters 72–82), originally yet another separate work, and still bearing its title *Book on the Motion of the Luminaries of Heaven*, Enoch receives astronomical knowledge from the angel Uriel, the angels' leader. Although the astronomy is elementary, not to say rudimentary (the basis of a calendar of 364 days), it is presented with an air of mystery.[5] Still another work, conventionally called *2 Enoch*, surviving only in Slavonic, and of uncertain provenance, describes Enoch's journey upward through the seven heavens, one by one.[6]

The portrayal of Enoch in these apocryphal books as a prophet who visited the celestial heights had ramifications also in Arabic literature.[7] It was evidently well known enough to the followers of Muḥammad that they identified Enoch with Idrīs, whom God "raised to a high place" according to the Qur'ān, as one finds, for example, in the chronicle of the Andalusian ʿAbdalmalik ibn Ḥabīb (d. 853), one of the earliest annalistic historical works to survive in Arabic, written around the time of Abū Maʿšar's *Book of the Thousands*.[8] Once the Qur'ānic Idrīs was identified with Enoch, it became possible for Muslims to use the ancient tales about Enoch's ascent and visions as a part of their own accepted tradition. It is uncertain whether Books of Enoch were in circulation among early followers of Muḥammad, and, if so, which ones and in what form. There is no direct trace of them in Arabic, as far as is now known, so one may want recourse to a hypothesis of oral transmission or folklore, or to suppose that the stories were introduced by converts to Islam.[9] Nevertheless, the sources prove that a certain amount of Enochic lore was definitely available to the early Muslim scholars of Jewish and Christian traditions about the Hebrew Bible and its related texts, experts in the stories called *isrā'īlīyāt*.[10] Arabic historians of the tenth century and later, like their Greek- and Syriac-writing predecessors, came to cite this material side-by-side with chronographic traditions such as those discussed in the previous chapter.

One typical example of a Muslim author treating the legends of Enoch under the name of Idrīs is Muṭahhar ibn Ṭāhir al-Maqdisī, author of *Kitāb al-Bad' wa-t-ta'rīḫ*

4. Trans. M. Black 1985: 67. The Ethiopic text (Knibb 1978) has, *malā'əkt aḫazani ədēya za-yamān wa-anšə'ani wa-awḍə'ani ḫaba kʷəllu xabu'ata məḥrat wa-xabu'ata ṣədq wa-ar'ayani kʷəllo xabu'ata aṣnāfa samāy wa-kʷəllo mazāgəbta kawākəbt wa-bərhānāt kʷəllomu əm-xaba waḍ'u la-gaṣṣa qəddusān wa-kabato manfas la-Hēnok wəsta samāya samāyāt.*

5. A mathematical analysis of the astronomical chapters of the *Book of Enoch* is presented by Neugebauer in M. Black 1985: 386–414.

6. Andersen 1983 provides an English translation and detailed commentary for *2 Enoch*.

7. On aspects of the mythology of heavenly ascent in the Qur'ān and among the early followers of Muḥammad, and for bibliography on heavenly ascent in general, see van Bladel 2007b.

8. Ibn Ḥabīb 36 (67).

9. The exception is the legend of the Watchers and their fall, related by the chronographic tradition of Annianus, and preserved in the works of al-Bīrūnī and al-Ǧuzǧānī, as seen in the last chapter. In any case, this was not transmitted through early Islamic tradition but by astrologers and chronographers.

10. A general introduction to "The Reception of Biblical Materials in Early Islam" is presented by Adang 1996: 1–22. On Idrīs as Enoch in aṭ-Ṭabarī and his sources (though not on Enoch's heavenly ascent), see Alexander 2000.

(*The Creation and the Chronicle*), which he wrote in Bust, Siǧistān, in 966. In the section of his work dealing with the ancient prophets, Idrīs figures third, after Adam and Seth, the place held by Enoch in Christian and Jewish accounts.[11] Al-Maqdisī begins his account of Idrīs with the biblical genealogy of Enoch derived from *Genesis* 5, followed by reports going back to anonymous early Muslim tradents. Idrīs was the first to write with the pen after Adam and the first to sew clothing and wear it. This is close to the report, already mentioned, attributed to Wahb ibn Munabbih. God revealed to Enoch astronomy and medicine (*anzala ʿalayhi n-nuǧūm wa-ṭ-ṭibb*). Next, al-Maqdisī tells of a disagreement among early Muslim authorities about the specifics of Idrīs' heavenly ascent. According to his citation of the *Book of Abū Ḥuḏayfa*,[12] Idrīs asked the angel of the sun to take him into heaven (the sky, *as-samāʾ*) so that he could worship God together with the angels. God raised him up to the Fourth Heaven.[13] Another version of the same story, allegedly going back to ʿAbdallāh ibn al-ʿAbbās (d. ca 687), says rather that the angel of the sun told Idrīs the name by which one could ascend to heaven, and it was by this name that he went up to the Fourth Heaven. Al-Maqdisī relates a different report attributed to Zayd ibn Arqam (d. 685/686), a Companion of the Prophet, saying that Idrīs was raised to Paradise (*al-Ǧanna*). Among other details in his ensuing discussion, al-Maqdisī insists that it is possible for bodies to enter heaven, and that the prophets ascended bodily, not spiritually. All the sources cited here agree that Idrīs (Enoch) did in fact ascend into the heavens in some way. Al-Maqdisī goes further in an attempt to provide a synchronism between Idrīs and ancient Iranian kings, in particular Hōšang (*Hūšank*), combining thereby the concerns of Iranian chronographers with early Muslim traditions in a way typical of tenth-century Muslim Arabic historiography.[14] As this example shows, stories about the heavenly ascent of Enoch, under the name of Idrīs, were well known and the subject of some discussion among Muslims since at least the eighth century.

When Hermes was identified with Idrīs, then Hermes, too, became the subject of similar stories of heavenly ascent originally proper to Enoch. An epistle of al-Ǧāḥiẓ, composed sometime from 842 to 845, implies the existence of an ongoing discussion over whether Hermes was indeed Idrīs. Al-Ǧāḥiẓ mocks the Rāfiḍī Aḥmad ibn ʿAbdalwahhāb with difficult historical and philosophical questions. Among other things, he asks whether Hermes is in fact Idrīs and, in another passage, what Hermes had to say about the nature of the celestial sphere (*ṭabīʿat al-falak*). Although uncertainties remain about the precise context in which al-Ǧāḥiẓ' addressee had

11. al-Maqdisī, 3.11.8–13.17.

12. Abū Ḥuḏayfa Isḥāq ibn Bišr (d. 821, GAS 1.293–294) wrote several books, but the one mentioned here is almost surely his *Kitāb al-Mubtadaʾ*. If the title is any indication, it was one of the models for al-Maqdisī's work. Huart supposed in his translation (al-Maqdisī 1.138) that this Abū Ḥuḏayfa was rather Mūsā ibn Masʿūd an-Nahdī al-Baṣrī (d. 835, GAS 1.41), the author of a *Tafsīr*.

13. The Fourth Heaven is that of the sun, the heaven in the middle of seven planetary heavens.

14. al-Maqdisī, 3.11–13.

employed references to Hermes, giving rise to such a response, the passage does at least locate the myth of Hermes' heavenly ascent, an immediate product of his identification with Idrīs-Enoch, within discussions of the views of the ancient philosophers on the nature of the heavens, and, importantly, it does so at this relatively early date.[15] When al-Ǧāḥiẓ wrote this work, the identification of Hermes and Idrīs was still a matter of debate, not yet perfectly decided or common knowledge. Soon afterward, however, Abū Maʿšar's biography of Hermes, written approximately between 840 and 860,[16] would establish it as common knowledge. Later authors mentioning Hermes in Arabic regularly add to his name the remark, "he being Idrīs," *wa-huwa Idrīs*.

The early Ismāʿīlīs, already from the late ninth century, seem to be the first organized group with an official doctrine about Hermes. The Ismāʿīlīs were in the late ninth century a secretive Šīʿite movement aimed at overthrowing the ʿAbbāsid dynasty and replacing it with the leadership of their imām. They maintained that their imām, Muḥammad ibn Ismāʿīl, was in hiding; that he represented the true line of succession from Muḥammad; that their imām alone was the source of correct interpretations of the divine, and particularly of the symbolic expressions allegorically interwoven in the Qurʾān; that their imām and his elect would initiate the

15. *Kitāb at-Tarbīʿ wa-t-tadwīr*: "Inform me about Hermes: is he Idrīs?" (al-Ǧāḥiẓ 1955: 26 §40). James Montgomery (2007) suggests that one passage in this work of al-Ǧāḥiẓ (1955: 87–88 §§165–166) alludes to al-Ǧāḥiẓ' having publicized that his addressee, the Šīʿite Aḥmad ibn ʿAbdalwahhāb, was the real author of some pseudepigraphical work, to Aḥmad's consternation. This is possible, but one must be wary of over-reading the passage. It is nevertheless worthwhile to consider Montgomery's hypothesis. The very same passage refers to Aḥmad's familiarity with Hermes, Plato, and Aristotle. Was Aḥmad—a secretary for the vizier Muḥammad ibn ʿAbdalmalik, a major patron of translations from ancient Greek ca 833–847 (Montgomery 2007, 446)—actually writing spurious works under one or more of these philosophers' names? An earlier passage in al-Ǧāḥiẓ' epistle (1955: 45 §83) refers to the same philosophers in the same order: *fa-ḫabbirnī ʿammā ǧarā baynaka wa-bayna Hirmisa fī ṭabīʿati l-falaki wa-ʿan samāʿika min Aflāṭūna wa-mā dāra fī ḏālika baynaka wa-bayna Arisṭāṭālīsa*, "Now inform me about your conversation with Hermes on the nature of the celestial sphere, and about the lectures you heard from Plato, and your discussions with Aristotle on that subject." These words imply that Aḥmad ibn ʿAbdalwahhāb indeed considered himself to know in a peculiarly personal way what these ancient philosophers taught. That in turn might be taken as support for Montgomery's suggestion that al-Ǧāḥiẓ is alluding to a pseudepigraphical work or works that Aḥmad wrote, perhaps even under these philosophers' names. Unfortunately it was not the purpose of al-Ǧāḥiẓ to inform us directly about what was commonly known in his own social circle at the time, and the tantalizing allusions here may rather be designed merely to mock Aḥmad's claim to know about the views of these ancients, or perhaps specifically to mock Aḥmad's belief in reincarnation (by which means he might claim accurately to know the views of ancient teachers). (For evidence of his view on reincarnation see al-Ǧāḥiẓ 1955: 43, §84 and 73–74 §130.) It remains an open question.

16. Abū Maʿšar is supposed to have lived from August 10, 787 to 886 (Pingree 1968: 2 and 1962: 487n6). He is said to have come to the practice of astrology only at the age of 47 (lunar years, hence in 833–834; cf. Pingree 1968: 20). After considering evidence suggesting that Abū Maʿšar was already casting horoscopes as early as 826, Pingree supposed anyway that 840–860 would be a reasonable period in which to expect the *Thousands* to have been written (Pingree 1968: 20). Without any better indication, one may assume the Hermes biography must have become known around the time of al-Ǧāḥiẓ or soon after.

Muslim community into true and higher, actually salvific grades of understanding; he would abolish previous religious laws and restore unity to the divided Muslim community and justice to its discontents. The Ismāʿīlī movement can be discerned in the historical sources as early as the 870s, from which time they had an active mission. Based in their secret headquarters at a mansion in the Syrian desert town of Salamīya until 902, the hidden leaders, who claimed to be in contact with the imām, sent missionary agents (*dāʿīs*) guarding their conspiratorial network throughout the ʿAbbāsid caliphs' territories to gain sympathizers and more active participants in their anticipated revolution. Their community was riven in 899–900 when the supreme leader was revealed to be someone other than the expected imām Muḥammad ibn Ismāʿīl. Soon afterward this leader, Saʿīd ibn al-Ḥusayn, was located by the authorities in Baġdād and forced to flee Syria to North Africa. From there he would achieve independent control of North Africa and create the successful Fāṭimid caliphate (909–1171) under the name al-Mahdī. Under his great-grandson al-Muʿizz, this dynasty conquered Egypt and was based from 969 in their newly founded capital of Cairo. Yet even in the early stages of the movement, two early tenth-century Ismāʿīlī *dāʿīs*, working far apart from one another in North Africa and northern Iran, use Hermes to further the very same argument in discussions of the relative merits of revelation and reason. Both insist that the principles of true philosophy and the sciences are part of divine revelation, and that human reason can tinker with these revelations but never really surpass them. In both cases Hermes is their special example of a prophet of science.

One of these was Abū ʿAbdallāh Ibn al-Hayṯam, a native of Qayrawān and already a Šīʿite, who in 909 joined the Ismāʿīlī movement through discussion with and initiation by Abū l-ʿAbbās Muḥammad, a longtime *dāʿī* who knew the emerging imām al-Mahdī personally. Ibn al-Hayṯam testified to him that the prophet Idrīs revealed astrology and arithmetic, and that his name in Greek was Hermes. He mentions it as the first point in an argument for the legitimacy of ancient Greek philosophy alongside revelation but definitely subordinate to it. He says that the Greek philosophers were "the followers of prophets" such as Hermes, although they did not found the sciences themselves.[17] The *dāʿī* Abū l-ʿAbbās is portrayed as seeming satisfied with these answers to his questions.

Far away in northern Iran, the Ismāʿīlī *dāʿī* Abū Ḥātim ar-Rāzī (d. 934/935), writing independently ca 919–924,[18] makes the very same argument more elaborately and contentiously. He says that Hermes is Idrīs, a true prophet and the source of the science of astrology as well as "one of our imāms, not one of the imāms of the heretics; between him and Adam were five patriarchs." By heretics he means here the

17. Madelung and Walker 2000: 137–139 (English), 87–90 (Arabic). This Ibn al-Hayṯam was writing in 946 or soon thereafter about a conversation, reported in detail, that had happened in 909. For an account of the Ismāʿīlī initiations by oath, see Halm 1996b.

18. Halm 1996a: 289.

Greek philosophers, and their followers—such as his opponent in his book, the self-described philosopher Abū Bakr ar-Rāzī—who reject prophecy and innovate erroneously in developing the revealed sciences. Abū Ḥātim furthermore quotes the Qur'ān's testimony that God raised Idrīs to a high place, which he interprets as the top of the mountain at the navel of the earth (*surrat al-arḍ*), where an angel taught him astrology.[19] The implication of these missionaries' arguments is, of course, that their imām, God's living interpreter, is the final arbiter of all philosophical as well as scriptural issues and that even sophisticated reasoning cannot refute him. These two examples illustrate the early Ismā'īlī doctrine regarding their appropriation of ancient teachings: the founders of the sciences are prophets, who establish the principles (*uṣūl*) on which each science is based, whereas the followers of those founders, the philosophers, often go astray and introduce errors.[20] For these early Ismā'īlīs, revelation is superior to reason and is itself the ultimate basis of science.[21] Moreover, it appears from the exact agreement of these *dā'īs*' deployment of Hermes in their homologous arguments that they were relating the views of their common source: the hidden leaders of the *da'wa* before the schism of the Syrian and Northern Iranian branches after 899–900.[22] We can therefore assign this particular use of Hermes to about 875–899, the most secretive, obscure, and earliest period of the Ismā'īlī movement.

In the Ismā'īlī discussions of science and revelation just described, both *dā'īs* also directly mention, and regard highly, the book of Pseudo-Apollonius called *Sirr al-ḫalīqa*, a work composed in Arabic ca 813–833, in which Apollonius discovers, and then explains, the secrets of nature he found inscribed on Hermes' Emerald Tablet hidden in a long-lost subterranean tunnel.[23] This is not the place to attempt an

19. Abū Ḥātim ar-Rāzī, *Kitāb A'lām an-nubūwa*, 210–211 = *bāb* 7, *faṣl* 1. He also insists that Muḥammad's reported admonition against astrology was not an absolute proscription but rather a warning that the exclusive interest in astrology comes at the expense of study of revealed laws (*Kitāb A'lām an-nubūwa*, 211).

20. On the impossibility of humans finding the *uṣūl* of a science without divine aid: Ibn al-Hayṯam in Madelung and Walker 2000: 137 (English), 87 (Arabic); Abū Ḥātim ar-Rāzī, *Kitāb A'lām an-nubūwa*, 221–226 = *bāb* 7, *faṣl* 2.

21. As Rowson, forthcoming, shows, later Muslim philosophers, some of them actually anti-Ismā'īlī, would make similar arguments for the revealed character of science.

22. On this stage of the *da'wa* see Halm 1996a: 5–78; Daftary 1990: 105–143; J. H. Kramers and F. Daftary, "Salamiyya," *EI*², 8.921a–923b (922a). On the one hand, Abū l-'Abbās Muḥammad was the courier between the Egyptian *dā'ī* and the headquarters in Salamīya, and, "as far back as 278/891, intimately linked to the central activities of the *da'wa*" (Madelung and Walker 2000: 8 and 37–38; also Halm 1996a: 39 and 58). On the other hand, Abū Ḥātim was an important follower of Ġiyāṯ, a student of the *dā'ī* Ḫalaf who had first been dispatched to Rayy from the Syrian headquarters (Stern 1983d: 189–195; Halm 1996a: 289); Abū Ḥātim may well also have studied under Ḫalaf himself earlier on, because the latter took up residence in Abū Ḥātim's home district of Pašāpūya, near Rayy, upon his arrival from Syria for his mission.

23. Abū Ḥātim says that Abū Bakr ar-Rāzī knew the *Sirr al-ḫalīqa* to be a composition by an author in the time of al-Ma'mūn (r. 813–833), but unfortunately he does not relate for us the author's name. Abū Ḥātim admits it is true (*Kitāb A'lām an-nubūwa*, 208 = *bāb* 7, *faṣl* 1; text also in Kraus 1936: 372–373). Weisser 1980: 54 and 69 (following Kraus 1942–1943: 2.275 and 2.278), considers the *Sirr*

interpretation of the notoriously obscure text of the Emerald Tablet, copied out in full in the last pages of the *Sirr al-ḫalīqa*.[24] Suffice it to say here that the Emerald Tablet describes an ascent to heaven and a return to earth in the context of a mysterious revelation of cosmic significance, and ends with the statement, "That is my boast, and because of that I am called Hermes Triplicate-in-Wisdom." Precisely who or what is supposed to have ascended to heaven and returned is unclear.[25] Perhaps it was Hermes himself. But what is clear is that the *Sirr al-ḫalīqa* was highly esteemed by the early Ismāʿīlīs because it exemplifies the notion that natural science is *revealed knowledge*, with its source in Hermes, the prophet Idrīs. It is plausible, if not likely, that the *Sirr al-ḫalīqa*, along with its peculiar representation of Hermes, was also disseminated by the hidden leaders of the Ismāʿīlīs in Syria through their *daʿwa* in the last quarter of the ninth century.[26] More generally, these examples show how some Muslims, by about 900, wholeheartedly regarded Hermes as a legitimate prophet. The Arabic Hermetica, and here his astrological works in particular, are understood to constitute revealed knowledge on a par with the knowledge derived from any other true prophet. It can only have helped the argument that Hermes was already identified during the ninth century as the prophet of an ancient religion, that of the Ṣābi'a.

---

al-ḫalīqa to be a translation from Greek, and supposes that only the longer recension of the work was from the time of al-Ma'mūn. Zimmerman 1981 considers the work probably to be a new composition in Arabic, albeit exploiting earlier materials (apparently accepted also by Burnett 2001a: 179n2). Rudolph 1995 presents convincing arguments, on the basis of the theology clearly expressed in the work, pinning it to the time of al-Ma'mūn. It is precisely the book's distinctive theology that was pointed out already by ar-Rāzī. Ullmann 1980: 93–94 argues that a number of the Arabic terms and usages found in the work were unknown before the ninth century. In any case, everyone agrees that the *Sirr al-ḫalīqa* incorporates material from earlier sources including, in its longer recension, adapted parts of Nemesius of Emesa's *On the Nature of Man* (Kraus 1942–1943: 2.275–280; see also Weisser 1980: 55–68, and Zimmerman 1981).

24. Weisser 1979: 524–525 (book V, section 33.2); text also printed by Ruska 1926: 112–113; Plessner 1927: 99–100. See also Ullmann 1972: 170–172.

25. The tenth- or eleventh-century author of the Hermetic *ar-Risāla al-falakīya al-kubrā* (*Great Treatise of the Spheres*, ed. in Vereno 1992) evidently thought it was Hermes himself who did so. Other interpreters take the ascent and descent as an allegorical reference to an alchemical operation on material substances, not a personal ascent.

26. Abū Ḥātim ar-Rāzī (*Kitāb Aʿlām an-nubūwa*, 90, *bāb* 3, *faṣl* 4, *fī bāb al-maṯal wa-l-maʿnā*) summarizes the introduction of *Kitāb Sirr al-ḫalīqa*, where Apollonius receives the emerald tablet from Hermes (ed. Weisser 1980: I.1.2), and says that this passage is a metaphor, but he does not provide us an interpretation of the metaphor. Ibn al-Hayṯam of Qayrawān apparently had a copy of the work in North Africa before joining the Ismāʿīlī movement, and his teacher, the *dāʿī* Abū l-ʿAbbās, knew the work already, expressed his admiration of it (*šarrafahū*), and asked to see his copy: Madelung and Walker 2000: 140–141 (English), 91 (Arabic). Significantly, the addressee of al-Ǧāḥiẓ in his *Kitāb at-Tarbīʿ wa-t-tadwīr*, Aḥmad ibn ʿAbdalwahhāb, whom al-Ǧāḥiẓ mocked some decades earlier with the question about Hermes' identity (see note 15), was also a Šīʿite holding views widely considered heterodox. In the context of the questions on nature, matter, and the occult properties of substances posed by al-Ǧāḥiẓ in the epistle, this could be taken to suggest that the *Sirr al-ḫalīqa*, through which Hermes acquired some fame, was esteemed among certain Šīʿites such as Aḥmad ibn ʿAbdalwahhāb already in the first half of the ninth century. The popularity of the book among Ismāʿīlīs would in that case be an ongoing development of a tendency already attested.

The idea of Hermes' ascent on high—mentioned only in passing by Abū Ḥātim—thereafter became closely linked with his role as a prophet of science. Already in the eighth century, Hermes had been famous for his astrological knowledge, known to astrologers from the Hermetic books circulating under his name in Middle Persian, Arabic, and other languages. Once Hermes-Idrīs' heavenly ascent was also accepted, this was evidently soon understood as *the means by which* he had discovered astrology. The Greek *Corpus Hermeticum* does refer in different ways to a kind of heavenly ascent of Hermes, but there is no definite evidence at present for the transmission of these particular Greek passages into Arabic.[27] Rather, the idea of Hermes' heavenly ascent came about more probably though the identification of Hermes with Idrīs-Enoch. This explains, among other things, the regular association of Hermes with heavenly angels, from whom he learned the sciences, as seen in the traditions reported by al-Maqdisī, Abū Ḥātim the Ismāʿīlī, and the *Kitāb at-Tuffāḥa* (discussed below).

It appears, then, that it was primarily the combination of ancient Judaean lore about Enoch with a Hellenistic astrological tradition that gave Hermes the character of a prophet of philosophy and science. An important testimony to this development is provided by Agapius (Maḥbūb), Bishop of Manbiǧ, called the Byzantine (*ar-Rūmī*), one of the earliest known Christians to write history in Arabic (rather than in Syriac or another language), sometime in the second quarter of the tenth century.[28] Firmly in the tradition of Greek and Syriac Christian world chronography, he treats the antediluvian patriarchs early in his work, and he therefore has something to say about Enoch. After discussing the variant periods for Enoch's life given in the Greek and Hebrew Bibles, and explaining that Enoch was raised to Paradise where he could worship God, he turns to what a certain, unnamed author has to say about him.[29]

وزعم بعض الحكما ايضا ان اخنوخ الذي هو ادريس عرف واظهر ودل على الكتاب والحروف والنجوم والحساب وزعم ماناثون الحكيم المصري المنجم ان الله رفع اخنوخ حتى ماس الفلك الداير وعرف البروج التي فيه والكواكب الثابتة والجارية †والاشياه† والحدود والوجوه من العشرات والصور التي فيها وغير ذلك من اسرار علم النجوم ولذلك يقال ان كتابه في النجوم يسما كتاب المعاني وجميع الحرانيين الذين يعبدون الاصنام والكواكب متمسكون بقول ماناثون المصري

27. The *Poimandres* (*CH* 1.24–26) describes a kind of psychic ascent to the eighth, outermost sphere of heaven by way of a progressive shedding of the body and base attributes. The *Korē Kosmou* (*SH* 23.7) says that Hermes went off to heaven after giving his teachings. Cf. Rowson 1988: 242. For an example of Hermes' more mundane and earlier fame merely as an astrologer, see Gutas 1998: 100–101, where a courtier reportedly flattered the caliph al-Ma'mūn (r. 813–833) saying, among other things, "If [you take up] astrology, you are like Hermes in your calculations." No hint of Hermes' prophetic character occurs in this earlier instance; indeed Hermes' expertise in astrology is explicitly contrasted in this passage with the expertise of others in other subjects, such as medicine and religion.

28. Graf 1944–1953: 2.39–41.

29. Agapius 1.35.9–36.3 = *Patrologia Orientalis* 5.591–592.

> One of the philosophers (*al-ḥukamā*) also has claimed that Enoch, who is Idrīs, discovered, revealed, and taught books, writing, astrology, and calculation. Manetho, the Egyptian sage and astrologer, claimed that God raised Enoch until he strode the turning celestial sphere and discovered the signs of the zodiac that are in it, the fixed and moving stars, the †lots†,[30] the *termini*, the decans and the images in them, and other mysteries of astrology. For that reason it is said that his book on the stars is called *The Book of Meanings.* All the Ḥarrānians, who worship the idols and the planets, adhere to the doctrine of Manetho the Egyptian.

Agapius does not mention Hermes by name. "One of the philosophers" evidently refers to an unnamed authority, and the references to Manetho and to the Ḥarrānians are difficult to interpret. In any case, the text clearly attributes to Manetho the notion that Enoch ascended to the heavenly spheres, and then returned to reveal not just any astral science, but astrology in particular. The question then is what that work of Manetho might be, or what the reference to Manetho can mean. Several writings attributed to the ancient Egyptian priest Manetho survive in Greek, though mostly in fragments. They include pieces of an Egyptian chronicle, along with a few brief fragments of other works thought to be authentic,[31] some extracts of a different chronicle called the *Book of Sothis* (studied in the previous chapter), and a poem on astrology in six books of epic meter, called *Apotelesmatika*. This poem and the *Book of Sothis* are both considered today to be falsely attributed to Manetho.[32] The poem does refer to Hermes as a source of astrological teachings, but these are from Hermes' inscriptions in Egyptian *charaktēres* on hidden stelae, not from an alleged heavenly ascent.[33] One might propose that the reference to Manetho points to *The Book of Sothis* as interpreted by Annianus, but this is purely speculative. The extant works of Manetho thus offer no concrete help.

There is another clue about the origin of this passage that may prove to be more fruitful. A substantial scholarly discussion has focused on the lost Syriac chronicle of Theophilus of Edessa (d. 785), court astrologer of the caliph al-Mahdī,[34] and the

30. The text of the edition here is obscure: *al-ašyāh*. It is uncertain what the correct reading should be. Charles Burnett has tentatively suggested to me *al-sihām*, the astrological lots, and I have no better suggestion. In any case, the intended word is surely an astrological term. Vasiliev's French translation to his edition has "horoscopes" with no comment.

31. The fragments are presented in English translation by Verbrugghe and Wickersham 1996: 129–171.

32. Partially presented by Verbrugghe and Wickersham 1996: 173–182. For an edition and English translation of the *Apotelesmatika*, with commentary, see Lopilato 1998.

33. Manetho 5.1–11: "From the books of the temple sanctuaries, O King Ptolemy, / and the hidden stelae, which all-wise Hermes devised / and inscribed in *charaktēres* with the appropriate forecasts of the heavenly stars, / having found Asclepius as an adviser of prudent wisdom, / I, having made impressions with figure-retaining wax, have recovered / my flower-gathering song, a gift sweeter than (that of) the bees..." (translation adapted from Lopilato 1998: text 99, translation 263, commentary 394–401).

34. On Theophilus in general see Hoyland 1997: 400–409 and Pingree 2001: 13–20.

consensus appears to be that Theophilus was a common source for several later historical works: the chronicle of Agapius (in Arabic) under discussion here, the Byzantine chronicler Theophanes (in Greek),[35] and Michael the Elder and the *Anonymous Chronicle of 1234* (both in Syriac). Theophilus is their common source at least for their histories of the years 630–750.[36] The full extent to which Agapius relied on Theophilus remains undetermined. Modern historians have focused on reconstructing the portion of Theophilus' text dealing with the period of Muḥammad's life and the Arab kingdom until the coming of the ʿAbbāsids.[37] They have not yet attempted to determine whether Theophilus' history of more ancient times can be reconstructed on the basis of these later works.[38] Nevertheless, it seems that Agapius' history may represent a good deal more of Theophilus than previously discerned. I think that the matter is fairly clear from Agapius' own statement: "Theophilus the Astrologer, from whom we have taken these narratives, said: 'I have myself been a continuous eyewitness to these conflicts [the ʿAbbāsid revolution], and I set matters down in writing so that nothing pertaining to them escaped me.' He wrote many books on such subjects, but we have condensed this book from them, adding what we knew could not be dispensed with, while avoiding prolixity as much as we could."[39] Agapius' reference to his source comes late in his chronicle, in the midst of reports on the ʿAbbāsid revolution. Notably, he does not say that he condensed *only this part* of his book from the works of Theophilus, but rather that *his book*, as such, was condensed from Theophilus' work, although he added to it what he considered indispensable. The astrological relevance of the report about Enoch's ascent, knowledge about the Ḥarrānians—who are very significantly *not* called Ṣābians, probably indicating a source prior to *circa* 830—and the reference to "Manetho the Astrologer," otherwise unknown in Arabic, all point to an early source familiar with both chronography and astrology as well as Greek literature. Given

35. Theophanes' chronicle was really a continuation of the chronicle of George Syncellus, and probably it was the latter who assembled the materials Theophanes used, including a translation of the chronicle of Theophilus of Edessa. See Mango 1978, Mango and Scott 1997: lxxv.

36. Conrad 1990; Conrad 1992: 322–348; Palmer and Hoyland in Palmer 1993: 95–98; Mango and Scott 1997: lxxxiii–lxxxvii; Hoyland 1997: 400–409. Bar ʿEbrāyā (Ibn al-ʿIbrī) tells us that Theophilus wrote his chronicle in Syriac (Syriac chronicle: Budge trans. 116, facsimile of Bodleian MS Hunt. 52, 43v, column b; cf. his Arabic chronicle 220.2, where the language of the chronicle is not specified).

37. Hoyland 1997: 631–671 provides the fullest outline to date of the reconstructed Syriac Common Source, accepted as Theophilus, for the years 589–760.

38. Theophilus may well be the common source for some of these later historians' information on ancient history. For example, Drijvers 1966: 116–118 points out that there is an unknown common source of information on the cosmology of Bar Dayṣān for Michael the Elder, Agapius, and Bar ʿEbrāyā (Ibn al-ʿIbrī). Theophilus could be this source.

39. Agapius (ed. Vasiliev 2.265.5–7 = *Patrologia Orientalis* 8.3.525.5–7; ed. Cheikho 369.2–6) *wa-qāla Tawfīlu l-munağğimu llaḏī aḫaḏnā ʿanhu hāḏihī l-aḫbāra innī lam azal mušāhidan li-hāḏihī l-ḥurūbi bi-nafsī wa-kuntu aktubu ašyāʾa ḥattā lam yašiḏḏa ʿannī minhā šayʾun wa-lahū min ḏalika kutubun kaṯīratun illā annā ḫtaṣarnā minhā hāḏā l-kitāba wa-alḥaqnā fīhi mā ʿalimnā annahū lā ġanāʾa ʿanhu fīhi wa-tağannabnā t-taṭwīla ğahdanā*. I have used the translation of Conrad 1992: 331, with modifications.

that Theophilus, the astrologer from Edessa, was a major source for Agapius, he is also the obvious candidate for the source of this particular passage, which is hardly likely to be something "indispensable" that Agapius himself would have added. If this assumption is correct, then this passage becomes an important early testimony to the astrological significance of Enoch's heavenly ascent, without direct reference to Hermes by name, composed by Theophilus originally in Syriac and in the court of al-Mahdī. It is conceivable that Enoch had come to be regarded already by Theophilus' time as the revealer of astrology in its entirety, instead of merely the simpler astronomy of the *Book of Enoch*. In fact, a Syrian contemporary of Theophilus claimed that the early Syrian philosopher Bar Dayṣān (d. 222) held that Enoch was the inventor of astrology, the Chaldaean art (*kaldāyūtā*).[40]

Agapius' source's description of the Ḥarrānians requires some comment and is harder to understand. One possible interpretation is just that the Ḥarrānians are thought to follow the *astrological doctrines* of Manetho. This could refer merely to his *Apotelesmatika*, mentioned above, or to something similar. Another interpretation is that the Ḥarrānians follow Manetho in asserting that Enoch went to heaven and returned. In that case, the reference would be probably to Hermes identified with Enoch, because the Ḥarrānians became famous, albeit only in the ninth century, for holding Hermes as their prophet. Again, the problem of specifying the source of this passage is critical. If the source is Theophilus, then the reference to the Ḥarrānians, and the testimony about their idolatry and astrolatry, precede both Theodore Abū Qurra's work (discussed in section 3.4.2) and the birth of Ṯābit ibn Qurra by a few decades. Perhaps Agapius added the comment on his own, based on his tenth-century knowledge of the legend of the heavenly ascent of Hermes, which had become widespread by that time. Unfortunately, there are too many questions to know for certain how to interpret the passage exactly. In any case, the Christian Agapius' testimony about the heavenly ascent of Enoch matches descriptions of Hermes from some of his tenth-century Muslim contemporaries.

Reports in later Arabic philosophical works sometimes omit the name Enoch entirely, preferring instead just the name Hermes for the prophetic figure who ascended to the heavens to learn celestial secrets. An influential description of the ascent of Hermes appears as an accepted fact in the little book known as *Kitāb at-Tuffāḥa* (*The Book of the Apple*). Like Plato's *Phaedo*, the purpose of the *Kitāb at-Tuffāḥa* is to exhort the student philosophically not to fear death and to teach the immortality of the soul. This work was widely read over several centuries, cited in Arabic by tenth-century authors, translated into Persian by Bābā Afḍaluddīn Kāšānī in the thirteenth century,[41] translated into Hebrew in Barcelona in 1235, and from

40. Theodore Bar Koni (wr. 791–792), 11.3 (trans. Hespel and Draguet 2.213).

41. First edited and translated by Margoliouth 1892; also published in Kāšānī 1952: 1.113–144. The same Kāšānī translated the work of Hermes *Kitāb fī Zaǧr an-nafs* (KZN) into Persian as *Yanbū ʿ al-ḥayāt dar muʿātabat-u naṣīḥat-i nafs* (Kāšānī 1952: 1.331–385). One of the two earliest known Arabic copies of

Hebrew into Latin in 1255 in Sicily, whereafter it became popular among European scholars under the name *Liber de pomo*.[42] A few remarks on the text are required before discussing the references to Hermes within in it. All modern studies of the work agree that the work is definitely modeled on Plato's *Phaedo*, which related Socrates' last dialogue before his death. The *Kitāb at-Tuffāḥa* is a dialogue between a dying sage and his disciples on the same subject—the afterlife and fear of death—but there are lingering questions about the origins of the Arabic adaptation. Several versions of phaedonic texts still exist in Arabic manuscripts today, and no published study has sorted them out adequately.[43] An Arabic version of Plato's *Phaedo* appears to be cited by Isḥāq ibn ʿAlī ar-Ruhāwī (later ninth century) in his *Adab aṭ-ṭabīb*, by al-Bīrūnī in his work on India, and by al-Mubaššir in his anthology of ancient wisdom.[44] The philosopher al-Kindī is reported to have written a *Risāla fī Ḫabar mawt Suqrāṭ* (*Treatise on the Story of the Death of Socrates*), from its title evidently having contents related to the *Phaedo*.[45] As for the *Kitāb at-Tuffāḥa*, it exists in different recensions in at least six known manuscripts, some published, some not, but never studied all together.[46] Of the five Arabic manuscripts about which details are currently available, two name Socrates as the principal interlocutor, whereas three have not Socrates but Aristotle as the dying teacher. The Persian and Latin versions also both have Aristotle as the teacher. Dimitri Gutas has made the commonsense suggestion that this situation indicates that the work first appeared in Arabic featuring Socrates, as in the original *Phaedo*, but later was changed in Arabic to make Aristotle the main speaker.[47] The history of the book may, however, be more complicated, as I will show.

In the examples of the text available to me, Hermes is said by the teacher (Aristotle or Socrates) to have brought philosophy down from heaven after learning it from the angels there:[48]

---

this Hermetic work occurs in MS Nuru Osmaniye 4931, 38b–57b, dated 1275–1276 (674 AH); this manuscript also contains the earliest known Arabic copy of the *Kitāb at-Tuffāḥa* as well as some of the Persian works of Kāšānī, including also a copy of Kāšānī's Persian translation of the *Kitāb at-Tuffāḥa* (126b–136a).

42. The Latin text is studied, edited, and translated into German by Acampora-Michel 2001, with a reprint of Margoliouth's English translation of the Persian version.

43. The standard studies are those of Jörg Kraemer 1956 and Bielawski 1974, to be supplemented by Gutas 1986: 31 and 36n61. The following information is taken from these authors.

44. Bielawski 1974: 123 and Endress 2007: 332 (mistakenly citing Ayyūb ar-Ruhāwī rather than Isḥāq ibn ʿAlī ar-Ruhāwī; on the latter see Ullmann 1970: 223–224).

45. *Fihrist* 319.12 (ed. Flügel 260.5–6); Bielawski 1974: 134.

46. Gutas 1986: 36n61. Everett Rowson, who has been studying this text, has kindly supplied me with a list of these manuscripts as well as other notes on the text, from which I have benefited in some matters of detail.

47. Gutas 1986: 31 and 36n61.

48. Besides the Persian version already mentioned, I have consulted MS Nuru Osmaniye 4931, 76b–85a, dated 674 AH (i.e., 1275–6 CE); the Arabic edition of Ḫayrallāh 1919–1920, published on the basis of a manuscript he happened upon in the service of Gregory IV, Greek Orthodox Patriarch of Antioch (1906–1928), in which Socrates is the speaker (text noted by Jörg Kraemer 1956, 488–489); and

> Diogenes:[49] Whence did the philosophers arrive at this view?—The sage answered: The preachers of the ages have always preached this up to the horizons of the earth. The first of the people of our land[50] to whom this science was revealed (*ūḥiya ilayhi hāḏā l-ʿilm*) was Hermes.—Diogenes asked: Whence did it come to Hermes?—The sage answered: He ascended in his spirit (*ʿaraǧa bi-rūhihī*) into heaven and heard it from the heavenly host (of angels: *al-malaʾ al-aʿlā*) who took it from the Noble Record (*aḏ-ḏikr al-karīm*).[51] [Then he returned to the earth and the sages learned it from him.][52]—Diogenes said: How do I know that he took it from the inhabitants of heaven?—The sage answered: If this matter is the truth, then its advent was from on high.

Just before dying at the end of the *Kitāb at-Tuffāḥa*, the sage gives his final advice, recommending certain books to his students that should guide them when he is gone. The list begins with a book on the first principles (*uṣūl*) called "the book of Hermes" (*kitāb Hirmis*),[53] and goes on to include the four books of Aristotle's logic, named as the *Categories*, *On Interpretation*, *Analytics*, and *Apodeictics*. Strangely, even in Ḫayrallāh's text, in which Socrates is the teacher, he says that he wrote these four books.[54] This appears to argue against Gutas' suggestion that Socrates was original to the *Kitāb at-Tuffāḥa*, for who would have attributed Aristotle's well-known books on logic to Socrates? A similar indication is in the sage's reference at one point to his own teacher, Plato.[55] The speaker in the original *Kitāb at-Tuffāḥa* must have been Aristotle; a later copyist must then have substituted Socrates, knowing something of the true story of the *Phaedo*, as, for example, al-Bīrūnī knew it, but when this copyist that I am supposing did so, he neglected to change or omit these famous books of Aristotle, so that now Socrates

---

two Arabic versions published by an-Naššār and aš-Širbīnī 1965: (1) a "*muḫtaṣar*" with Socrates as speaker, 337–352 (noted by Jörg Kraemer 1956: 488), and (2) an incomplete version with Aristotle as speaker, 355–363. The latter is said to be from the notes of D. Santillana, who lectured in Egypt in 1910–1911, and the editors do not know whether he found it in an unknown manuscript or translated it from Margoliouth's Persian version! Clearly a critical edition of the Arabic text is needed. The passage translated here is from Ḫayrallāh 1919–1920: 220 compared with the Nuru Osmaniye manuscript.

49. Or Diochus.

50. The reference to Hermes as being from "our land," *min ahli bilādinā*, appears to refer to ancient Greece. The Persian text has *az zamīn-i mā*, but Margoliouth (1892: 251, line 555; cf. Kāšānī 143.3) mistranslates it as "on earth."

51. MS Nuru Osmaniye 4931, 84b10: *aḏ-ḏikr al-ḥakīm*.

52. Words in brackets from MS Nuru Osmaniye 4931, 84b10: *ṯumma ʿāda bihī ilā l-arḍi fa-qtabasahū l-ḥukamāʾ*.

53. MS Nuru Osmaniye 4931, 85a1–2, has instead: *kitābī l-maʿrūf fīmā baʿda ṭ-ṭabīʿa*, "my well-known book on metaphysics."

54. Ḫayrallāh 1919–1920: 221.

55. MS Nuru Osmaniye 4931, 83b5–6: *qad nafaʿaka Aflāṭūnu bi-mā aslama ilayya min nafʿi ʿilmihī llaḏī aslamtu ilayka*, "Plato has benefited you through the benefit of his knowledge that he granted to me, which I have granted to you."

claims to have written them.[56] The reference to the sage's teacher Plato also remained unchanged. As odd as this may seem, such a manipulation of the text could account for the two different traditions, one having Aristotle, the other Socrates as teacher, but in both cases named as the author of Aristotle's famous books. A critical edition is needed to decide, and in either case the question is irrelevant to the myth of Hermes.

Earlier in the dialogue, the teacher had already referred to a book of Hermes when he asked whether his student agrees with the statement that Hermes registered in the "Book of the Natures of Creation" (*Kitāb Ṭabā'iʿ al-ḫalq*), that "natures become strong only by combination with their likes, and they become weak only by receiving that which is unlike them."[57] As recognized by Jörg Kraemer, this very teaching, in almost the same words, is found near the beginning of the *Sirr al-ḫalīqa* (*The Secret of Creation*) of Pseudo-Apollonius of Tyana, and it is described there as "knowledge of the first principles," the word being *uṣūl* in both works.[58] Although the *Sirr al-ḫalīqa* is not a work *by* Hermes, it has contents in common with one known Arabic work of Hermes, the *Kitāb al-Isṭamāṭīs*, and it is the source of the famous Hermetic *Tabula Smaragdina*, the alleged basis of the whole work.[59] Hermes' book on "first principles," mentioned at the end of the dialogue in the list of books, is therefore probably also to be identified as the "Book of the Natures of Creation" mentioned earlier in the *Kitāb at-Tuffāḥa*. Further investigation may be able to

56. Gutas 1986: 31 regards it as "difficult, if not impossible, to maintain that it was subsequently also attributed to Socrates" if the original form had Aristotle as its teacher. Yet it is equally difficult to think the original would have attributed Aristotle's books to Socrates. It *is* remotely possible that Ḫayrallāh himself "corrected" the name of Aristotle, found in his manuscript, to Socrates; he shows in the foreword to his edition that he had been studying the story of Socrates in Arabic translation (1919–1920: 475–476). The Persian, Hebrew, and Latin versions have Aristotle as the teacher.

57. MS Nuru Osmaniye 4931, 82b5, and Ḫayrallāh 1919–1920: 107.12–13: *aḫbara anna ṭ-ṭabā'iʿa lā taqwā illā bi-ttiṣāli* (not *bi-nifiṣāli* as in Ḫayrallāh; cf. Kraemer 1956: 503) *aškālihā wa-lā taḍʿufu illā bi-liqā'i ḫilāfihā* (*iyyāhā*). Cf. the Persian version (Kāšānī 1952: 136.18–19 and Margoliouth, text and translation 1892, lines 440–441): "He has stated that no nature acquires strength except by union with its like; and that none acquires weakness except by union with what is unlike it" *hīč ṭabʿ nīrū na-gīrad illā az payvand-i hamsān-i ḫwīš va-sustī na-yābad illā az payvand-i muḫālif-i ḫwad*. The "*Muḫtaṣar*" of the work, in which Socrates is the main speaker, provides the same citation of Hermes (an-Naššār and aš-Širbīnī 1965: 350.15–16).

58. The two Arabic recensions of the text published by an-Naššar and aš-Širbīnī do not include the list of books. The Nuru Osmaniye manuscript and the Persian give the list of books, including the book of Hermes in the Persian (*kitāb-i Hirmis*, Kāšānī 1952: 143.20–21), and describing the contents of the book of Hermes more specifically as dealing with metaphysics and theology. MS Nuru Osmaniye 4931, 85a1–4: *fīmā baʿda ṭ-ṭabīʿa*; Margoliouth 1892: 571–572, and Kāšānī 1952: 143.20: *ʿilm-i avval-u ḥikmat-i rubūbīyat*.

59. Kraemer 1956: 504; Weisser 1980: 68–70. In the edition (Weisser 1979: 4.4–6: book I, section 1.1.3) the text describes the *ǧawāhir*, "essences," as *mustaʿdiyatu aškālihā bi-ttiṣālihā bihā wa-dāfiʿatun aḍdādahā bi-ḫilāfihā lahā. hāḏā sirru l-ʿālami wa-maʿrifatu uṣūli ṭ-ṭabā'iʿi*, "deriving assistance from their likes by combination with them and repelling their opposites by difference from them. This is the secret of the world and knowledge of the first principles (*uṣūl*) of the natures." On the *Tabula Smaragdina* Ruska 1926; Ullmann 1972: 171; Weisser 1980: 44–46.

discover whether the *Kitāb at-Tuffāḥa* refers to the *Sirr al-ḫalīqa* or to another Arabic Hermetic text.

Any definitive conclusion about the authorship of the *Kitāb at-Tuffāḥa* would require a careful study of all the manuscripts and, ideally, a critical edition. Previous suggestions about its author include al-Kindī[60] or a Ḥarrānian (because of the reference to Hermes).[61] It would seem uncharacteristic of al-Kindī to create a work such as this, especially when he evidently had access to the *Phaedo* itself in some form, as seen above. As for the other hypothesis, it can no longer be maintained automatically that Ḥarrānians are responsible for anything connected with Hermes. Moreover, as just shown, the doctrine of Hermes in the *Kitāb at-Tuffāḥa* is related to a passage in Pseudo-Apollonius' *Sirr al-ḫalīqa*; in the same section of that work the Ṣābians and planet and idol worshippers are accused of error in worshipping created things.[62] This might make it seem rather *anti*-Ḥarrānian. In the *Kitāb at-Tuffāḥa*, Aristotle, on his deathbed, cites and recommends a book of Hermes known to his disciples. A similar chain of transmission of knowledge from Hermes to Aristotle is also assumed in the unpublished works I have dubbed Talismanic Pseudo-Aristotelian Hermetica (TPAH); these include the *Kitāb al-Isṭamāṭīs* just mentioned. The *Kitāb at-Tuffāḥa* was composed by someone who had in mind such a chain of transmission of knowledge, and one may therefore suppose that he knew of the TPAḤ Whoever it was also had knowledge of the *Phaedo* through a more faithful translation or adaptation of the Greek, such as that known to al-Bīrūnī, which served as the basis of his adaptation; he knew that Hermes was Idrīs-Enoch, who had experienced a heavenly ascent and communicated with angels; and he believed that Hermes had revealed the principles of physical science. The earliest allusion in Arabic to Aristotle's holding an apple while on his deathbed occurs in the famous *Kitāb al-ʿIqd al-farīd* of the Andalusian Ibn ʿAbd Rabbih (d. 940).[63] Here it appears in a letter, cited in its entirety, supposedly sent by a slave girl along with the gift of an apple to her master the caliph al-Ma'mūn; this is not, however, strictly an explicit reference to the *Kitāb at-Tuffāḥa*. Direct references to the *Kitāb at-Tuffāḥa* itself occur only in the 980s, when it received widespread attention all at once in far-flung lands: by Ibn Ǧulǧul in al-Andalus (wr. 987),[64] by the Iḫwān aṣ-Ṣafā' (Brethren of Purity), who wrote probably in Baṣra around 980,[65]

60. GAL S1.373; Bielawski 1974: 133–134.

61. Margoliouth 1892: 190. An-Naššār and aš-Širbīnī 1965: 231 discuss Margoliouth's hypothesis inconclusively.

62. Weisser 1979: 35–36 (book I, 2.3.6).

63. Ibn ʿAbd Rabbih 1940–1968: 6.288.7–8 (near the end of Book 23, i.e., *az-zabarǧada aṯ-ṯāniya*, on the natures of humans and animals).

64. Ibn Ǧulǧul 27.5–6.

65. For a brief introduction, including a balanced discussion of the dating of the text, see Paul Walker, "Ekwān al-Ṣafā'," *EIr* 8.294b–297b. The fundamental articles are Stern 1946 and 1983b. Joel Kraemer 1986: 165–178 summarizes the evidence and discusses their place in the intellectual milieu of their time. *Pace* Abbas Hamdani, who argues in numerous instances, unconvincingly, that the *Rasā'il*

by al-ʿĀmirī when he composed his *Kitāb al-Amad* in Buḫārā in 985–986.[66] These attestations all argue for a tenth-century composition. Interestingly, the mostly comprehensive book catalog, the *Fihrist* of an-Nadīm, completed 987/988, does not mention the *Kitāb at-Tuffāḥa*, at least not under that name.[67] As shown above with a couple of examples, early Ismāʿīlī authors held that the principles (*uṣūl*) of the sciences were revealed knowledge, and that astronomy and astrology were first revealed to the prophet Hermes, whom God raised up in order to teach him these sciences. The insistence that the *uṣūl* or first principles of a science are revealed knowledge and its connection with the *Sirr al-ḫalīqa*, a work known and highly appreciated by early Ismāʿīlīs, are both typical of early tenth-century Ismāʿīlī writings. One should look in this direction to find the author of the *Kitāb at-Tuffāḥa*.

Later philosophers writing in Arabic, including the readers of the *Kitāb at-Tuffāḥa* just mentioned, accepted it as a fact that the ancient Hermes ascended to the heavens to gain knowledge. The eastern Iranian philosopher al-ʿĀmirī (wr. 985–986) reports on Hermes in his treatise on the afterlife: "Hermes ascended in his spirit to (the celestial sphere of the zodiacal signs) and became well-informed about what was in them. Then he told us about it."[68] By al-ʿĀmirī's time, the information in the *Kitāb at-Tuffāḥa* was reinforced by the generally circulating traditions of the *isrāʾilīyāt* and Muslim chronicles about the ascent of Idrīs.

The anonymous philosophers of Ismāʿīlī tendency called Iḫwān aṣ-Ṣafāʾ, "the Brethren of Purity," wrote their encyclopedia of the sciences in the form of collected epistles.[69] In these they cite the *Kitāb at-Tuffāḥa* and make further use of the idea that Hermes ascended through the planetary spheres. In the second epistle of the mathematical section of their collection, there is a subsection (*faṣl*) entitled "On the abstraction of the soul (from the body) and its yearning for the world of the heavenly spheres." Here they are concerned explicitly with spiritual ascent to heaven and insist that they are not talking about bodily ascent. As a part of their program to integrate and harmonize all the philosophical systems known to them, including their understanding of Islam and its sects, they provide, as a testimony to the salvation available to philosophers and pious people, what amounts to a list of great ancient sages and prophets who experienced or who testified to the reality of heavenly ascent. These include Ptolemy, Hermes, Aristotle, Pythagoras, Jesus, and

---

Iḫwān aṣ-Ṣafāʾ are a ninth-century work. A citation of the *Kitāb at-Tuffāḥa* occurs at Iḫwān aṣ-Ṣafāʾ 4.120.16–17 (in the fourth *risāla* of *al-ʿulūm an-nāmūsīya wa-š-šarʿīya*), where Socrates refers to his noble brethren awaiting him after death; the Iḫwān aṣ-Ṣafāʾ sometimes call themselves by the same term: *al-iḫwān al-kirām*.

66. See below on al-ʿĀmirī's discussion of the ascent of Hermes.

67. It does, however, refer to the *Phaedo* in a list of Plato's works ascribed to Theon (*Fihrist* 306.24).

68. al-ʿĀmirī, *Kitāb al-Amad ʿalā l-abad* XVI.4 (Rowson 1988: 144–145): *wa-inna Hirmisa qad ʿaraǧa bi-rūḥihī ilayhi* (scil. *ilā falak al-burūǧ*) *fa-ṭṭalaʿa ʿalā mā fīhi fa-aḫbaranā bihī*.

69. On the authors' identity see Stern 1946 and 1983b.

Muḥammad.[70] For Hermes, they say, "It is related about Hermes the Triplicate in Wisdom, who is Idrīs the prophet—peace be upon him—that he rose to the sphere of Saturn and turned together with it for thirty years until he witnessed all the states of the heavenly sphere. Then he descended to the earth and informed the people of the science of astrology. God said, 'We raised him to a high place.'"[71] The sphere of Saturn is the seventh and outermost planetary sphere, and thirty years was the generally accepted approximate tropical period of Saturn, a parameter commonly occurring in the astrological literature.[72] In the view of the Brethren of Purity, Hermes could learn all the states of the heavenly sphere by abiding with the outermost planet through one of its cycles. Here the Qur'ānic reference to Idrīs, the statement of the *Kitāb at-Tuffāḥa*, the equation of Hermes and Idrīs with Enoch, and the view that Hermes revealed astrology are all used together to testify that the purified spirit of the sage can ascend through the heavens without the body in order to experience a divine revelation of the foundations of science.

The conclusion reached above, that the myth of the heavenly ascent of Hermes is a product of the identification of Hermes with Idrīs-Enoch in the ninth century, subsequently developed and popularized—at first by Ismāʿīlī missionaries, it seems, and then by philosophers—in the tenth century, provides criteria helping to determine whether some Hermetic texts in Arabic are original compositions in Arabic or not. An example from an Arabic alchemical treatise of Hermes, edited by Ingolf Vereno, illustrates this. Known from at least eight manuscripts, this short work, attributed specifically to Hermes of Dendara (ad-Dandarī), is called *ar-Risāla al-falakīya al-kubrā* (*Great Treatise of the Spheres*, abbreviated RFK), and begins with the following preface: "Hermes said: for the one whose service to the highest light is lasting, things go as he wishes. I am the master of wonders, who ascended the seven celestial spheres! I mastered (*malaktu*) the brilliant sun and the illuminating moon. I planted the luminous tree of wisdom, whoever eats of the fruit of which does not hunger and needs no food and drink; he becomes spiritual, divine (*rūḥānīyan ilāhīyan*), has inexhaustible knowledge and unceasing good."[73] Here Hermes, the author, says that he learned secret knowledge through a heavenly ascent. The passage is clearly inspired partly by the early ninth-century Emerald Tablet of Hermes already mentioned,[74] but the ascent to the seventh sphere is similar to the view of the Iḫwān aṣ-Ṣafā'. A few lines later, Hermes of Dendara elaborates further

70. The reference to Aristotle here is really to Plotinus, parts of whose *Enneads* were known in Arabic as the *Theology* of Aristotle. For Pythagoras, a translation of his *Golden Verses* is cited.

71. Iḫwān aṣ-Ṣafā' 1.92.

72. This period was known in Greek since Eudoxus (fourth cent. BCE): Neugebauer 1975: 391, 436–437, and 605–606; Evans 1998: 309.

73. Vereno 1992: 161.8–10 (RFK sections 2–4).

74. The "wonders" (*ʿaǧā'ib*) here, together with references to sun and moon, evokes the "working of wonders (*ʿamal al-ʿaǧā'ib*) from One, as all things are from One by one procedure—its father is the sun, its mother the moon," the expression of the Emerald Tablet of Hermes (*Sirr al-ḫalīqa*, Weisser 1979: 524.6–7 in VI.33.2).

on his deeds: "I am the one who constructed the Egyptian temple (*birbā*) and filled it with divine wisdom, and erected in it the image (*ṣanam*) of *rṭym'ws*,[75] endowed with four faces that indicate with their signs what happens in the four directions of the world."[76] In the course of the text, Hermes pursues a mysterious project with the assistance of the instructions of the arch-deacon (*ra's aš-šammāsīn*) Ūwīrūs. The priests are said to be "the spiritual being of the celestial sphere" (*wa-hum rūḥānīyat al-falak*). Evidently the text is an alchemical allegory, in which the sun and the moon stand for sulfur and mercury, respectively; the contents are difficult to understand.[77] The end of the text explains its alleged rediscovery: "End of the treatise of Hermes of Dendara on the spiritual art and the divine wisdom, which was recovered from the subterranean tunnel (*as-sarab*) in the temple of Dendara, under the image of *'rṭms*,[78] in the time of king *Lq'mn*."[79] The motif of a text discovered in an ancient ruin is fairly common in Arabic literature of the ninth and, especially, the tenth century.[80] As for the supposed find-site, the temple of Ḥathor at Dendara, a construction of late Ptolemaic times, is to this day one of the best preserved in Egypt, and it does have tunnels beneath it. The capitals of the four-sided columns of the temple are decorated on each side with a face of Hathor.[81] Hermes is thus imagined to have erected these pillars along with the rest of the temple at Dendara. Hieroglyphic inscriptions and illustrations in relief cover the walls, ceilings, and columns of this temple. A famous, elaborately designed circular relief in the temple portrayed the heavens and the zodiac. This zodiac was visible at Dendara until it was removed by Sébastien-Louis Saulnierin in 1820; it now resides in the Louvre. The beautiful inscriptions, combined with the obvious astrological significance of the Dendara zodiac, must have inspired the author of the text, who took them as likely proof of the wisdom with which he thought the legendary Hermes filled the place.[82]

75. Vereno's reading of Artemis, *'rṭms*, is not warranted. His critical apparatus shows that the manuscripts have *rṭym'ws* and *rṭm'ws*.

76. Vereno 1992: 161.9–11 (RFK sections 7–8).

77. Vereno's commentary, dealing also with another alchemical text of Hermes, is a mixture of important insights with unlikely speculation and some irrelevant details (1992: 182–338).

78. It is not clear from the edition whether this is the reading of a manuscript or the conjecture of Vereno.

79. Vereno 1992: 181.7–9 (RFK sections 192–193). *tammati r-risālatu li-Hirmisa d-Dandarīyi fī ṣ-ṣan'ati r-rūḥānīyati wa-l-ḥikmati r-rabbānīyati stuḫriǧat mina s-sarabi llaḏī fī birbā Dandarata min taḥti ṣanami 'rṭms fī zamāni lq'mn al-maliki.*

80. Examples include the Pseudo-Aristotelian *Sirr al-asrār* (on which see van Bladel 2004) and the *Sirr al-ḫalīqa* of Pseudo-Apollonius (ed. Weisser 1979: 5–7 (book I, section 1.2); cf. Weisser 1980: 74–75 and English translation of the passage in Rosenthal 1975a: 246). See also Weisser 1978 and van Koningsveld 1998: 347–355.

81. Vereno 1992: 308–310.

82. Images of the temple at Dendara, and the "four-faced Hathor" (*Ḥt-ḥr... m ḥr fdw*) can be seen in the famous *La Description de l'Egypte*, 1809, IV, plates 2–34. Complete plates, including images of the crypts beneath the temple, appear in Chassinat 1934–. Photographs of the site, including the four-faced capitals, are also easily found on the Internet.

This treatise of Hermes of Dendara is cited by aṭ-Ṭuġrā'ī, a secretary of the Salǧūqs and a renowned alchemist (d. 1121),[83] providing a *terminus ante quem*. Vereno suggests therefore a date of composition in the tenth or eleventh century. The next step is to find a *terminus post quem*. The preceding discussions have shown that the notion of Hermes as an Egyptian sage who carved his wisdom into monuments was popularized by Abū Ma'šar's *Kitāb al-Ulūf*. The identification of Hermes with Idrīs-Enoch, spread through the same work, also prompted the development of a legend of Hermes' heavenly ascent, known among philosophers through the *Kitāb at-Tuffāḥa*, which was referred to only in the late tenth century. Moreover, Hermes declares at the beginning of the work that he ascended through the celestial spheres. It seems, then, that Vereno must be right in pointing to the tenth or eleventh century for the date of composition of this work of Hermes. The motif of Hermes' heavenly ascent, which became widely known only in the late tenth century, might argue for a composition later than the 980s. Given that the author must have seen the temple at Dendara in Egypt with his own eyes, and his high regard for Hermes as a prophet who experienced heavenly ascent, it is likely that he lived and wrote under the Ismā'īlī regime of the Fāṭimids, who ruled Egypt from 969 to 1169 and under whom scholarship thrived with extraordinary vigor.

Vereno's further suggestion that there must have been a Greek *Vorlage* to the text, dating to the Hellenistic period and a time when the cult of Hathor-Isis persisted in Dendara, must be rejected.[84] Any tourist, medieval or modern, could see the four-faced capitals at the temple of Hathor and the marvelous inscriptions there. The only possible hint of truly ancient knowledge in the text is the name of figure in the four-faced image, which manuscripts give as *rṭym'ws* and *rṭm'ws*. Vereno proposes the reading of *'rṭms*, in his opinion to be read as Artemis, with the idea that Hathor was identified with Isis, who was identified with the Greek goddess Artemis. Even if his interpretation of this name is correct—the reading is uncertain—one can hardly take a reference to Artemis as proof that the text is in any way of early Hellenistic origins. Although Vereno's *terminus ante quem* is surely correct, the *terminus post quem* must be Abū Ma'šar's *Kitāb al-Ulūf* (ca 840–860) at the very earliest. As already argued, the developed motif of Hermes' heavenly ascent in the text suggests the work was written at least a century later.

The foregoing examples show that the identification of Hermes with Enoch had far-reaching consequences for Arabic scholarship concerned with the ancient sciences. Hermes became a prophet not only for the Ḥarrānians but for philosophers and seekers of knowledge in general, indeed a Qur'ānic prophet. The teachings of Hermes, including astrology, were based on revelations received in the highest celestial spheres and even from conversation with angels. In Arabic, it was thought that

83. F. C. de Blois, "al-Ṭughrā'ī," *EI*², 10.599a–600a. Vereno 1992: 163, 173, 179–181 and 333. The work of aṭ-Ṭuġrā'ī in which the citations occur is his *Kitāb Tarākīb al anwār* (Ullmann 1970: 230).

84. Vereno 1992: 333–334.

Aristotle himself, who described his own spiritual ascent in his book of *Theology* (actually an Arabic adaptation of *Enneads IV-VI* of Plotinus),[85] depended on the teachings of Hermes. Aristotle explains this on his deathbed in the *Kitāb at-Tuffāḥa* and he describes the teachings of Hermes on the occult sciences in great detail in the several Talismanic Pseudo-Aristotelian Hermetica. It is true that not all Arabic scholars were interested in the sciences of the ancients, and of those who were, not all of them accepted the validity of alchemy and even of astrology. However, the many who did accept such sciences found legitimation for their practices in the prophetic character of Hermes. From the survey above, it appears that this aspect of Hermes appealed especially to those Šīʿites for whom a basis in revelation was a criterion of truth. Simultaneously, many Arabic philosophers from al-Kindī onward held that the soul could undergo heavenly ascent, or mental contact with the higher spheres, when purified.[86] In this respect Hermes came to constitute a model for aspiring philosophers, as attested by the citations of the *Kitāb at-Tuffāḥa* seen above, in the same way that the prophet Muḥammad provided a more general model of conduct, a *sunna*. The medieval development of this myth of Hermes will be discussed further in section 5.4.

For historians today, the motif of Hermes' heavenly ascent offers an approximate *terminus post quem* for establishing the dates of texts and a tool for locating their historical contexts. It was possible for an author to employ this motif to good effect in a text, as in the treatise of Hermes of Dendara examined briefly here, only *after* Hermes was securely identified with Idrīs-Enoch. The myth of Hermes' heavenly ascent would be meaningful only after this was widely known to many scholars, and part of general knowledge, in the tenth century, and especially the later tenth century.

5.2 Hermes the Lawgiver: Hermes as Presented by al-Mubaššir ibn Fātik

Al-Mubaššir ibn Fātik (already introduced in section 3.4.6) is the author of one of the largest and most important collections of wise sayings (*ḥikam*) extant in Arabic,[87] which he compiled in Egypt in 1048/1049, when books in all branches of learning were abundantly available in the libraries sponsored by the Fāṭimids.[88] Along with the sayings of the philosophers, al-Mubaššir relates a great amount of biographical information about the ancients who are said to have uttered them. Thus his work is

85. Ed. Badawī 1948: 3–164; the standard monograph on this work is Adamson 2002.

86. Al-Kindī 1950: 278 (*On the Soul*) explains that the soul is purified as it ascends through each planetary sphere, beginning with Mercury. Cf. Genequand 1987–1988: 8–9. The idea of the heavenly ascent of the philosophers was also popularized by the Pythagorean *Golden Verses*, a text known in Arabic.

87. Franz Rosenthal, "al-Mubashshir b. Fātik," *EI*², 7.282a–283a.

88. P. Walker 1997.

effectively a gnomological history of ancient Greek philosophy. The first four teachers in the book are Seth, Hermes, Ṭāṭ,[89] and Asclepius (followed by Homer, Solon, etc.), and each of the other three is linked to Hermes through internal references: Seth is the first "*Ūrānī*" "among the Greeks"[90] (an obscure name discussed below), whereas Agathodaemon and Hermes are the second and third "*Ūrānīs*;" Ṭāṭ, also called Ṣāb, is the son of Idrīs-Hermes; Asclepius is a disciple of Hermes. Each of the four is made the author of wise sayings, but Hermes has the fullest treatment of all of them.

Al-Mubaššir's biography of Hermes, which treats the Egyptian sage primarily as the founder of a religion, is quite different from that presented by Abū Maʿšar two hundred years earlier. It has an entirely different source, as will be shown. Later authors, however, sometimes mix al-Mubaššir's account of Hermes together with Abū Maʿšar's account. This is so with all the major Arabic historians of philosophy of the thirteenth century: Ibn al-Qifṭī, Ibn Abī Uṣaybiʿa, Ibn al-ʿIbrī, and aš-Šahrazūrī. Although Abū Maʿsar's *Book of the Thousands* is not preserved, al-Mubaššir's much less technical compendium of ancient wisdom was extremely successful and became widely known, and it is extant in many languages. Through translations of his book into Spanish, Latin, French, Provençal, and English, it came to be popular in Western Europe.[91] I begin with a translation of the story of Hermes in the Arabic text as given by al-Mubaššir, with numbered paragraph divisions that I have added.

> **1.** Hermes of the Hermeses was born in Egypt, in the city of Memphis there. In Greek he is "Irmīs," and then it was pronounced "Hirmīs." The meaning of "Irmīs" is Mercury (*ʿUṭārid*).[92] He was also named, upon him be peace, "Ṭrismīn"[93] among the Greeks; among the Arabs, "Idrīs;" among the Hebrews, "Enoch." He is the son of Jared, son of Mahala'il, son of Cainan, son of Enosh, son of Seth, son of Adam, upon them be peace.[94]
>
> **2.** He was before the great deluge that inundated the world, that is, the first deluge. After it there was another deluge that inundated the people of

89. The edition has only Ṣāb, but Ṭāṭ was the reading determined by Rosenthal on the basis of the manuscripts (handwritten notes).

90. Al-Mubaššir 4.7: *wa-smuhū ʿinda l-Yūnānīyīna Ūrānī l-awwal.*

91. Rosenthal 1960–1961 presents a bibliography of all these texts including their modern editions. The Latin translation of the maxims of Hermes from this text is not taken into account in the recent catalog of Latin Hermetica (Lucentini and Perrone Compagni 2001).

92. The name Hermes is translated here with the Aramaic and Arabic name for the planet Mercury.

93. This is the reading of Rosenthal's handwritten notes to his unpublished edition, and is clearly a corruption of "Trismegistus."

94. I give the names derived from the Hebrew Bible in their English, not their Arabic, forms. It has already been shown that Hermes was identified with Enoch by the time of Abū Maʿšar, in the middle of the ninth century, and that this identification might have been made already by Annianus (or even perhaps by Manichaeans). The biblical genealogy of Hermes given here is an immediate result of this identification. (Interestingly, it would also make Hermes an ancestor of Jesus according to Luke 3:37.) As seen above, the surviving versions of Abū Maʿšar's account do not give this genealogy.

Egypt only. In the beginning of his career he was a student of Agathodaemon (*Ġūṭādīmūn*) the Egyptian. Agathodaemon was one of the prophets of the Greeks and the Egyptians; he is for them the second Ūrānī, and Idrīs is the third Ūrānī, upon him be peace. The explanation of the name Agathodaemon is "he of happy fortune."[95]

3. Hermes left Egypt and went around the whole earth. He returned to Egypt and God raised him to Himself there. God the Exalted said, "And We raised him to a high place."[96] That was after [he had lived] eighty-two years. In seventy-two languages he called the people of the entire earth's population to worship the Creator, the Mighty and High. God granted him wisdom (*al-ḥikma*) so that he spoke to them in their different languages, taught them and educated them. He built for them a hundred and eight great cities, the smallest of which is Edessa (*ar-Ruhā*). He was the first who discovered astrology, and he established for each region (*iqlīm*) a model of religious practice (*sunna*) for them to follow which corresponded to their views. Kings were his servants, and the whole earth's population and the population of the islands in the seas obeyed him. Four kings served him; each one of them, by his order—upon him be peace—was in charge of the whole earth. The first of them was Īlāwus, meaning the Merciful; the second his son Āwus; the third Asclepius; the fourth Ammon, who is Ibsīlūḫus.[97]

4. He preached God's judgment (*dīn*), belief in God's unity, mankind's worship [of God] (*ʿibādat al-ḫalq*), and saving souls from punishment. He incited [people] to abstain piously (*az-zuhd*) from this world, to act justly, and to seek salvation in the next world. He commanded them to perform prayers that he stated for them in manners that he explained to them, and to fast on recognized days of each month, to undertake holy war (*al-ǧihād*) against the enemies of the religion, and to give charity from [their] possessions and to assist the weak with it. He bound them with oaths of ritual purity from pollutants, menstruation, and touching the dead. He ordered them to forbid eating pig, ass, camel, dog, and other foods. He forbade intoxication from every type of beverage, and stated this in the most severe terms. He established many feasts for them at recognized times, and prayers and offerings in them. One of [these times] is that of the entry of the sun into the beginnings [i.e., the first degree] of the signs of the zodiac. Another is that of

95. *as-saʿīd al-ǧadd*. In the short treatise *Tafsīr asmāʾ al-ḥukamāʾ* (Explanation of the Names of the Sages) attributed to al-Fārābī (d. 950), the name Hermes, rather than the correct Agathodaemon, is explained as *as-Saʿīd al-ǧadd* (Rosenthal 1942: 73). A common source must underlie their etymologies.

96. Qurʾān 19:57.

97. The identity of these names, distorted in the manuscript tradition, is quite uncertain. As for the last name, we have for example al-Mubaššir 8.15 *ʾbsylwḫš*, 23.3 *bsylwḫys*; in Ibn al-Qifṭī 7.6 *bsylwḫs*; in aš-Šahrazūrī 138.3 *ʾbwsylwḫs*. Grignaschi 1975: 212n5 supposes that the four kings are Asclepius-Imouthes and three personifications of Ammon; this is not likely to be correct.

the sighting of the new moon and that of the times of astrological conjunctions. And whenever the planets (*al-kawākib*) arrive at their houses and exaltations or are aspected with other planets, they make an offering. The offerings for what he prescribed include three things: incense, sacrificial animals, and wine. Of the first fruits of aromatic plants they offer roses. Of grains, they offer wheat and barley, of fruit, grapes, and of drink, wine. He promised them that a number of prophets would come after him, and he informed them that some of the characteristics of the prophet to be sent are that he will be free of all causes of blame and ailments, perfect in all praiseworthy virtues, will not fail to answer correctly questions asked about anything in the heavens and the earth (*al-araḍīn*), and that he will indicate cures for every pain, and that his prayers will be answered in everything he asks for, including the sending down of rain, the lifting of ailments, and other sorts of requests. His doctrine (*maḏhabuhū*) and his preaching (*da'watuhū*) would be the doctrine by which the world becomes well and by which its prosperity increases.

5. He ordered people into three classes: priests (*kahana*), kings, and subjects. The rank of priest is above the rank of king, because the priest prays to God for himself, his king, and his subjects, while it is not for the king to pray to God the Exalted for anything other than for himself and his subjects, and it is not for the subjects to pray to God for anything that is not for themselves alone.

6. He was, upon him be peace, a man of dark[98] complexion, of full stature, bald, of handsome face, thick-bearded, of pleasant lineaments and perfect arm-span, broad-shouldered, big-boned but of little flesh, with flashing, dark-lined eyes, unhurried in his speech, often silent, his limbs at rest; when he walked, he mostly kept his gaze toward the earth; he thought much; he was serious and stern. He moved his index finger when he talked. His period on the earth was eighty-two years.

7. There was on the bezel of his seal-ring that he wore every day: "Patience combined with faith in God bequeaths victory." And on the bezel of the seal-ring that he wore at religious feasts was, "Perfect joy at religious feasts is good works." And on the bezel of his seal-ring that he wore when he prayed for a dead person, "The time of death is the harvest of hope; death is a watchman never heedless." And on the belt that he always wore, "Consideration of the next life bequeaths security to body and soul from harmful accidents." On the belt that he wore to religious feasts, "Keeping religious duties and law is the fulfillment of religion, and the fulfillment of religion is the fulfillment of valor." On the belt that he wore at the time of prayer for the dead, "Whoever considers his soul is victorious, and his intercession with his Lord is his good works."

98. Or white, or ruddy, complexion! Cf. Lane 1863–1893: "Ādamu," 1.37a–b.

> **8.** His religious law, the *ḥanīfī* community (*al-milla al-ḥanīfīya*), also known as the Right Religion (*dīn al-qayyima*), reached the eastern and western ends of the earth, and the north and the south, and spread throughout the earth in its entirety until there remained no human on the face of the earth who did not practice this religion.[99] Their direction of prayer was towards true south along the line of the meridian.[100]

Although some sections of this account are not organically linked with the others and were probably originally separate, such as the physical description of Hermes (6), the divisions of the other sections are not so clear. They are divided only to facilitate discussion.

The introduction (1) simply relates the names of Hermes in different languages. Hirmīs. *Irmīs* reflects the Greek pronunciation of Ἑρμῆς of late antiquity and later; al-Mubaššir's source thinks that this is the original Greek form, and that *Hirmīs*, with an *h* or *spiritus asper*, is a later development. *Hermīs* is also the normal way to render the name in Syriac. Interestingly, the author of this source has heard of the Greek title Trismegistos. Enoch's biblical genealogy is trivial information easily acquired. The next three paragraphs (2, 3, and 4) offer more interesting and new material. Hermes is characterized here as an antediluvian Egyptian who founded a religion presented in terms very similar to those used for the religion of Ḥarrān in the account of al-Kindī related by as-Saraḫsī (section 3.4.3). Although Agathodaemon is the second Ūrānī and Hermes is the third, the previous section of the book has already stated that Seth is the first Ūrānī. The earliest known reference to Ūrānī appears in as-Saraḫsī's description of the Ḥarrānian religion, where the luminaries of the Ḥarrānians (*inna mašhūrīhim wa-aʿlāmahum*) are said to be Urānī (with short u), Agathodaemon, and Hermes (here also *Hirmīs*).[101] No one has ever successfully explained this name Urānī/Ūrānī, nor how Seth, Agathodaemon and Hermes could be the first, second, and third Ūrānī, although there is no shortage of haphazard guesses.[102] The connection with Ḥarrān, however, through as-Saraḫsī's

99. We encounter here again the notion that the religion of Hermes was the universal, primordial religion. See chapter 3 in this volume, on the Ṣābians.

100. Al-Mubaššir, *Muḫtār*, 7.8–10.19.

101. Rosenthal 1943: 42; *Fihrist* 383.12–13 (ed. Flügel 318.20).

102. Chwolsohn 1856: 1.800–801, after reviewing the many highly corrupt renderings of this name in the Arabic script, considered it to represent Orpheus. Several scholars, beginning with de Goeje (cf. de Boer 1912, Scott 4.249n3), suggested οὐράνιος (followed by Ullmann 1972: 374). But Scott rightly asked who this Οὐράνιος could be; however, he then unfortunately suggested instead that the name was invented by the Ḥarrānians as their eponymous ancestor (I suppose something like *Ḥarrānī—this suggestion must be rejected on phonetic grounds alone). Jolivet and Monnot, in their notes their French translation of aš-Šahrastānī, follow M. Tardieu's strange suggestion that it be read as an abbreviation of the name Arnebeskhēnis (a name of Egyptian Horus mentioned once in SH 26.8) (Jolivet and Monnot 1986: 2.170n17). I will not add my own speculation about this name here.

Ibn Ḥazm al-Qurṭubī (d. 1064) cited similar information about the ancient prophets of the Ṣābians (Scott 4.257) about the same time that al-Mubaššir was writing. The name also appears in al-Masʿūdī's *Tanbīh*, 161.17 (with references to the first and the second Urānī) and *Murūǧ aḏ-ḏahab* (ed. Pellat §1234 =

description of the Ḥarrānian Ṣābians, is unmistakable in this passage, for Ūrānī is never mentioned in any other context.

In paragraph 4 of the translation above, where the religious laws of Hermes are explained in detail, there is an unspecified "they" whom Hermes instructed. "They" might be understood to mean just the people of Hermes' time, but at one point we are told about the things that "they offer" in sacrifice (*yuqarribūna*), the imperfect/*muḍāriʿ* verb here implying ongoing activity. Because the sacrifices, purity rules, and the astronomically regulated times of sacrifice prescribed by Hermes here are also somewhat like those of Ḥarrān as described by as-Saraḫsī, on al-Kindī's authority, it can be inferred that "they" here are in fact the Ḥarrānian Ṣābians. When this account was excerpted from its source, direct reference to the Ḥarrānians was removed, but the verbal traces of the original context remain. The reference to Edessa, a city in Syria very near to Ḥarrān, named as one of the cities founded by Hermes, also probably points to an origin of the text in a description of the Ḥarrānians.

Hermes is said to have been raised to heaven after eighty-two years of life. The Qur'ān is cited to emphasize the truth of his heavenly ascent. In al-Mubaššir's account, however, this ascent was at the *end* of his life. Hermes is not said to have returned to earth in order to teach, as in the other accounts examined in the previous section. Moreover, the eighty-two-year life span conflicts with that of the biblical Enoch, who lived three hundred and sixty-five years. This shows an origin of the account of this Egyptian Hermes somewhat at odds with attempts to harmonize Hermes with the biblical Enoch, although the passage does admit that Hermes is Enoch in part 1. It lacks any reference to science brought down from heaven by Hermes, a theme that became popular only in the 980s. This probably implies an earlier composition of the source.

The previous chapter showed that the third Hermes of Abū Maʿšar's synthesis was borrowed from al-Kindī. He said that the third Hermes was Egyptian and that "he traveled around in different countries, wandering in them, knowing the foundations of cities, their natures, and the natures of their peoples." Here, in al-Mubaššir's account, "Hermes left Egypt and went around the whole earth. He returned to Egypt... He built for the (peoples of the earth) a hundred and eight great cities..." The Arabic words used are different, but the similarity in meaning is too close to be fortuitous. I have suggested tentatively that al-Kindī derived his idea about Hermes from a Ḥarrānian tradition. The context of al-Mubaššir's description of Hermes also

2.308.15–309.2; trans. 2.465 = *bāb* 52 *fī ḏikr al-kihāna*, etc.; Pellat emends the name to Orpheus) and in aš-Šahrazūrī, 136.2, as *'wry'*. The damaged page 1a of Chester Beatty Arabic MS 5153 (copied in 1360 CE), introducing the Hermetic *Kitāb Zaǧr an-nafs* under the title of *The Codex of Agathodaemon* (*Muṣḥaf Ġūṯāḏīmūn*), offers another interpretation of the name. Using information derived from paragraph 2 of al-Mubaššir's account of Hermes, translated here, it says that Seth, Agathodaemon, and Hermes were the first, second, and third *Ūrānīs*, but it adds (1a6) that the meaning of *Ūrānī*, [*'wr'ny*] is *nūrānī* ("luminous" in Arabic). This interpretation is apparently based on the Hebrew word *ōr*, [*'wr*], "light."

appears to be Ḥarrānian. In light of the coincidences of content and context, I assume that al-Mubaššir's and al-Kindī's accounts derive from a source common to both of them. This might have been an older Ḥarrānian tradition; it is also possible that the common source was a chronographic work known to both al-Kindī and al-Mubaššir's source. There are nevertheless major differences between al-Mubaššir's account and that of al-Kindī. Hermes is explicitly antediluvian in the former, explicitly postdiluvian in the latter.[103] In the former, Hermes founds a primordial, worldwide, revealed religion; in the latter, Hermes is merely a scholar whose books are still known.

The physical description of Hermes (6) is a part of al-Mubaššir's general method. He includes similar physical descriptions of other philosophers elsewhere in the book, which was originally illustrated.[104] The illustration of Hermes unfortunately does not survive, but it is reasonable to suppose that the physical descriptions are related to the illustrations. The paragraph dealing with the inscriptions on Hermes' rings and belt buckles (7) is clearly part of a pre-Islamic genre: records of maxims inscribed on rings of famous individuals are attested in Arabic texts of Iranian origin.[105] This portion is surely an excerpt from a larger collection of wise maxims adapted for the present purpose of describing Hermes. We will return to the discussion of such adaptations in section 5.3 in this volume.

The last paragraph (8) describes the religion founded by Hermes as *al-milla al-ḥanīfīya*, "the *ḥanīfī* community."[106] Although the word here means primarily "pagan," it is a term that meant different things to different communities at different times. As F. de Blois has shown with a decisive analysis, the Aramaic word *ḥanpā*, together with its Arabic equivalent *ḥanīf*,[107] went through a complicated evolution in different religious communities. To explain what this word signified in the time

103. To be precise, the statement that Hermes was after the Flood is not present in al-Kindī's words as related by Ibn Nubāta. It is found only in Ibn Ǧulǧul's version.

104. One of the Arabic manuscripts of al-Mubaššir (Istanbul Topkapı Saray Ahmet III, 3206, copied thirteenth cent.) is accompanied by depictions of the philosophers; unfortunately, the miniature of Hermes is missing, apparently "the result of the nefarious activities of an art lover who tore out the page containing the picture" (Rosenthal 1975b: 213; also Rosenthal 1961: 142). Another manuscript of the same text (Chester Beatty Arabic 3027, copied 1221 CE) contains blank spaces where the miniatures were supposed to be added but never were. The blank space for the depiction of Hermes has the caption, "The picture of Idrīs, peace be upon him, while he is speaking to and exhorting King Ammon." Rosenthal suggests (1975b: 213) that the illustrations intended for these early manuscripts were based on copies of original miniatures commissioned by the author, al-Mubaššir himself.

105. In *Kitāb Ādāb al-falāsifa*, al-Anṣārī's collection of aphorisms, one of the same bezel inscriptions of Hermes occurs in a list of the sayings on the rings of many ancient philosophers (al-Anṣārī 48.6). The origins of and the relationship between al-Mubaššir's list of Hermes' ring and belt inscriptions and al-Anṣārī's list of the different philosophers' ring inscriptions are not clear. In any case, the motif of wise sayings inscribed on the seal rings of kinds is a well attested in Iranian literature but not in Greek (Shaked 1986: 85–87; Zakeri 2004: 180).

106. The Ḥarrānians are called *al-ḥanīfīya* also in as-Saraḫsī's treatise: *Fihrist* 383.12 (ed. Flügel 318.19).

107. The Aramaic and Arabic words *ḥanpā* and *ḥanīf* are formed from the same etymological root but are not, strictly speaking, cognates. On the relationship between the two words, awkward to resolve, see the best available analysis: de Blois 2002: 23.

of al-Mubaššir, or for his later readers, it is worthwhile here to recapitulate his argument in brief.[108] By the time of the Syriac Aramaic translation of the Bible in the second century CE, Aramaic *ḥanpā*, meaning originally something like "deceiver," came to signify in a derogatory way "gentile" (ἐθνικός), or non-Judaean. Christian speakers of Aramaic continued to use the term to refer to pagans, and the Greek ethnonym "Hellene" (Ἕλλην) became synonymous with it. Among later Christians using Arabic, the Arabic equivalent *ḥanīf* kept the same negative meaning, referring to pagans or non-Christians. Meanwhile, another interpretation of the term had developed: legends about Abraham, progenitor of both the Jews and Arabs, described him as a gentile (Arabic *ḥanīf*), or as the one through whom the gentiles (*ḥunafāʾ*) would be saved. The latter theme was elaborated already in Paul's letters to the Galatians and Romans. Abraham there is not a polytheist, but he is not a Jew either. Now, the Arabs were thought to be descendants of Abraham, through Hagar, although they were gentiles. The Qurʾān reflects this tradition about Abraham when it describes him as a *ḥanīf*. This probably meant here just "gentile," or at least a person not bound by Jewish law, but as terminology developed by which Muḥammad's followers identified themselves as members of a confessionally separate group, the word describing Abraham's religion took on a new, more positive meaning based on the Qurʾānic usage. In this new context, the *ḥanīfī* religion of Abraham, Muḥammad's predecessor, came to be regarded as the ancient monotheistic antecedent of Islam. Moreover, because Abraham is qualified in the Qurʾān as a *ḥanīf muslim*, a "gentile submissive (to God)," the term *ḥanīf* was understood by some exegetes, after Muḥammad's death, to be practically equivalent to *muslim*.[109] The different histories of the word in the different communities had two results. Christians used the word in a negative way to describe both Muslims and pagans, but Muslims used the term in a decidedly positive way to describe the religion of Muḥammad and the prophets who were his forerunners, and even as a synonym for Muslim. For a fuller discussion, I refer the reader to de Blois' study.

The description of the religion of Hermes as *ḥanīfiya* in section (8) of al-Mubaššir's account is therefore potentially quite confusing. Inasmuch as the religion allegedly founded by Hermes was paganism, however that was understood, there can be no doubt that *ḥanīf* here was meant in the sense of "pagan," like Aramaic *ḥanpā*. For a Christian, the term is a negative one, but for pagans of Ḥarrān, it bore, as for Muslims, positive connotations. Thus in a famous Syriac passage attributed to Ṯābit ibn Qurra, the religion of the *ḥanpē* (pagans), that is, *ḥanpūtā* or "paganism," is praised and defended against Christianity.[110] It must have been convenient for the Ḥarrānian pagans, like Ṯābit, that the Arabic word *ḥanīf* had by the ninth century

108. de Blois 2002: 16–25.

109. On the problem of the development of the term *muslim*, see Donner 2002–2003, and also Hoyland 1997: 545–550 to temper Donner's view of a totally indistinct confessional identity. Hoyland 1997: 414n88 provides early non-Muslim attestations of the term *muslim*.

110. Bar ʿEbrāyā, Syriac chronicle, trans. Budge 153, Bodleian MS Hunt. 52 facsimile 55r.b39–56v.a19.

taken on such positive connotations in the Arabic of their Muslim rulers, for now they could claim not only to be the Ṣābians mentioned in the Qurʾān as a licit sect, but also to be following the *ḥanīfī* religion, like Abraham, who, according to *Genesis* 11:31, had indeed dwelled in Ḥarrān.[111] Al-Mubaššir's source was thus exploiting the positive evaluation of the term *ḥanīf* among Muslims for the promotion of the status of the Ḥarrānian (pagan) religion and the teachings of Hermes in the eyes of Muslims.[112] The direct implication of this conclusion, combined with the foregoing evidence, is that the source of the passage is a Ḥarrānian Ṣābian author.

The end of the passage insinuates the religion of Hermes further into the imagined prehistory of Islam with a clear Qurʾānic reference. The religion of Hermes and his *ḥanīfī* community (*al-milla al-ḥanīfiya*) is described as *dīn al-qayyima*, roughly, "the Right Religion." The phrase is from the Qurʾān 98:2–5, in which the term describes the religion brought by "an apostle from God reciting purified scrolls in which are right scriptures. Those who received the book went their separate ways only after the clear proof came to them. They were ordered only to worship God, sincerely practicing his religion as *ḥunafāʾ*, and to practice prayer and to bring alms. That is the Right Religion (*dīn al-qayyima*)." The precise interpretation of the phrases in this passage, like so many Qurʾānic passages, is open for discussion. However, al-Mubaššir's source could not have connected the religion of Hermes with the *ḥunafāʾ* of the Qurʾān more clearly. The implication is that this short Qurʾānic *sūra* is really referring to the ancient revelation of Hermes. The remaining verses of *sūra* 98 tell how the unbelievers and the polytheists will go to hell, whereas those who believe rightly are saved. If one were to pursue the implication of the reference to the Right Religion, found in al-Mubaššir's description of Hermes' teaching, to its conclusion, by reading it in its entire context, we find the true, monotheist "pagans," or *ḥunafāʾ*, distinguished sharply from the actual polytheists (*mušrikūn*), the former being saved, and the latter, who went astray from the message of Hermes, destined for hell.

Compared with the biographies of Hermes presented by Abū Maʿšar, the treatment of Hermes here is much more concerned with Hermes' religious teaching. Moreover, Hermes is not the name of three individuals but only one ancient figure (though there are at least three *Ūrānīs*). The source of the passage was evidently primarily concerned with Hermes' role as a lawgiver for the ancient religion. This again points to a Ḥarrānian background, for the Ḥarrānian Ṣābians were the only group known to regard Hermes as a prophet of their own whose religion they followed. The wise sayings of Hermes related by al-Mubaššir immediately after this biography, discussed at length in the next section (5.3), are to a great extent in the form of religious and ethical commandments, demonstrating Hermes' prophetic

111. Tardieu 1986: 8–9, also makes this point.

112. Levy-Rubin 2003: 214–222 provides examples of the confusion about the term *ḥanīf* in Arabic in the tenth, eleventh, and twelfth centuries, and authors asserting or denying that the Ḥarrānian Ṣābians could rightly be called *ḥunafāʾ*. As noted above, the Ḥarrānian religion was called *al-ḥanīfiya* as early as as-Saraḫsī (*Fihrist* 383.12; ed. Flügel 318.19) in the later ninth century.

character. Although the precise sources of these passages remain unknown, it nevertheless seems likely that a Ḥarrānian background underlies much of the account. The analysis in section 5.3 below will strengthen this hypothesis further.

As I have argued, the treatment of the prophet Seth, the "first Ūrānī," and that of Hermes, the "third Ūrānī," immediately following are from the same source. A glance at maxims of Seth fortifies the conclusion that this source was a Ḥarrānian Ṣābian. They begin with a list of the sixteen virtues that must be in "believers and *ḥanīfs*," *fī l-mu'min wa-l-ḥanīfī*. Most of these virtues are ordinary: one must do good, protect the poor, speak the truth, and so forth, but the fourteenth virtue is "animal sacrifices and offerings in thanks to God the Exalted for the benefactions he has given to His creation."[113] Both Seth and Hermes teach the *ḥanīfī* religion involving animal sacrifices and offerings to God. It seems unlikely that a Muslim, Christian, or Jew would want to promote Seth and Hermes as a teacher of such rites.

Al-Mubaššir's account of Hermes was the only successful alternative to that of Abū Ma'šar. It described only one Hermes, not three, and it emphasized much more strongly the role of Hermes as the founder of a primordial religion. Besides its career in Arabic, al-Mubaššir's collection as a whole was translated into the languages of Western Europe. The numerous vernacular versions attest to its popularity. Franz Rosenthal has provided a succinct overview of these translations, all of which lack the attribution of authorship to al-Mubaššir but were transmitted anonymously.[114] A somewhat abbreviated version of the text circulated in a Latin translation of the Spanish, made near the end of the thirteenth century, under the title *Liber philosophorum moralium antiquorum*. The portion of this text relating the brief biography of Hermes and his wise sayings has gone unnoticed in the recent and very useful catalog of the medieval Latin Hermetica.[115] But al-Mubaššir's treatment of Hermes is important for the Latin tradition, for it seems likely that it, together with the maxims of Hermes, was known to the Italian scholars of the late fifteenth century, such as Ficino, who took a new and special interest in the Greek *Corpus Hermeticum*. This description of Hermes' teachings testified that Hermes was an antediluvian sage and teacher of holy rituals regulated astrologically. The Latin text relates a shortened version of the biography of Hermes as follows:

> Hermes in Egipto natus fuit, et Hermes in greco dicitur, Mercurius latine, et in ebrayco Enoch; qui fuit filius Jared, filii Macthalaleb, filii Quenam, filii Enoy, filii Sed, filii Adam. Et fuit ante magnum diluvium quod mundum submersit; post quod fuit aliud diluvium quod submersit Egiptum tantum. Et recessit Hermes de Egipto et ambulavit totam terram octoginta et duobus annis, invitans omnes ad obediendum Deo cum septuaginta et duabus

113. al-Mubaššir 5.3–4: *aḍ-ḍaḥāyā wa-l-qarābīnu šukran li-llāhi ta'ālā 'alā mā awlā ḫalqahū mina n-ni'ami.*

114. Rosenthal 1960–1961: 133–135 and 149–155.

115. Lucentini and Perrone Compagni 2001.

linguis, et construxit centum et octo villas et instruxit eas scientiis. Et fuit primus inventor scientie stellarum; et stabilivit omni populo cuiuslibet climatis legem pertinentem et convenientem suis opinionibus. Cui obedierunt reges et tota terra, et habitantes insulas maris. Et invitavit omnes ad legem Dei et ad confitendum veritatem, et ad horrendum mundum et observandam iusticiam et querendam salvationem alterius mundi. Et mandavit fieri orationes et ieiunare die sabbati quolibet mense et expugnare inimicos fidei, et dare hiis qui Dei sunt peccuniam ut auxiliarentur per eam debiles et impotentes; et prohibuit eos comedere carnes porcinas et zebrarum et camelorum, et alios similes cibos. Et inhibuit eos quovis inebriari vino, et stabilivit festivitates multas certis temporibus, et offerre sacrificia, aliqua ex eis in introitu solis, in principio signorum, et aliqua in prima visione lune et in coniunctione planetarum, et etiam quando planete intrabunt domos suas proprias et exaltationes suas, aut in aspectibus planetarum. Et offerebant sacrificia de rebus omnibus, de floribus scilicet rosas, et de granis ordeum et triticum, et de fructibus uvam et de potibus vinum.[116]

Hermes was born in Egypt, and he is called Hermes in Greek, Mercury in Latin, and in Hebrew, Enoch. He was the son of Jared, son of Machtalaleb, son of Quenam, son of Enoy, son of Sed, son of Adam. He was before the great flood that submerged the world, after which there was another flood that submerged just Egypt. Hermes withdrew from Egypt and traveled the whole world for eighty-two years, calling all, with seventy-two languages, to obey God. He built one hundred and eight towns and taught them in the sciences. He was the first discoverer of the science of the stars, and he established for every people of each climate a law applying to and fitting for their views. Kings, and the entire world, and the inhabitants of the islands of the sea obeyed him. He called all to the law of God and to confess the truth, to dread the world, to observe justice, and to seek the salvation of the other world. He ordered that there be speeches and fasting on the day of the sabbath in each month, fighting of the enemies of the faith, giving of money to those who are of God in order that the weak and powerless be assisted through it. He forbade them to eat the meat of pigs, asses, camels, and other similar foods. He kept them from being drunk on any wine, and established many festivals for particular times. [He ordered them] to offer sacrifices, some from them at the entrance of the sun in the first degree of the zodiacal signs, and some at the first sight of the moon, and at the conjunction of the planets, and also when the planets entered their own houses and exaltations, or in the aspects of the planets. They offered sacrifices of all things, of flowers such as roses, and of grains, barley and wheat, and of fruits, grapes, and of drinks, wine.

116. Ed. Franceschini 1931–1932: 402–403.

The European translations, beginning with the Spanish version from which this Latin text was made, notably leave out the part about prophets after Hermes, and the characteristics of the true prophet, as well as the rest of the account thereafter. The eleven extant Latin manuscripts have been dated to the fourteenth and fifteenth centuries; four of them are in Italy today, further supporting the notion that the text would have been available to those around Ficino in the late fifteenth century.[117] In any case, it can be taken for granted that some readers of the Greek *Corpus Hermeticum* in its first Latin translation (originally published in 1471) would have been familiar already with the story of Hermes presented here, not to mention the wise maxims and commands of Hermes. These maxims of Hermes in the Spanish and Latin versions also include the short text of the *Testament of Hermes to Ammon*, analyzed below, Ammon being a name of one of Hermes' students also found in the better known work *Asclepius*.[118] It is likely therefore that al-Mubaššir's text, perhaps ultimately of Ḥarrānian Ṣābian derivation, was a factor conditioning the reception of the Hermetica in Renaissance Europe. It is for specialists in Italian scholarship of fifteenth century to determine whether the *Liber philosophorum moralium antiquorum* had any importance for this region and period, and in particular for the first printing of the *Corpus Hermeticum* in Latin translation. The text in its European versions should in any case be taken into account in studies of European Hermetica.

It is certain at least that some English readers of Ficino's translation of the *Corpus Hermeticum* (entitled *Pimander*) were also aware of al-Mubaššir's account of Hermes, which appeared in Caxton's English translation (printed 1477). We see this from William Baldwin's *Treatise of Morall Phylosophie* (first printed 1547, expanded and rearranged by Thomas Palfreyman ca 1635), which cites "approximately a quarter" of the sayings of Hermes from Caxton right alongside material from Ficino's preface to the *Pimander* (first printed 1491), without distinguishing clearly between these two sources of information about Hermes.[119] Baldwin had two biographical reports about Hermes: one from Ficino, who took it from Augustine, and the other ultimately from al-Mubaššir. The conflict between the two accounts arouses Baldwin's comment:

> Hermes, otherwise called Mercurius Trismegistus, was not only the most excellent of the philosophers, but also the most ancient... Howbeit some which write of him, hold opinion, that he was Enoch... which opinion (although it be not to be utterly reiected) yet it is not sufficient without proofe to bee beleeved, for Enoch whom they take for Hermes, was before Noes floud, in which all the works which were written, if they had at that

117. Franceschini 1931–1932: 395–396.

118. Hermes addresses Ammon also in several fragments of lost Hermetica appearing in the *Stobaean Anthology* (*SA* 12–17).

119. Gill 1984.

time any use of letters were drowned, but the works of this Hermes of whom we treat, are yet appearing in divers languages...[120]

5.3 The Wisdom of the Arabic Hermes

It is reported that Hermes ascended into heaven, and many wise maxims were related in his name.

—Ibn Kammūna (thirteenth cent.) [trans. Perlmann 1971: 38]

Al-Mubaššir's anthology was only one of many such collections of the wise maxims (*ḥikam*, sing. *ḥikma*) of the philosophers assembled on the basis of earlier works, many of them translations into Arabic from other languages. The genre of the philosophical gnomologium, intended to provide the reader with a moral education and a storehouse of wit and wisdom, practical advice for living, and sometimes also a protreptic and propaedeutic to more advanced philosophy, was an ancient and popular one. Collections of such maxims circulated widely in Greek, Middle Persian, Aramaic, Sanskrit, and other languages already many centuries before the advent of Islam. Wise sayings also constituted part of the folklore of the pre-Islamic Arabs, such as those attributed to the Arab sage Luqmān.[121] In fact, the *ḥikma* has remained an important genre of Arabic literature until today. These maxims were from the earliest times a part of oral literature, readily repeated at appropriate occasions, and were used within poetry and song and in different kinds of written literature, but they also formed the special subject of popular and copious anthologies written in Arabic from at least the eighth century onward.[122] Like several other genres of classical Arabic literature, this one also was enriched by the inheritance of antiquity through translations.

Hermes is one of the many ancient philosophers to whom wise sayings were attributed in such Arabic collections. In his article on Hermes in the *Encyclopaedia of Islam*, 2nd ed., Martin Plessner wrote that a comparative study of the Arabic maxims attributed to Hermes was "badly needed."[123] The sense of urgency is no doubt due to the interest and value in locating the precise connections, if any could be found, between the Greek *Corpus Hermeticum* and the Arabic words attributed to Hermes. Franz Rosenthal had already discovered one passage, about a paragraph in length, in the Arabic collection of wise sayings compiled by the philologist Ibn Durayd (d. 933), that was in fact derived, albeit indirectly, from the *Corpus Hermeticum*.[124] If other scattered excerpts of an Arabic translation of the Greek

120. Palfreyman ca 1635: 16r.

121. A useful introduction to wisdom literature in Arabic is Gutas 1981, where the meaning of the term *ḥikma*, as well as its sources, genre, and literary function are treated. For Luqmān, see B. Heller and M. A. Stillmann, "Luḳmān," *EI*² 5.811a–813b.

122. Books of proverbs (*amṯāl*) are reported to have been compiled already in seventh-century Arabic (Zakeri 2007: 1.310–311 for a list of early *amṯāl* works).

123. Plessner, "Hirmis," *EI*² 3.463b–465a.

124. Rosenthal 1958: 54 and 183. The passage is SH 1.

Corpus could be found in these anthologies, this might give indications of an Arabic translation of the Greek Hermetic dialogues. Therefore the next step was to collect and enumerate all of the extant Arabic *ḥikam* attributed to Hermes in an attempt not only to find their origin, but also to ascertain what the Arabic Hermes had to say. After they were thoroughly collated and compared, it became mostly clear which collections depended on which for this material. In the end, however, the results of this research were different from those anticipated. There were no more links to the Greek *Corpus Hermeticum*. Instead other findings of interest appeared.

This is not the place to present the nearly two hundred maxims of the Arabic Hermes in the original text or in translation. A critical edition of these sayings should involve all the Graeco-Arabic gnomologia together, including the sayings of not just one sage like Hermes, but of all the ancient figures, in all of the various collections. This is because they repeat sayings attributed to the different figures in different versions, drawing from each other in complicated, ramified, interconnected series.[125] In other words, the problem is that the sayings attributed in some sources to Hermes sometimes show up elsewhere under the names of different Greek philosophers, and it is, for now, impossible to determine the textual history or even the exact wording of each saying without an all-inclusive study of the maxims not only of Hermes but also those of the other ancient philosophers.[126] Some of the examples below will clarify this situation.

These reservations notwithstanding, a detailed discussion of the maxims attributed to Hermes is in order. The Arabic Hermetic maxims are an important part of the history of the Arabic Hermes as he was imagined to be and as his story was told. From a sample of his sayings, even based on the imperfect editions available, the reader can also see what kind of figure Hermes, the prophet Idrīs, appeared to be in Arabic wisdom literature. The gnomological collections in which these sayings are found seem to have been more popular than the works on astrology and talismans attributed to Hermes, which required of their readers at least a basic technical knowledge of their subjects. The maxims of Hermes were often accompanied by a version of the biography of Hermes (or the Hermeses) that established his character as a prophet. Thus they more or less represent the stories and ideas that a typical literate medieval scholar of Arabic would have associated with Hermes.

Outside of technical treatises on astrology, alchemy, and the like, the earliest citations of Hermes in Arabic literature in general are mostly not strictly gnomological but doxographical. That is, they do not convey wisdom but rather philosophical doctrine. They appear not in large collections, as do the wise maxims of Hermes, but as individual citations of authoritative views, much like the citations of ancient Hermetica by the Greek church fathers, who used them to further their own

125. Gutas 1975: 436–441.

126. The difficulty of the enterprise is demonstrated by the monumental labor of Zakeri 2007, vol. 2, which shows parallels for more than two thousand maxims attributed to many different figures in various permutations.

arguments. For example, the early Arabic pseudepigraphical correspondence between Aristotle and Alexander called *as-Siyāsa al-ʿāmmīya* represents Alexander as citing Hermes in such a way; in one instance, Alexander says, "Hermes the preeminent in knowledge spoke well when he said, 'Man is a microcosm and the celestial sphere is the macrocosm.'"[127] Pseudo-Apollonius of Tyana also cites Hermes for such doctrines.[128] Another substantial citation of Hermes in the gnomologium of Ibn Durayd (d. 933) is actually doxographic in character. This is the one that Rosenthal discovered to be derived, albeit indirectly, from the Greek Hermetica:

وقال هرمس إنّه لصعب أن يوقف على حقيقة أمر الخالق وغير ممكن أن يوصف وذلك أنّه
غير ممكن أن يصف جسم مدرك ما ليس بمدرك ولا يدرك التامّ ما ليس بتامّ ويصعب أن يقرن
الأزليّ بما ليس بأزليّ فإنّ الأزليّ باق أبداً وغير الأزليّ فانٍ والفاني خيال وظلّ فعلى قدر ما بين
الضعيف القويّ وما بين الدون والأشرف فكذلك بين الفاني وبين الإله الذي لا يموت

> Hermes said: It is difficult to know the truth about the Creator, and impossible to describe Him. For it is impossible for a comprehensible body to be described as something incomprehensible,[129] nor can what is perfect be comprehended by what is imperfect. It is difficult to combine the eternal with something that is not eternal, for the eternal is everlasting, while the non-eternal is transitory, and that which is transitory is a phantom and shadow. The relationship of things transitory to the immortal deity is like that of weakness to strength, or meanness to nobility.[130]

In its original Greek form, this text is best known today as it appears in the *Anthology* of Stobaeus (fifth cent.), in which a quantity of excerpts from lost Hermetic discourses survives.[131] The same Hermetic passage was also particularly popular among Christian theologians. It was cited by Pseudo-Justin (before 302),[132] Lactantius (ca 250–ca. 325), and Cyril of Alexandria (ca 375–444). It was probably also known by the Carthaginian bishop Cyprian (ca 200–258) and the Emperor Julian (r. 361–363).[133] Lactantius stated that this passage was taken from the beginning of a book of Hermes writing to his son, and Scott plausibly suggested that the popularity of this passage was due to its appearing at the beginning of a collection of discourses from Hermes to his son, Tat.[134] It is not surprising to find such a well known passage also in Arabic. It must have been translated either as part of a Greek gnomologium or as

127. Grignaschi 1967: 75.1–2, 76.10–11; 1975: 141.5–7, 144.1–2.
128. Weisser 1980 (index pp. 245–246).
129. On the error in this transmitted Arabic version, see Rosenthal 1958 n1.
130. Rosenthal 1958: 54 and 183. The translation here is that of Rosenthal, slightly altered.
131. SH 3.2–3. For commentary, see CH 3.xiii–xiv.
132. Relevant references to Pseudo-Justin's *Cohortatio ad gentiles* and its dates can be found in Scott 4.6.
133. Scott 3.301.
134. Ibidem. Lactantius, *Epitome* 4.5: *Huius ad filium scribentis exordium tale est…*

cited in one of the Christian works just reviewed.[135] Rosenthal argued that the passage must have been transmitted into Arabic through a Greek gnomologium now lost.[136] In any case, the appearance of the passage in Arabic is an exception that has nothing to do *directly* with the ancient Greek Hermetica. Moreover, it does not seem to have been repeated in later florilegia by other Muslim authors; this particular Hermetic citation is ultimately a rare one in Arabic.

Rosenthal found two more sayings of Hermes in an unpublished Arabic translation of a Greek gnomologium on music attributed to a certain Paulus, who is said to have compiled the statements of the philosophers on music at the Byzantine emperor Heraclius' (r. 610–641) bidding.[137] The work is entitled *Kitāb ʿUnṣur al-mūsīqā wa-mā ftaraqat ʿalayhi l-falāsifa min tarkībihī wa-māʾīyatihī* (*The Element of Music and the Differences of the Philosophers Concerning Its Composition and Essence*). The translator is named Isḥāq; Rosenthal surmised this to be the famous translator Isḥāq ibn Ḥunayn (d. 910/911). One of these two sayings was subsequently cited, together with its attribution to Hermes, by Ibn Ḫurradāḏbih (ca 820–ca 911), a companion of the caliph al-Muʿtamid (r. 870–892), in his work *Kitāb al-Lahw wa-l-malāhī* (*On Musical Entertainment and Instruments*), establishing a rough *terminus ante quem* for the translation. Here is an example: "Hermes Triplicate-in-Wisdom, when he sat for drinking with friends, said to the musician: Let fly the soul from its tethers!"[138] Evidently the reference is both to the strings of the instrument as well as to the psychic effects of music. Perhaps the Byzantine Paulus adapted his two sayings of Hermes on music from the Greek *Corpus Hermeticum* 18.1–8, a discussion of musical harmony and the effects of music on the soul;[139] however, no direct resemblance between the two texts is readily apparent.

Mostly Hermes is cited in Arabic as an authority in numerous works of astrology, magic, and alchemy. Outside of these genres, and apart from the accounts of the life of Hermes-Idrīs, the scraps of Hermetic material cited in Arabic literature down to the late tenth century are fairly meager. Hermes appears to have remained an author known largely for technical treatises used by specialists in the ancient sciences until the end of the tenth century. This changed with the compilation of the two most important Arabic gnomologia of ancient philosophers, very popular from the eleventh century onward: the *Ṣiwān al-ḥikma*, or *The Bookcase of Wisdom*, and the compendium of al-Mubaššir ibn Fātik, *Muḫtār al-ḥikam*. These two works are

135. Cyril's *Contra Julianum*, where the citation occurs, was known in Syriac translation, a part of which survives (Baumstark 1922: 161). Fragments of it were also known through chronographic extracts in Arabic (Stern 1972: 439–443).

136. Rosenthal 1958: 156–158.

137. Rosenthal 1966.

138. Ibn Ḫurradāḏbih 1961, 22.6–7; Paulus in Manisa Genel Kütüphane MS 1705, 123v–130v (Hermes on 125r1–2), which I have seen in Franz Rosenthal's microfilm collection. Cf. *Muntaḫab Ṣiwān al-ḥikma* (ed. Dunlop 1276).

139. If true, this would constitute an important testimony for the presence of Hermetica in seventh-century Byzantium.

undoubtedly the most important Arabic collections of the maxims of the Greek philosophers. They remained fundamental sources for the history of philosophy in Arabic tradition. Both share the same format: biographies of the ancient sages are provided, where available, followed by sometimes lengthy lists of their sayings. Both works discuss Hermes.

The *Ṣiwān al-ḥikma*, or *The Bookcase of Wisdom*, composed between 985 and 1030, does not survive in its original form, which must have been long. According to one estimate, it contained about twenty-five hundred sayings attributed to at least one hundred thirty-seven Greek and thirty-three Muslim philosophers.[140] It consisted of biographies for each philosopher followed by that philosopher's wise maxims. The author of the *Ṣiwān* is unknown, but he was surely someone acquainted with the prominent philosophers and scholars of Baġdād and northern Iran of the late tenth century.[141] The original text is not extant, but several abridgements and selections of the original do survive, not to mention supplements extending the treatment to cover more Arabic and Persian philosophers in later times.[142] Two of the surviving derivative works are particularly important here, because they relate material on Hermes: (1) the *Muḫtaṣar Ṣiwān al-ḥikma* (*Abridgment of the Bookcase of Wisdom*), made by ʿUmar ibn Sahlān as-Sāwī (d. 1145), a philosopher who flourished in Nīšāpūr, and (2) the *Muntaḫab Ṣiwān al-ḥikma* (*Selection from the Bookcase of Wisdom*), made by an unknown scholar between 1191 and 1241. Both of these are now published.[143] The *Muḫtaṣar* omits the biographies of the philosophers, but from the *Muntaḫab*, which includes them, we can see that the life of Hermes in the original *Ṣiwān* was taken from Abū Maʿšar's well-known account (as mentioned in section 4.1). Between the two derivative works, it is possible to collect forty-six different *ḥikam* attributed to Hermes, ranging in length from short sentences to paragraphs.

The *Muḫtār al-ḥikam* of al-Mubaššir ibn Fātik has already come under discussion (in sections 3.4.6 and 5.2). The biography of Hermes contained in it, the origins of which I have supposed above to be connected with a Ḥarrānian, is entirely different from that of Abū Maʿšar, and the wise maxims accompanying this biography are also mostly different from those related in the remnants of the *Ṣiwān*. Depending on how one divides them from one another, there are about one hundred and twenty-five sayings of Hermes in al-Mubaššir's collection. Only fourteen of them are found also in the *Ṣiwān*-materials just described, and in several of these instances they are worded there quite differently. Clearly, therefore, al-Mubaššir used at least

140. Kartanegara 1996: 35.

141. Modern scholars formerly attributed it to Abū Sulaymān as-Siǧistānī (d. ca 985), the influential Iranian philosopher of Baġdād and teacher of the famous litterateur Abū Ḥayyān at-Tawḥīdī. However, Wadād al-Qāḍī (1981: 116–119) argued convincingly that Abū Sulaymān could not have written it, and she suggested instead as its author Abū l-Qāsim al-Kātib, a student of al-ʿĀmirī.

142. For a survey of all the texts derived from *Ṣiwān al-ḥikma*, see Gutas 1982 and Daiber 1984.

143. Two editions of the *Muntaḫab*, by Dunlop 1979 and Badawī 1974, are listed under the title in the bibliography. On these see Gutas 1982. The *Muḫtaṣar* was edited in a PhD dissertation: Kartanegara 1996. I have also consulted MS Istanbul Fatih 3222, 35a–36a, for the *Muḫtaṣar*.

one major source unknown to the compiler of the *Ṣiwān* for his treatment of Hermes. In addition, two other related types of texts, besides the individual maxims, are included in the section on Hermes in the *Muḫtār al-ḥikam*: the short list of wise inscriptions on the rings and belt buckles of Hermes (already translated in the previous section as a part of his biography) and Hermes' testimony addressed to King Ammon, a coherent set of maxims for kings running a few pages in length (discussed below). We also find the sayings of Ṭāṭ or, as his name is given in some manuscripts, Ṣāb, immediately following those of Hermes. Although these references to Ṭāṭ and to Ammon bring immediately to mind the ancient Greek Hermetica, wherein Hermes addressed both Tat and Ammon, there is no evident textual connection between this part of al-Mubaššir's work and the Greek Hermetica. The sources that *have* appeared are quite different.

Together these two compilations are the sources of the great majority of the known Arabic sayings of Hermes and were drawn upon freely by later authors of gnomologia and similar works, such as Miskawayh (d. 1030),[144] al-Anṣārī (eleventh cent.),[145] aš-Šahrastānī (fl. early twelfth cent.),[146] and aš-Šahrazūrī (fl. late thirteenth cent.).[147] Each of these authors (aš-Šahrazūrī excepted) adds a few sayings of Hermes not known elsewhere; it is difficult to say whether these derive from a lost, more complete version of the *Ṣiwān*, perhaps even the original *Ṣiwān* itself, or from other sources entirely. A thorough and complete study and translation of all the wise sayings of Hermes in Arabic could form the subject of an entire monograph. Here several representative examples of Hermetic maxims in Arabic will have to suffice instead. Collectively these sayings convey a kind of general morality common to wisdom literature in many countries: be kind and humble, keep your promises, control your passions in word and deed, don't go to extremes, find trustworthy friends, respect learning, esteem the next life more highly than this one and the spiritual more highly than the bodily, and obey God and the king. Occasionally the sayings are in the form of riddles or even jokes. Here is a brief sample from the *Muntaḫab Ṣiwān al-ḥikma*:

> He said: The oppressed, the cheated, the tortured and the afflicted are fortunate next to the oppressor, the cheat, the torturer and the afflicter. For it is God's obligation to repay the oppressed with relief and the oppressors with affliction.[148]

144. Miskawayh, *Kitāb al-Ḥikma al-ḫālida*, ed. Badawī 214.3–216.4. Obviously Miskawayh was too early to use al-Mubaššir's book.

145. *Kitāb Ādāb al-falāsifa*, ed. Badawī, 133.2–135.15 and 48.6. This work used to be attributed to Ḥunayn ibn Isḥāq (d. 873). Mohsen Zakeri 2004 (also 2007: 1.59–73) showed that only a small section of the *Ādāb* as we know it can be ascribed to Ḥunayn. Rather, the *Ādāb al-falāsifa* is really just the work of a certain al-Anṣārī, who compiled it from a variety of sources. This moves the date of this collection from the ninth to the eleventh century and the sayings of Hermes it presents now appear to be highly abridged from al-Mubaššir's collection.

146. aš-Šahrastānī, *Kitāb al-Milal wa-n-niḥal*, 241.7–292.6.

147. aš-Šahrazūrī, *Kitāb Nuzhat al-arwāḥ*, 139.3–160.3.

148. *Muntaḫab* (Dunlop 1315–1317; Badawī 188.9–11): *wa-qala l-maẓlūmu wa-l-maḫdū'u wa-l-mu'aḏḏabu wa-l-mubtalā fī ǧanbi ẓ-ẓālimi wa-l-ḫādi'i wa-l-mu'aḏḏibi wa-l-mubtalī su'adā'u fa-innahū ḥaqqun 'alā llāhi an yu'aqqiba l-maẓlūmīna rawḥan wa-ẓ-ẓālimīna balā'an.*

> He said: Practicing good habits is easier than stopping bad habits.[149]

> He said: Fire is extinguished by water, poison is held off by theriac, patience protects one from sadness, long absence distracts from infatuation, but rancor is an evil with no cure.[150]

These and the rest of the sayings of Hermes from the *Ṣiwān al-ḥikma* are typical of ancient wisdom literature in general and are not especially distinctive. By contrast, al-Mubaššir's collection has a different and peculiar character, that of commandments or even religious laws. According to their heading, they are "A Selection of the Exhortations and Lessons of Hermes, who—peace be upon him—is Idrīs, the prophet, triplicate in prophethood, wisdom, and kingship."[151] Clearly his prophetic character is foremost in the mind of the author of this source. Many of the sayings related in this section have the form of imperative sentences, which are rare in the Hermetic sayings from the *Ṣiwān*. Here are some examples:

> He said: Oh man! When you fear your Lord and guard against the ways leading to evil, you will not fall into it.[152]

> He said: Dress yourselves in wisdom. Follow religion. Make yourselves accustomed to a peaceful, dignified bearing. Adorn yourselves with fine and beautiful manners. Reflect on your affairs and do not be too hasty, especially in punishing an evil-doer. Fill your faces with shame and stuff your sides with fear of God. Conduct yourselves soundly and straightly.[153]

> Be on your guard against and avoid noxious foods; shy away from base acquisitions. For even if they fill your cups with money, nevertheless they will empty your hearts of faith.[154]

149. *Muntaḫab* (Dunlop 1317–1318; Badawī 188.13) = Kartanegara 1996: 170.16 (Istanbul Fatih 3222: 36a1–2): *wa-qāla ʿtiyādu l-ḫayri aysaru min qaṭʿi ʿādati š-šarri.*

150. *Muntaḫab* (Dunlop 1329–1330; Badawī 189.12–13): *wa-qāla n-nāru yuṭfiʾuhā l-māʾu wa-s-sammu yadfaʿuhū t-tiryāqu wa-l-ḥuznu yakuffuhū ṣ-ṣabru wa-l-ʿišqu yuslīhi ṭawīlu* (read *ṭūl*?) *l-ġurbati wa-l-ḥiqdu šarrun lā dawāʾa lahū.*

151. al-Mubaššir 11.1–2: *muḫtār mawāʿiẓ Hirmis wa-ādābihī wa-huwa ʿalayhi s-salām Idrīs an-nabī al-muṯallaṯ bi-n-nubūwa wa-l-ḥikma wa-l-mulk.*

152. al-Mubaššir 11.9 = aš-Šahrazūrī 140.10–11: *wa-qāla ayyuhā l-insānu iḏā ttaqayta rabbaka wa-ḥaḏirta ṭ-ṭuruqa l-muʾaddiyata ilā š-šarri lam taqaʿ fīhi.*

153. al-Mubaššir 12.19–23 = aš-Šahrazūrī 143.2–6: *wa-qāla stašʿirū l-ḥikmata wa-tbaʿū d-diyānata wa-ʿawwidū anfusakumu l-waqāra wa-s-sakīnata wa-taḥallaw bi-l-ādābi l-ḥasanati l-ǧamīlati wa-tarawwaw fī umūrikum wa-lā taʿǧalū wa-lā siyyamā fī muǧāzāti l-musīʾi wa-ǧʿalū l-ḥayāʾa malʾa wuǧūhikum wa-l-ḫīfata mina llāhi ḥašwa ǧunūbikum wa-tadabbarū bi-ṣ-ṣiḥḥati wa-l-istiqāmati.*

154. al-Mubaššir 13.7–10 = aš-Šahrazūrī 144.1–2; cf. Ibn al-Qifṭī 6.7: *wa-qāla taḥarrazū wa-hrubū mina l-maʾākili l-ḫabīṯati wa-ḥtašimū l-makāsiba d-danīyata fa-innahā wa-in malaʾat akyāsakum mina l-māli fa-innahā tufarriġu qulūbakum mina l-īmāni.*

> Obey God, obey your chiefs, submit to your ruler, honor your elders, revere your teachers, and let the love of God and truth master you.[155]

Not quite all of the sayings of Hermes related by al-Mubaššir are strictly commandments. Later in the collection the sayings become more typically gnomic in form:

> The good man is he whose good is the good of everyone and who considers the good of everyone to be good for himself.[156]

> Death is like a dart already thrown, and your lifespan is as much as its course toward you.[157]

> Intellect without education is like the barren tree. Intellect accompanied by education is like the fruitful tree.[158]

As seen in the previous section, al-Mubaššir's biography of Hermes, explaining how he founded a religion, is probably connected somehow with a description of the Ḥarrānians. At the same time, most of the sayings accompanying this biography, with their pious commandments, do not appear elsewhere in Arabic before al-Mubaššir. Therefore it seems likely that the biography and the commandments of Hermes (or at least a large portion of them) in al-Mubaššir are taken from a single, otherwise unattested source. Because the biography seems to be of Ḥarrānian Ṣābian origin (section 5.2), most of the sayings related by al-Mubaššir could have the same background, too. It is tempting to suppose that this section of al-Mubaššir's book is even an excerpt of the reported *Kitāb Nawāmīs Hirmis* (*The Book of the Laws of Hermes*), with its Arabic version due to the Ṣābian court physician Sinān ibn Ṯābit ibn Qurra (d. 943), though one cannot be in any way certain about it.[159] The exhortation to obey not only God but also "your chiefs" (*ru'asā'akum*) calls to mind the "chiefs" (with the same word) of the Ḥarrānian Ṣābians listed in the *Fihrist*, as well as the chief (*ra'īs*) of the Ḥarrānians mentioned in the anecdote about Ṯābit's arrival in Baġdād.[160] One cannot, however, make too much of this point, because the word can refer to any authority, and the office of *ra'īs* was common throughout towns of southwestern Asia.[161] One of Hermes' exhortations here advises people in how to

155. al-Mubaššir 14.5–7 = aš-Šahrazūrī 145.4–5; cf. Ibn al-Qifṭī 6.7–8: *wa-qāla aṭī'ū llāha wa-aṭī'ū ru'asā'akum wa-ḫḍa'ū li-sulṭānikum wa-akrimū kubarā'akum wa-burrū mu'addibīkum wal-taġlib 'alaykum maḥabbatu llāhi wa-l-ḥaqqi.*

156. al-Mubaššir 18.20 = aš-Šahrastānī 244.4–5 = aš-Šahrazūrī 151.8–9: *wa-ṣ-ṣāliḥu man ḫayruhū ḫayrun li-kulli wāḥidin wa-man ya'uddu ḫayra kulli wāḥidin li-nafsihī ḫayran.*

157. al-Mubaššir 19.4 = al-Anṣārī 134.11 = aš-Šahrazūrī 151.11: *al-mawtu ka-sahmin mursalin wa-'umruka bi-qadri masīrihī naḥwaka.* Same as Ibn al-Mu'tazz, *Kitāb al-Ādāb* 121.19–20.

158. al-Mubaššir 19.23 = aš-Šahrazūrī 153.1–2: *qāla l-'aqlu bi-ġayri adabin ka-š-šaǧarati l-'āqiri wa-l-'aqlu ma'a l-adabi ka-š-šaǧarati l-muṯmirati.* Same as Ibn al-Mu'tazz, *Kitāb al-Ādāb* 83.1–2.

159. See 3.4.5 above.

160. *Fihrist* 390.12–21 (ed. Flügel 326.7–19) gives a brief *Ta'rīḫ ru'asā' aṣ-Ṣābi'īn*. See also chapter 3, notes 4 and 23, in this volume.

161. A. Havemann, C. E. Bosworth, and S. Soucek, "Ra'īs," *EI*², 8.402a–403a.

deal gently and piously with those who argue against them rudely in matters of religion.[162] This could very well apply to Ḥarrānian Ṣābians in their status as members of a small and marginal religious group of questionable legitimacy surrounded by Muslims in Baġdād. These are no more than suggestions and definite proof is lacking for the moment.

Any conclusions about the origins of these sayings of Hermes have to take into account their parallels and antecedents in other languages. As expected, some of the sayings of Hermes in Arabic certainly derive from Greek gnomologia, where they were nevertheless *not* Hermetic, but attributed to various Greek philosophers. One of these in the *Muntaḫab Ṣiwān al-ḥikma* is a translation of a Greek maxim based on a passage in Plato's *Republic* (5.473c–d). Though it is attributed here to Hermes, at-Tawḥīdī elsewhere ascribes it to Diogenes.[163] The Arabic has the following: "He said: Things go as God the Exalted wishes only when the philosopher is king or when the king practices philosophy."[164] The Greek original, under the name of Plato, is found in the *Gnomologium Vaticanum*: "When he was asked, 'How may city-states (*póleis*) best be managed?' he said, 'If the philosophers are kings and the kings practice philosophy.'"[165] Although the Arabic translation is not strictly literal, it is still fairly close to the original Greek text. Aside from the Greek question becoming the Arabic apodosis, and the loss of the sense of the Greek "less vivid future" conditional, the principal changes are two, showing that the translation was made according to the cultural context of its recipients: the substitution of "by God's will" for the adverbial "best," and the reference to the management of the Greek *pólis*, inappropriate to the new context, replaced with the neutral "things go." In any case, there can be no doubt that this saying of Hermes is ultimately a translation of the Greek. Another example of an ancient Greek source behind a Hermetic maxim in Arabic is in the *Muntaḫab*, where Hermes advocates "voluntary death of the appetites."[166] A version of this maxim is

162. al-Mubaššir 14.20–15.2 = aš-Šahrazūrī 146.5–8: "He said: If sectarians debate with you about religion with coarseness and ill speech, do not requite them likewise, but gently, with right direction, guidance, and courteous conversation. Have recourse to God and say—all of you—'O God! Correct your creatures and bring them, according to your divine decree, what will led them to congeniality, peace, faith, and guidance.'" *qāla iḏā ǧādalakumu l-muḫālifūna lakum fī d-dīni fi-l-faẓāẓati wa-sū'i l-qawli fa-lā tuqābilūhum bi-miṯli ḏālika bal bi-r-rifqi wa-d-dalālati wa-l-hidāyati wa-luṭfi l-muḫāṭabati wa-'taṣimū bi-llāhi wa-qūlū bi-aǧma'ikumu llāhumma aṣliḥ barīyatakum wa-aǧri 'alaykum min qaḍā'ika wa-qadarika mā yaqūduhum ilā l-ulfati wa-s-silmi wa-l-īmāni wa-l-hudā.*

163. Gutas 1975: 467. The names Hermes and Diogenes can appear similar to one another in Arabic script; perhaps this is the basis for the confusion.

164. *Muntaḫab* (ed. Dunlop 1272–1273): *qāla innamā taǧrī l-umūru bi-mašī'ati llāhi 'azza wa-ǧalla iḏā kāna l-faylasūfu malikan awi l-maliku mutafalsifan* = Miskawayh 1915: 215.18–19.

165. Gnomologium Vaticanum no. 443 (Plato): ὁ αὐτὸς ἐρωτηθεὶς πῶς ἂν ἄριστα αἱ πόλεις οἰκοῖντο ἔφη, εἰ οἱ φιλόσοφοι βασιλεύοιεν ἢ οἱ βασιλεῖς φιλοσοφοῖεν.

166. *Muntaḫab* (ed. Dunlop 1275–1276): "He said: There are two kinds of death: voluntary death and natural death. Whoever makes his [appetitive] soul die the voluntary death will have a natural death that is for him life." *wa-qāla inna l-mawta mawtāni mawtun irādīyun wa-mawtun ṭabī'īyun fa-man amāta nafsahū mawtan irādīyan fa-inna mawtahū ṭ-ṭabī'īya lahū ḥayātun.*

attributed to Socrates, the speaker in the *Phaedo*, in another Arabic collection also derived from the *Ṣiwān al-hikma*.[167]

It has been easier to locate parallels and antecedents for sayings in al-Mubaššir. One of these, for example, is attributed here to Hermes, but al-Mubaššir gives it another time among the sayings of Plato, whereas the tenth-century philosopher al-ʿĀmirī attributes the same saying to Homer.[168] Another seems to be adapted from a Greek saying attributed in the *Gnomologium Vaticanum* to Anacharsis the Scythian.[169] This saying has a rendering closer to the Greek again with al-ʿĀmirī, who attributes the saying this time to Aristotle.[170] Another saying of Hermes is found with slight variants attributed to Pythagoras in an Arabic gnomologium and, with greater alterations, is attributed to Zeno by aš-Šahrastānī.[171] One more comparison will show a case of clearly Greek origins, though there is some change of the content of the saying in Arabic. Al-Mubaššir includes this saying: "He [Hermes] was asked: Why is it that scholars go to the doors of the rich more than the rich go to the doors of scholars? He answered: For the understanding of the scholars has no need of riches, and the ignorance of the rich has no need of science."[172] Something similar appears in Greek tradition both under the name of Antisthenes and in the *lógoi* of Aesop: "The tyrant asked him: Why do the rich not go to the wise, but rather the reverse? He said: Because, whereas the wise know what is fitting for them for their lives, the rich do not know, but are concerned with money more than with wisdom."[173] These two sayings are definitely related—the

167. This is the "Philosophical Quartet," edited by Gutas 1975: 102–103 (Socrates no. 32), commentary 309–311. For other parallels in Arabic literature see Rosenthal 1940: 409. The validity of this saying was debated by the vizier Ibn Saʿdān and Abū Ḥayyān at-Tawḥīdī in the 980s (summarized by Joel Kraemer 1986: 18 and 205–206).

168. al-Mubaššir 20.12 = aš-Šahrazūrī 153.10: "He said: The faults of the intelligent man do not let him rejoice in his readily apparent good qualities." *wa-qāla l-ʿāqilu lā tadaʿuhū ʿuyūbuhū yafraḥu bi-mā ẓahara min maḥāsinihī*. Cf. Gutas 1975, 130–131 (Plato no. 33), commentary 355.

169. al-Mubaššir 20.20 = al-Anṣārī 153.13 = aš-Šahrazūrī 154.2–3: "What saves you from the envious is that he is unhappy when you are happy." *wa-yašfīka mina l-ḥāsidi annahū yaġtammu waqta surūrika*. Cf. *Gnomologium Vaticanum* no. 19 (Anacharsis): "When asked by someone why the envious people are always distressed, he said, 'because not only do their own faults eat at them, but also the good things of others distress them." ὁ αὐτὸς ἐρωτηθεὶς ὑπό τινος, διὰ τί οἱ φθονεροὶ ἄνθρωποι ἀεὶ λυποῦνται, ἔφη· ὅτι οὐ μόνον τὰ ἑαυτῶν αὐτοὺς κακὰ δάκνει, ἀλλὰ καὶ τὰ τῶν πέλας ἀγαθὰ λυπεῖ. See also Gutas 1975: 403.

170. Gutas 1975: 172–174 (Aristotle 35), commentary 402–403.

171. al-Mubaššir 19.19 = aš-Šahrazūrī 152.10: "He was asked, What is it that destroys men? He answered, Anger and envy, but graver than either is worry." *wa-suʾila mā llaḏī yahuddu r-rağula fāqāla l-ġaḍabu wa-l-ḥasadu wa-ablaġu minhumā l-hammu*. Cf. Gutas 1975: 72–73 (Pythagoras no. 21), commentary 246.

172. al-Mubaššir 19.20–22 = aš-Šahrazūrī 152.11–13: *wa-suʾila mā bālu l-ʿulamāʾi yaʾtūna abwāba l-aġniyāʾi akṯara mimmā yaʾtī l-aġniyāʾu abwāba l-ʿulamāʾi fa-qāla li-maʿrifati l-ʿulamāʾi bi-faḍli l-ġinā wa-ğahli l-aġniyāʾi bi-faḍli l-ʿilmi*.

173. Gnomologium Vaticanum no. 6 (Antisthenes): ὁ αὐτὸς πυνθανομένου τοῦ τυράννου, τί δήποτε οὐχ οἱ πλούσιοι πρὸς τοὺς σοφοὺς ἀπίασιν, ἀλλ' ἀνάπαλιν, εἶπεν· ὅτι οἱ σοφοὶ μὲν ἴσασιν ὧν ἔστιν αὐτοῖς χρεία πρὸς τὸν βίον, οἱ δ' οὐκ ἴσασιν, ἐπεὶ μᾶλλον χρημάτων ἢ σοφίας ἐπεμελοῦντο. Aesopic *lógos*: Perry 1952: 252, no. 19.

Arabic is clearly a translation of some form of Greek—but the path of transmission into Arabic remains unclear.

Yet it is not only a Greek background that is relevant to the sayings of Hermes in Arabic. At least one saying of Hermes in al-Mubaššir is a word-for-word translation of a Middle Persian saying that occurs in a Zoroastrian compendium of wisdom literature, Book 6 of the *Dēnkard*:

> M. Pers.: *Ast saxwan ke a-passaxwīh passox ud ast kār ke a-kerdārīh pērōzīh.*
> Arabic: *Rubba kalāmin ǧawābuhū s-sukūtu wa-rubba ʿamalini l-kaffu ʿanhu afḍalu . . .*[174]

The Arabic can be translated as, "Often talk is best answered by silence. Often actions are better abstained from."[175] (The Arabic goes on to add, "Often avoiding a quarrel is more correct.") There is an almost perfect correspondence not only between the words of the Arabic and Middle Persian given above, but also in the word order:

Mid. Pers.	= Arabic
ast[176]	= *rubba*
saxwan	= *kalām*
ke . . . passox	= *ǧawābuhū*
a-passaxwīh	= *as-sukūt*
ud	= *wa-*
kār	= *ʿamal*
ke ākerdarīh	= *al-kaff ʿanhu*
pērōzīh	≈ *afḍal*

The Middle Persian words formed with privative prefixes (*a-*) are translated here in Arabic with simple antonyms (as opposed to phrases of multiple words), because Arabic has no analogous type of word formation and would require a separate negative word (e.g., *ġayr*) to render the Middle Persian word precisely. Yet if it were the other way around, that the Middle Persian was translated from the Arabic, then we would expect not to find those Middle Persian words with privative prefixes but rather more exact equivalents of the positive Arabic words. For example, we would expect a Middle Persian word for "silence" or the equivalent for Arabic *as-sukūt*, "silence." Therefore the original form was Middle Persian and the Arabic is its translation. The comparison also shows that the translator made an effort to obtain a word-for-word correspondence with the original. The first part of the saying is

174. al-Mubaššir 21.19–20 = aš-Šahrazūrī 155.2–3: The Middle Persian is in Shaked 1979: 212–213, E45a.

175. Shaked's translation of the Middle Persian is, "There are words to which lack of answer is the answer, and there are actions in which inaction is victory."

176. Middle Persian *ast* ("is") has one of the meanings of Arabic *rubba* in initial position: "Sometimes there is/are . . . ," "There are some . . . "

found already in the great gnomologium of the secretary of the caliph al-Ma'mūn, 'Alī ibn 'Ubayda ar-Rayḥānī (d. 834), a compiler of wise sayings translated from Middle Persian, and perhaps even a translator of such sayings himself.[177] Another instance of a wise maxim derived from Middle Persian reattributed to a Greek sage in Arabic literature has been shown elsewhere.[178] The practice of taking sayings of Iranian background and inventing Greek sources for them is probably due to the prestige of the ancient Greeks as authors and the demand for their books among Arabic scholars of the ninth and tenth centuries. In the case of the sayings of Hermes, in which one short passage is found to be a translation from Middle Persian, it seems likely that more Middle Persian antecedents are waiting to be discovered.[179]

A similar case involves sayings translated from Sanskrit into Arabic and then reattributed to Hermes, both in the *Ṣiwān* and in al-Mubaššir's *Muḫtār*. It may come as a surprise at first to find ancient Indian maxims attributed to Hermes Trismegistus, but the context of early Arabic literature, in which translations from several ancient languages abounded, makes it perfectly plausible. Here we are dealing with the political teachings (*nītiśāstra*) of the ancient and legendary Indian sage Cāṇakya, minister of king Candragupta Maurya (fourth century BCE). These are extant in countless different Sanskrit collections with complicated histories of their own, not to mention the derivative traditions in Tibetan, Burmese, and other languages. It appears that a collection of the teachings of Cāṇakya was translated into Arabic sometime during the late eighth century, when other translations of Sanskrit works were sought by the Barmakid vizier Yaḥyā ibn Ḫālid ibn Barmak (deposed in 803), whose family came from Ṭuḫāristān, a region in which Sanskrit was studied up to the time of the Arab conquest.[180] At least two teachings of Cāṇakya from this Arabic version found their way under the name of Hermes in al-Mubaššir's *Muḫtār*. The precise Sanskrit original, or something very close to it, in the traditional verse form, appears in a published collection of maxims of Cāṇakya:

dhanikaḥ śrotriyo rājā nadī vaidyas tu pañcamaḥ
pañca yatra na vidyante na tatra divasaṁ vaset

A creditor, a brahmin, a king, a river, and fifth, a physician:
Where these five are not known, he should not dwell there [even] a day.[181]

A saying of Hermes in the *Muḫtaṣar Ṣiwān al-ḥikma* corresponds almost exactly, leaving out the physician but adding another requirement: "[Hermes] said: Do not

177. Zakeri 2007: 1.317–326; parallel saying at 2.384, no. 742.
178. van Bladel 2004: 165–168.
179. On wisdom literature in Middle Persian, see Shaked, "Andarz. i. Andarz and Andarz Literature in Pre-Islamic Iran," *EIr*, 2.11b–16b. See also Shaked 1979 and 1987a.
180. On the social context of the early translations from Sanskrit into Arabic, see van Bladel, forthcoming.
181. Ludwik Sternbach 1963: 108 (no. 107).

make your home in a country in which there is not a resolute authority, a just judge, a functioning market, a flowing river, and a helping friend."[182] Al-Mubaššir provides this saying of Hermes, truer to the original in its list of requirements, but somewhat expanded and differing in its choice of words from the one in the *Muḫtaṣar*: "[Hermes] said: He who lives in a place where there is not a victorious ruler, a just judge, a learned physician, a functioning market, and a flowing river, has lost himself, his people, his possessions and his children."[183] The match is nearly perfect.

Arabic	Sanskrit
victorious ruler	= king
just judge (*qāḍī*)	= brahmin
learned physician	= physician
functioning market	= creditor
flowing river	= river

The Arabic must be derived from the Sanskrit, which is much older and in meter, and not the reverse. However, the Arabic translation of the Sanskrit original has been changed differently in the two different cases. In the *Muḫtaṣar*, the change is minor: the physician has been replaced by a helping friend, who is then introduced out of order. In al-Mubaššir, although the requirements of a habitable land are the same as in the Sanskrit, the consequences of living in a place not supplied with all the requirements are elaborated.

Another example confirms the hypothesis that the Arabic translation from Sanskrit was deliberately adapted for Hermes by al-Mubaššir's source. Although the Arabic translation of Cāṇakya's Sanskrit maxims appears to have received almost no attention after its early translation, a section of it survived to be cited by Abū Bakr Muḥammad ibn al-Walīd aṭ-Ṭurṭūšī (from Tortosa, d. in Cairo in 1126 or 1131) in his famous compendium for rulers *Sirāǧ al-mulūk* (*The Lamp of Kings*).[184] As it happens, this excerpt by aṭ-Ṭurṭūšī contains one of the passages adapted by al-Mubaššir's source for a saying of Hermes. In all likelihood, the preceding example also derives from another passage of the same translation from Sanskrit. First I give the Hermetic saying as related by al-Mubaššir: "[Hermes] said: When the king cannot conquer his senses and overcome his appetites, how can he restrain his aides? And if he cannot restrain his aides, how can he restrain his subjects and what is distant from his kingdom? The path of kings is to begin with rule over his self so that his rule over others is righteous."[185] The parallel and original passage, as preserved by aṭ-Ṭurṭūšī, is

182. Istanbul Fatih 3222: 36r5–7; Kartanegara 1996: 170.19. *wa-qāla lā tatawaṭṭan baladan laysa fīhi sulṭānun ḥāzimun wa-qāḍin ʿādilun wa-sūqun qāʾimun wa-nahrun ǧārin wa-wāddun musāʿidun.*

183. al-Mubaššir 23.1–2 = aš-Šahrazūrī 157.2–4: *wa-qāla man lam yaskun mawḍiʿan fīhi sulṭānun qāhirun wa-qāḍin ʿādilun wa-ṭabībun ʿālimun wa-sūqun qāʾimun wa-nahrun ǧārin fa-qad ḍayyaʿa nafsahū wa-ahlahū wa-mālahū wa-waladahū.*

184. A. Ben Abdesselem, "al-Ṭurṭūshī," *EI*², 10.739a–740b.

185. al-Mubaššir 26.1–3 = aš-Šahrazūrī 157.5–8; cf. Ibn al-Qifṭī 8.5–6: *wa-qāla iḏā lam yakunu l-maliku yaqdiru ʿalā qahri ḥawāssihī wa-ġalabati šahawātihī, fa-kayfa yaqdiru ʿalā ḍabṭi aʿwānihī*

attributed there specifically to Cāṇakya. The section is prefaced with the words: "Some of the wise sayings of Šānāq the Indian, from his book which he named *Collection of Jewels* for the king Ibn Qm'bṣ the Indian" (*min ḥikam Šānāq al-Hindī min kitābihī llaḏī sammāhu* Muntaḥal al-ǧawāhir *li-l-malik Ibn Qm'bṣ al-Hindī*).[186] The relevant portion is as follows: "Whoever does not restrain his self, it being just one, does not restrain his senses, they being five. When he does not restrain his senses, though they are few and base, it is difficult for him to restrain his aides, they being many and a rude bunch. The common subjects in the districts of the country and the outskirts of the kingdom are yet more distant from restraint. Therefore let the king begin with ruling his self, for there is no enemy more worthy than to begin with conquering his self."[187] Almost a century ago Theodor Zachariae wrote a brilliant article demonstrating that this passage, along with the entirety of this subsection of aṭ-Ṭurṭūšī's compendium, definitely derives from a Sanskrit original undoubtedly attributed to Cāṇakya.[188] Though the direct antecedent of the Arabic has not been identified, Zachariae was able to adduce many close parallels from various works of Sanskrit literature and, moreover, to discern the true Sanskrit proper names underlying the names of individuals, distorted in the Arabic script, mentioned later in the same passage.

In the excerpt above, clearly the basis of the Hermetic maxim reformulated in al-Mubaššir's *Muḫtār*, we are dealing with a fundamental concept in ancient Indian ethics: *indriyanigraha*, literally restraint of the sense faculties. This subject, referring to self-control in the face of desirable or repulsive sensory stimulus, often appears at the beginning of Sanskrit works on *rājanīti*, or king's conduct.[189] As it says near the beginning of one collection of such maxims of Cāṇakya, *sa rājā yo jitendriyaḥ*, "the king is he whose sense-faculties are controlled."[190] It is remarkable that Hermes Trismegistus is in effect advocating this Indian principle, yet the adaptor has slightly changed the words to naturalize the loan translation, which was evidently a bit strange to him. A literal translation of the Sanskrit expression *indriyanigraha* in

---

wa-iḏā lam yaqdir ʿalā aʿwānihī fa-kayfa yaqdiru ʿalā raʿīyatihī wa-mā baʿuda ʿan mamlakatihī fa-sabīlu l-maliki an yabtadiʾa bi-sulṭānihī ʿalā nafsihī li-yastaqīma lahū sulṭānuhū ʿalā ġayrihī. (Badawī's edition of al-Mubaššir omitted several words in a haplography; I have supplied the text from Rosenthal's handwritten edition.)

186. aṭ-Ṭurṭūšī 192.20–21. Haas 1876: 629 suggested that Sanskrit *Brahmaṇa* underlies this name. This cannot be so, as recognized by August Müller 1880: 479n. Sanskrit *h* is rendered with an Arabic *h*, not *q*.

187. aṭ-Ṭurṭūšī 192.25–28: *wa-man lam yaḍbuṭ nafsahū wa-hiya wāḥidatun lam yaḍbuṭ ḥawāssahū wa-hiya ḫamsun wa-iḏā lam yaḍbuṭ ḥawāssahū maʿa qillatihā wa-ḏillatihā ṣaʿuba ʿalayhi ḍabṭu l-aʿwāni maʿa kaṯratihim wa-ḫušūnati ǧānibihim fa-kānat ʿāmmatu r-raʿīyati fī qawāṣī* (read *nawāḥī*) *l-bilādi wa-aṭrāfi l-mamlakati abʿudu mina ḍ-ḍabṭi fal-yabdaʾi l-maliku bi-sulṭānihī ʿalā nafsihī fa-laysa min ʿadūwin aḥaqqu min an yabdaʾa bi-l-qahri min nafsihī*.

188. Zachariae 1914. Ludwik Sternbach 1981: 122b denies that aṭ-Ṭurṭūšī's passage from Šānāq is Indian in origin; this is a simple assertion without any argument.

189. Zachariae 1914: 189–190, with more parallels in Sanskrit.

190. Ludwik Sternbach 1963: 75.2 (no. 5).

Arabic is **ḍabṭ al-ḥawāss*, indeed two of the key words in the Arabic text. Looking at the translation copied by aṭ-Ṭurṭūšī, the same expression is found with a finite verb: *lam yaḍbuṭ ḥawāssahū*: "he does not restrain his senses." The same verb *ḍabṭ* is repeated also for the more physical restraint of the king's aides and subjects. It sounds, however, as unnatural in medieval Arabic as it does in modern English to speak of "restraining the senses" rather than restraining the passions.[191] The Arabic verb *ḍabṭ* similarly has a connotation of physicality, making it a literal rendering of the Sanskrit *nigraha*, "suppression," or the like. Yet this was a bit strange-sounding in Arabic—how can *senses* be held back?—so the adaptor of the text for Hermes therefore changed the verb in the first case to *qahr*, "conquering"; one could speak of conquering the senses more naturally than restraining them. He expanded it with "and overcoming his appetites," the word *appetites* taken from later in the original passage (thus in aṭ-Ṭurṭūšī) and suitable within the discourse of Hellenistic philosophy and medicine in Arabic translation. Yet the verb *ḍabṭ* remains in place in the rest of the Hermetic saying as a trace of its origin. One can see from this example how the unfamiliar Indian expression, translated literally, called for slight adaptation in the Arabic. Other adaptations in the saying of Hermes were appropriate rather to context: for example, the point of the Sanskrit original, that there are five senses but only one self, is eliminated in the Hermetic saying. This is because the original passage, as represented by aṭ-Ṭurṭūšī, expands the discussion of the dangers inherent in each of the five senses with five animal metaphors ("It is through the appetites that animals perish," *innamā yahliku l-ḥayawānu bi-š-šahawāti*). The adaptor, who wanted to leave these metaphors out, could also dispense with the reference to the number five. Beyond his purposes to eliminate foreign, less readily understood elements and to preserve its integrity as a short excerpt, the changes may also have been intended to hide somewhat the origins of the passage, something necessary for the Sanskrit wisdom of Cāṇakya to become appropriate to Hermes. It is likely that more sayings of Hermes are adapted from the collection of Cāṇakya known to aṭ-Ṭurṭūšī as *Muntaḥal al-ǧawāhir*.

Whoever adapted these various sayings for attribution to Hermes did not have recourse to the ancient originals in so many languages. He was using gnomologia already available in Arabic translation. As noted above, the maxims of Hermes related by al-Mubaššir begin largely with a list of commandments and imperatives, but eventually turn to more ordinary gnomic sayings without any section break. Mohsen Zakeri discovered that a substantial portion of this second group of sayings of Hermes are also found in the *Kitāb al-Ādāb* of the ʿAbbāsid prince and poet Ibn al-Muʿtazz (861–908), a work composed from earlier materials before 887.[192] In turn, Ibn al-Muʿtazz' source for some of these sayings, directly or indirectly, appears to be

191. The expression "restraint of the seneses" and the like will perhaps now be natural in English for those familiar with Indian learning in English translation.

192. Zakeri 2007: 1.39–43.

the *Ǧawāhir al-kilam* of the secretary ʿAlī ibn ʿUbayda ar-Rayḥānī (d. 834). Zakeri supposes that the common content of the sayings of Hermes and these two earlier works "insinuates the existence of an independent text of the 'Maxims of Hermes' before the time of al-Rayḥānī."[193] None of these earlier collections, however, makes any reference to Hermes. In light of the numerous cases in which maxims have clearly been adapted for Hermes from sources in which Hermes was not mentioned, as well as the fabricated and later character of the accompanying account of Hermes' career, it seems rather that the adaptor of these sayings simply took them from Ibn al-Muʿtazz, or from a source common to the two later than ar-Rayḥānī, as he took maxims for Hermes from many sources. There is no reason to suppose an earlier collection of maxims of Hermes as such. In fact, there is no direct evidence for any attempt to produce such a collection prior to the *Ṣiwān al-ḥikma* or al-Mubaššir's source. Clearly the latter had recourse to many different books in compiling the wisdom of the antediluvian prophet, including preexisting Arabic gnomologia.

What can one make of these maxims of Hermes with their various origins in Greek, Middle Persian, Sanskrit, and Arabic? In the history of wisdom literature, it is very common for sayings to be reattributed from one sage to another. This happened regularly already within the Greek manuscript tradition, and for a variety of causes, many of them unintentional. For example, sometimes the heading under which a group of sayings appeared might be lost or overlooked by a copyist, so that originally separate groups of sayings would come to be attributed to a single name. In Arabic, names as dissimilar as Diogenes and Hermes could be mistaken for one another in sloppy handwriting (ديوجنس and هرمس respectively). In the present case, however, it is impossible to escape the conclusion that someone has selected words of wisdom from various books and deliberately collected them to form something new, the wisdom of Hermes. There are too many different sources for these Hermetic sayings to suppose that every case is accidental. Our compiler has doubtless also added his own ideas and sayings to the ancient samples.

A conspicuous example of a new text under the name of Hermes, laced with ancient materials taken from other sources, is in Hermes' *Testament to Ammon*, related by al-Mubaššir in the midst of the other *ḥikam* of Hermes. The *Testament to Ammon* has never received special attention, yet it has been available to European scholars for a long time, not only through the medieval translations of al-Mubaššir's *Muḫtār* into Spanish, Latin, and other languages, but also, since 1903, in Lippert's edition of Ibn al-Qifṭī's *History of the Sages* (in the abbreviation of az-Zawzanī), a work that draws heavily from al-Mubaššir. This *Testament* is a remarkable document, demonstrating an interest in investing Hermes with all the characteristics of a prophetic lawgiver on the model of Muḥammad. Here is my translation from al-Mubaššir with section numbers added:

193. Zakeri 2007: 1.42.

1. He gave a testament to Basīlūḫīs,[194] who was King Ammon. He said: The first thing with which I charge you is a fear of God the Exalted and a love of obeying Him. **2.** Whoever is in charge of the populace must keep three things in mind: the first is that he has absolute power over many people; the second is that those over whom he has absolute power are free, not slaves; the third is that his authority lasts only a short time. **3.** Your path is to purify yourself with good intentions, O Ammon,[195] and speaking the truth. **4.** Take care not to delay battle and holy war (*al-ḥarb wa-l-ǧihād*) against those who do not believe in God—His name is most high—and who do not follow my custom and my law (*sunnatī wa-šarīʿatī*), because of your desire that they enter into obedience to God the Exalted. **5.** Guard against wishing to take their property or leaving them in their excess. For property is wanted only for spending it, and God—His name is most high—has no satisfaction in it. **6.** Know that the subjects trust in whoever treats them well. You will not improve the kingdom except by the subjects. Whenever the ruler has no subjects, the rule of [merely] himself is what occurs, when he is without them. **7.** You must not neglect to attend to their affairs and the affairs of your kingdom, and only then attend to yourself. **8.** Give priority to that by which your afterlife will be well and your present life becomes well. **9.** Your way, when you encounter war, is to be of resolute view in all of your affairs. **10.** Beware rout, for if it befalls an army, it cannot ready itself quickly. **11.** Use many spies so that you can have news of your enemies at every moment. **12.** Beware of a plot that works against you. **13.** When you give an order, inquire about it afterwards without failing, for then you will worry less. **14.** When you order that a letter be written for you, take care not to seal it and send it without first reading it yourself, because schemes befall kings. You are not the first king alarmed[196] at this matter. **15.** You must not keep confidence in anyone and reveal to him your secrets, but rather your nobles and subjects will keep confidence in you through your good governance of them. **16.** Make sleep for yourself just enough for your body's rest, and occupy yourself only with serious things, so that all your affairs are serious without jesting. **17.** When you have a concern, act. When you are victorious, continue. When you continue, be careful. **18.** Beware neglecting the greatest alchemy, governing its people, the inclination of their hearts, and pardoning them: they are the farmers, and alchemy is cultivating the earth with seeds and plants, for the subjects are supported by it, the army[197] becomes numerous out of it, houses of business prosper from it, and the dynasty is secure by it. Your way is not to ignore a matter with this

194. The name is totally corrupt in the manuscript, and this reading is far from certain. See discussion below.

195. *yā Āmmūn* supplied by Rosenthal's handwritten edition.

196. *wahila* (Rosenthal's handwritten edition).

197. *al-ǧund* (Rosenthal's handwritten edition).

> outcome. **19.** Your way is to be generous with the people of rank in treating each one according to the extent of his intellect[198] and science. **20.** Take the opportunity to be generous to them so that the subjects are not ignorant of the dues of the virtuous. **21.** Whoever seeks knowledge, be generous to him and know his due, shower him with benefaction to increase his ardor for it [*scil.* knowledge], to give his intellect grace, to let his mind be pure and his concern for this world be small. Take benefit from him, if God wills. **22.** Quickly punish the corrupters of the land (*al-mufsidīn fī l-arḍ*) after their offense is proven to you and their crime is evident. **23.** Whoever defames your rule, decapitate him and make it known so that others will beware. **24.** Whoever steals, cut off his hand. **25.** Who robs on a path, cut off his head and crucify him so that news of it spreads and your roads be safe. **26.** Whoever is found with a male like himself, fornicating with him, must be burned in fire. Whoever is found with a woman, committing adultery with her, strike him with fifty lashes and stone the woman with a hundred stones after establishing sure proof of it. **27.** Guard against listening to the words of a slanderer, but when you have proof that he is slandering, quickly punish him and make it known. **28.** Give your heart rest from concern with the unattainable. **29.** Take care not to ignore those in prison each month, so that none among them is oppressed. Whoever deserves to be released, you set him free after treating him well. And if he deserves punishment, do it quickly. He whose case warrants taking time to clarify his situation, send him back into it [*scil.* into prison]. **30.** Beware of being vain in your opinion. You must take counsel with the man of fine intellect, advanced in years, because of the experiences he has had, and obtain his opinions. For if you see in any of them the right thing to do, [do it]; and if not, then you yourself draw together the right view from all of them. Success is by God.[199]

The terminology of the passage shows that it was written by someone familiar with the Qur'ān and also the developed regulations of Islamic law, so it cannot have been written before the ninth century. Hermes is like Muḥammad in having a *sunna*—his conduct taken as a model of righteous behavior—and a *šarīʿa*—laws of his religion that he has revealed in his role as prophet.[200] Hermes commands *ǧihād*, holy war, to increase the population submissive to God's law, and opposition to "the corrupters in the earth" (*al-mufsidīn fī l-arḍ*), a frequent expression in the Qur'ān.[201] Hermes' punishments for several crimes are practically identical with the *ḥadd* punishments required by Islamic law: adultery punished with a fixed

198. *ʿaqlihī* (Rosenthal's handwritten edition).

199. al-Mubaššir 23.3–25.4 = aš-Šahrazūrī 157.5–160.3; cf. Ibn al-Qifṭī 7.10–8.4.

200. The term *šarīʿa* was employed by Jews and Christians, as well as Muslims, by the tenth century. N. Calder and M. B. Hooker, "Sharīʿa," *EI*², 9.321a–326a. Cf. Qur'ān 45:18 for the collocation of *ittibāʿ* and *šarīʿa*.

201. For example, Qur'ān 2:205, 5:64, 7:74, 11:85, 26:183.

number of lashes or stoning; highway robbery with crucifixion; and theft with the cutting off of the hand. The *ḥadd* punishment for *qaḏf*, slandering the legitimacy of someone's parentage, is not present, but *siʿāya*, slander in general, has taken its place.[202] In Islamic law, the imām, or rightful leader of the religious community, is technically responsible for enforcing the *ḥadd* punishments; Hermes here addresses king Ammon as if he were also that imām. Ammon is therefore implicitly treated as a caliph: imām and king in one.

Who is Ammon, and how would Arabic authors have known about him? As seen in the previous section, the biography of Hermes also related by al-Mubaššir explicitly makes King Ammon one of Hermes' students. Ammon is actually found very little in Arabic literature, but it is easy to see how a reference to Ammon as a disciple of Hermes could have been transmitted from Greek into Arabic. The name Ammon (also rendered Amun) referred to an ancient Egyptian god of Thebes, *Jmn*, whose importance and popularity rose and fell with the changing fortunes of Egyptian dynasties, particularly from the middle of the second millennium BCE onward.[203] Ammon is later mentioned a number of times in ancient Greek literature as an Egyptian god. In Hellenistic Greek he was best known as the god of the oracle in the western Egyptian oasis of Siwa, made especially famous by Alexander's visit there. His name occurs in the *Alexander Romance* of Pseudo-Callisthenes in its Syriac version; probably it is present in an Arabic version of that work, recently rediscovered in manuscript.[204] In his *Phaedrus* 274c–275b, Plato has Socrates relate an Egyptian story about Theuth (Thoth) instructing Ammon, described as the king of Egypt, in the sciences. This bears some resemblance to the basic idea of Hermes (Thoth) instructing King Ammon. Ammon's prophet Bitys is mentioned in the *De mysteriis Aegyptiorum* attributed to Iamblichus. Bitys is said there to have revealed "the Hermetic path."[205] Importantly, the Greek *Corpus Hermeticum* itself includes several discourses in which Hermes instructed Ammon, but these are for the most part concerned with Providence and Necessity and do not contain advice for a governor.[206] Neither is the Greek iatromathematical text addressed from Hermes to Ammon of direct assistance here.[207] In any case, the contents of the Hermetica connected with Ammon are so different from the Arabic *Counsel of Hermes to Ammon* that they cannot be directly related. Rather, it must be merely that the idea of Hermes' teaching King Ammon was transmitted into Arabic through an unidentified intermediate account, and a later author wrote these instructions in awareness of this idea that Ammon, a king, was Hermes' student.

202. Carra de Vaux, Schacht, and Goichon, "Ḥadd," *EI²*, 3.20a–22a.

203. E. Otto, "Amun," *Lexicon der Ägyptologie*, 1.237–248.

204. On the Alexander Romance in Arabic, see Doufikar-Aerts 2003. For the following and more references to Ammon in Greek literature, see Fowden 1993a: 32n115.

205. *De mysteriis* 8.4 (267).

206. SH 12–17.

207. Ideler 1841: 387–396.

At least one section of the *Testament of Hermes to King Ammon* is directly drawn from an ancient Greek source in translation. Sentence 2 comes from a maxim attributed in many Greek gnomologia to Agathon, a friend of Plato:[208] "Agathon said that the ruler must remember three things: first, that he rules men; second, that he governs according to laws; third, that he will not always rule."[209] Again, from the Arabic of the *Testament of Hermes to Ammon*:

> **2.** Whoever is in charge of the populace must keep three things in mind: the first is that he has absolute power over many people; the second is that those over whom he has absolute power are free, not slaves; the third is that his authority lasts only a short time.[210]

The Arabic translator was careful to render the nuances of the Greek original. For example, where the Greek sentence plays on three slightly different uses of the verb ἄρχειν, the translator has rendered these with several different words according to each idea.

There are also a few elements reminiscent of pre-Islamic Persian wisdom literature in the *Testament*. One of these is found in part 18: "Beware neglecting the greatest alchemy, governing its people, the inclination of their hearts, and pardoning them: they are the farmers, and alchemy is cultivating the earth with seeds and plants, for the subjects are supported by it, the army becomes numerous out of it, houses of business prosper from it, and the dynasty is secure by it. Your way is not to ignore a matter with this outcome." I can only suppose that the clumsy likening of agriculture to alchemy has the purpose of making Hermes, legendary as a teacher of alchemy, into a counselor for kings. The idea, however, that all parts of the kingdom are interrelated, and that the king's attention to the lowliest laborers is necessary for the prosperity of the realm, is found in several similar figures recorded elsewhere as attributed to pre-Islamic Iranian kings. Scholars today know such figures generally as "Circles of Justice."[211] For example, aṯ-Ṯaʿālibī (wr. before 1021) attributes these words to the Sasanid Ardašīr in his history of Persian kings, written

208. The same line is found in two late antique florilegia, Stobaeus 4.5.24 (fifth century) and the *Loci communes* of Ps.-Maximus the Confessor, the earliest form of which was compiled around 650 (Ihm 2001: 9.61/64, pp. 232–233; on the date, p. xxix). It was repeated, on the basis of these earlier collections, in many medieval Greek florilegia, such as the eleventh-century "florilegium of Patmos" (Sargologos 1990: 28.45), the twelfth-century florilegium of Antonius Melissa (*Patrologia Graeca* 136.1008), the thirteenth-century *Gnomologium Parisinum* (ed. Leo Sternbach 1894: 149, no. 124; cf. Searby 1999), the *Gnomica Basilensia* (Kindstrand 1991: 439), and the unedited *Florilegium Rossianum* (Ihm 2001: xvi and 233).

209. Ihm 2001: 9.61/64, pp. 232–233. Ἀγάθων ἔφη τὸν ἄρχοντα τριῶν δεῖ μεμνῆσθαι· πρῶτον μὲν ὅτι ἀνθρώπων ἄρχει, δεύτερον ὅτι κατὰ νόμους ἄρχει, τρίτον ὅτι οὐκ ἀεὶ ἄρχει.

210. *Testament of Hermes to Ammon* 2 (al-Mubaššir 23.4–7): *wa-man tawallā amra n-nāsi fa-qad yağibu ʿalayhi an yakūna ḏākiran ṯalāṯata ašyāʾa awwaluhā anna yadahū muṭlaqatun ʿalā qawmin kaṯīrin wa-ṯ-ṯānī anna llaḏīna yaduhū muṭlaqatun ʿalayhim aḥrārun lā ʿabīdun wa-ṯ-ṯāliṯu anna sulṭānahū innamā yaṯbutu muddatan yasīratan.*

211. For numerous examples of the history of such figures see Darling 2002.

a few decades before al-Mubaššir's *Muḫtār*: "Rulership is only through men. Men are only through wealth. Wealth is only through cultivation. Cultivation is only through work and good government."[212] The similarity to the doctrine of Hermes on the "alchemy of agriculture" is obvious. A more developed parallel with the Hermetic passage appears in the *Sirr al-asrār* of Pseudo-Aristotle, a work definitely drawing on pre-Islamic Iranian literature. In its edited, long recension we find a diagram meant to illustrate the harmonious interrelationship between the dynasty and the state, the subjects and the army, and the natural order of the world. This diagram is supposed to be the summation of the entire book. Arranged on the page in a geometrical design, a circle or an octagon, the aspects of this relationship are connected in an endless cycle.

> The world is a garden, whose fence is the dynasty.
> The dynasty is authority, which lives by law.
> Law is governance, governed by the king.
> The king is a shepherd supported by the army.
> The army is assistants fed by wealth.
> Wealth is sustenance gathered by the subjects.
> The subjects are slaves served by justice.
> Justice is a harmony and it is the life of the world.[213]

Although the *Testament of Hermes to King Ammon* does not arrange the vital components of the state in a cyclical pattern, we see a similar chain of causes and effect. The farmers must be cultivated, because they support all the subjects, the army, houses of work, and finally the dynasty. Thus the harmony of the kingdom depends on the king's care directed toward the humblest subjects. This is also one of the main points of the Circle of Justice. Interestingly, the *Sirr al-asrār*, in which the Circle of Justice occurs, also includes references to Hermes, as well as at least one passage adapted from another work and attributed spuriously to a Greek sage with an invented name.[214]

Who, then, composed this *Testament of Hermes to Ammon*? There are now many clues to the answer all pointing in one and the same direction. Whoever it was, he lived after Islamic law was well established and normative, but before al-Mubaššir: in the late ninth or the tenth century. He was a scholar, for he had extensive

212. Translation from the text as cited by Manzalaoui 1974: 214. *lā sulṭān illā bi-r-riǧāl wa-lā riǧāl illā bi-l-māl wa-lā māl illā bi-l-ʿimāra wa-lā ʿimāra illā bi-ʿamal wa-ḥusn as-siyāsa*.

213. Trans. from text in Badawī 1954: 127–128. Cf. the same passage in Ibn Ḫaldūn, *Muqaddima* (trans. Rosenthal) 1.82n30. In the same work, 1.80–81, another parallel is attributed to the Sasanian emperor Ḫusrō Anūširwān himself.

214. For more examples of parallels with and sources of the *Sirr al-asrār*, see van Bladel 2004: 160–168. There I suggested that the Testament of Hermes to Ammon was perhaps translated from Middle Persian; now it is clear that it cannot have been so, although its author had the benefit of translated literature including material translated from Middle Persian.

knowledge of wisdom literature in translation from Greek, Middle Persian, and Sanskrit, from which he carefully adapted numerous passages and put them under the name of Hermes. He very much wanted to convince his readers that Hermes was a prophet and the founder of a religion, and to present that founder and that religion after the accepted models of Muḥammad and Islam. Yet he wanted it to have the scent of ancient Greek philosophy.

As seen, the *Testament of Hermes to King Ammon* is transmitted in the midst of the other sayings of diverse origins, and was also adapted by someone for attribution to Hermes, someone with these same goals: surely the same person. The *Testament* and most of the maxims together accompany a biography of Hermes clearly portraying him again as a prophet with a religion, one described very much as the religion of Ḥarrānian Ṣābians was described by as-Saraḫsī. Both texts, the biography and the accompanying *Testament*, must have a common origin insofar as they share the same goals and both make Hermes the teacher of a ruler named Ammon. But, interestingly, this biographer makes no reference to the more famous account of Hermes by Abū Maʿšar, which was written ca 840–860. Given the foregoing, it is unlikely that he wrote before Abū Maʿšar; in that case, he must have deliberately ignored it, or even wanted to replace it. For this author, Hermes was not three, but one, and his name is not a title, as Abū Maʿšar claimed, but he is indeed definitely a prophet. He is explicitly connected with figures known from the ancient Greek Hermetica, Tat and King Ammon. The maxims of Ṭāṭ are introduced together immediately after those of Hermes, with the statement that Ṭāṭ is Ṣāb, for whom the Ṣābians are named.[215] Seth and Hermes, the first and third *Ūrānī*s, taught a religion called *ḥanīfī*, or pagan, involving animal sacrifices and offerings to God. As already explained, the name Ūrānī is first known from as-Saraḫsī as a teacher of the Ṣābians of Ḥarrān.

In chapter 3 the reader has seen good reasons to be skeptical of the existing claims concerning the alleged doctrines of Ḥarrānian origin. Despite such reservations, the evidence here strongly suggests that a certain Ḥarrānian Ṣābian living among Muslims wrote this text for a general Arabic audience to argue for the legitimacy of his religion: not only the material copied by al-Mubaššir on Hermes, but also the chapters on Seth, Ṭāṭ, and Asclepius. The author of these passages endowed Hermes with the characteristics expected of a prophet by Muslims, a prophet whose law is in remarkable harmony with Islamic law. To do so, he plagiarized a number of gnomologia, some of which may well already have become obscure by his own time. Although other Arabic Hermetica are of unknown authorship, here we seem to have found at last an example of the "Hermetism" of the Ḥarrānians in Baġdād so much speculated about. Instead of cosmic sympathies and revelatory initiations, we find corporal punishments, holy war, practical advice on managing a kingdom, and, in general, total conformity to the idealized mores of medieval kings. But one may not

215. On Tat as Ṣāb, see section 3.4.6 in this volume.

assume that this work is the product of "the Ḥarrānians" as a group. It is just one author's work. There is no reason to suppose that this text represents the views of the entire sect of Ḥarrānian Ṣābians, in which case it would be misleading in a sense actually to call it "a Ḥarrānian text" as such.[216]

Is there an alternative to supposing the author was a Ḥarrānian Ṣābian living in Baġdād? There is no reason to think that a Christian, Jew, or Magian would have taken such pains to present a picture of Hermes and the ancient religion of his followers that would be appealing to Muslims. Nor is it likely that a Muslim wrote the text. One might try to make the case that this work was composed by an Ismāʿīlī; as seen, Hermes was explicitly regarded by some early Ismāʿīlī authors as a prophet, and al-Mubaššir was working in Egypt under Ismāʿīlī Fāṭimid rule. But there are no especially Ismāʿīlī characteristics in Hermes' teaching here, which one would expect to find if this were the case. There are no references to esoteric (*bāṭinī*) or allegorical exegesis, nor is Hermes a part of the Ismāʿīlī account of the succession of the prophets that was to become normative. In this account Hermes promised that prophets would come after him, but there is no evident hint of anything special from Ismāʿīlī prophetology here. As for the Ṣābians, those in Ḥarrān itself produced no known authors, but the Ḥarrānian Ṣābians in Baġdād, specifically Ṯābit and his family, well-known scholars employed by the caliphs, would have a powerful motive to make their religion appear, and even to define their religion, in generically Islamic terms, and to make their prophet into a teacher of kings, as they themselves were advisors to kings. Two suggestions as to the author's identity come to mind. These are merely informed guesses. The first is Sinān ibn Ṯābit around the year 933. As told in section 3.5, in the reign of al-Qāhir, Sinān was put in the awkward position of having to resist the caliph's strongly stated wish that he convert to Islam, and had to face the immediate threat of the slaughter of his community. He might have elaborated these teachings of Hermes, including advice on just and moderate kingship, in an attempt to satisfy a violent caliph like al-Qāhir, or indeed any other caliph concerned about the legitimacy of the Ṣābians. Sinān had close dealings with three caliphs and had to live his last years, at least outwardly, with the conversion to Islam imposed upon him. He might have wished to safeguard the legitimacy of the religion of the rest of his family, and it would be in his interest, as a Muslim expected at the mosques for prayer, but whose family nevertheless remained Ṣābians, to emphasize the harmony between the two religions. Al-Masʿūdī, writing in 956, had already heard of Ṣāb, for whom the Ṣābians are named,[217] which might suggest that al-Mubaššir's source was somewhat earlier than that; again, Sinān would fit the chronology. The second suggestion is based on a reference by the Jewish scholar

216. Grignaschi 1975: 80–81 supposes this text to show that the "Hermétistes de l'époque musulmane" have adopted *ǧihād* as a requirement of a king who follows Hermes. As already shown, there is no reason to posit the existence of such an ill-defined group as Muslim Hermetists.

217. al-Masʿūdī, *Tanbīh*, 91.3–4.

Mūsā ibn Maymūn to "the book by Isḥāq the Ṣābian and his defense of the Ṣābian religious community, and his great book of the laws of the Ṣābians, the particulars of their religions, their feasts, offerings, and prayers."[218] Although no other reference to these two books is known, it seems plausible from the descriptions that these could be the sources exploited by al-Mubaššir. This Isḥāq aṣ-Ṣābi' could be the son of Ibrāhīm ibn Sinān ibn Ṯābit. Almost nothing else is known about this individual, save that he was a physician,[219] and that he flourished in the middle of the tenth century (his father's dates are 908–946). In any case, one must readily admit that other explanations are possible. It is conceivable that an anonymous author of other Arabic Hermetica in the tenth century wanted to assure his pseudonym a religious legitimacy. Further speculation is not warranted, but it appears likely that al-Mubaššir's texts on Hermes are part of one Ṣābian's apologetic, and that it was composed in the tenth century.

Subtle differences between al-Mubaššir's description of the religious teaching of Hermes and as-Saraḫsī's earlier description of the Ḥarrānian Ṣābian doctrines of the *ḥanīfīya* (section 3.4.3) support the interpretation of the text as implicit apologetic. The latter text, probably based in part on his acquaintance with Ṯābit ibn Qurra, says that the Ṣābians sacrificed to the planets, and *not* to the Creator, because the planets, and not God directly, actually administered the world.[220] In other words, they were really star worshippers. By contrast, in al-Mubaššir's description, Hermes teaches a religious calendar regulated by the planetary movements, but with no direct sign of planet worship as such. Hermes' teaching is presented as emphatically monotheistic. This subtle change in the description of the religion of Hermes can be interpreted as a response to the charge against the Ṣābians that they were illegitimate because they worshipped the stars, as in the *fatwā* of al-Isṭaḫrī against the Ṣābians under the caliph al-Qāhir.

5.4 The Synthesis in the Twelfth and Thirteenth Centuries and Beyond

So far this study has documented the development of the representation of Hermes in Arabic literary traditions from about 750 to 1050, and the genesis of each component of the Arabic myth of Hermes. By the end of this period, Hermes was known to learned men in different ways from the different sources described so far: as a sage or a prophet who had ascended to the celestial spheres and returned, who revealed an ancient religion or the science of the stars, or both, whose technical works on a variety of subjects were in circulation, whose teachings were represented

218. *Dalālat al-ḥā'irīn* 3.29 (ed. 588.13–16): *wa-kitāb Isḥāq aṣ-Ṣābi'ī* [sic ed.] *fī ḥtiǧāǧ al-milla aṣ-ṣābi'a wa-kitābuhū l-kabīr fī nawāmīs aṣ-Ṣābi'a wa-ǧuz'īyāt dīnihim wa-a'yādihim wa-qarābīnihim wa-ṣalawātihim.*

219. Chwolsohn 1856: 1.578.

220. *Fihrist* 384.2–4.

in wisdom literature, and whose life appeared, if only in brief, in works of history. In later centuries, Arabic scholars continued to weave all of these strands together.

The later reception of the wise maxims of Hermes was varied. Although they were copied into a number of subsequent collections and obviously were widely read, there are only a few concrete indications of any direct impact that the maxims of Hermes might have had. As already seen, they were at least authoritative and instructive enough to be preserved in many later wisdom anthologies. But there is at least one known instance, in the twelfth century, in which an important author makes use of a *ḥikma* of Hermes, along with information about the legend of Hermes, to further a theological argument. In his *Kitāb al-Milal wa-n-niḥal* of 1127/1128, the eastern Iranian heresiographer aš-Šahrastānī wrote a detailed refutation of the Ṣābians, as he understood them, excluding them from the ranks of the *ḥunafāʾ*—whom he treats as Muslims rather than as pagans—and focusing instead on the role of Ṣābians as "adherents of the spiritual beings" (*aṣḥāb ar-rūḥānīyāt*).[221] Again, "spiritual beings" is a term related to the talismanic literature, where the powers of the *rūḥānīyāt* are harnessed for a variety of magical purposes.[222] Guy Monnot has tentatively suggested that this anti-Ṣābian section of aš-Šahrastānī's book was based on an earlier work containing arguments derived from the early-tenth-century debate between the philosopher Abū Bakr ar-Rāzī and the Ismāʿīlī *dāʿī* Abū Ḥātim ar-Rāzī.[223] In any case, the legend of Hermes and his maxims inform aš-Šahrastānī's argument. Aš-Šahrastānī characterizes the Ṣābians as doubting that an angel could descend to teach humans, but the response to his imagined Ṣābian interlocutor includes the following: "You [Ṣābians] hold that the great Hermes ascended to the spiritual world (*al-ʿālam ar-rūḥānī*) and entered into their [the angels'] ranks (*fa-nḫaraṭa fī silkihim*). When you conceive of the ascent of humans, why can you not conceive of the descent of angels? When you verify the doffing of the garment of humanity,[224] why is it impossible that angels don the garment of humanity?" Several pages later he cites the *ḥikam* of Hermes, whom he introduces now as "the Great Hermes (*Hirmis al-ʿaẓīm*) ... who is counted as one of the greater prophets, who is said to be the prophet Idrīs, peace be upon him. He is the one who established the names of the zodiacal signs and planets; he arranged them in their houses; he determined for them exaltations, dejections, apogees, depressions; trine, sextile, and quartile aspects; and oppositions, conjunctions, retrograde and direct motions; he explained the equations and the true positions of the planets."[225] Although aš-Šahrastānī hastens to add that the prognostications made on the basis

221. Studied by Monnot 1986.

222. On *rūḥānī* as a loan word from Aramaic (Syriac), see Stern 1983a: 27–28.

223. Monnot 1986: 12–13.

224. This is a reference to the Arabic *Theology of Aristotle* (*Theologia* 1.21, Badawī 1948: 22.1–4) where Plotinus' description of his soul's leaving his body (*Enneads* IV.8.1) is adapted in Arabic in these terms. Zimmerman 1986: 138–141 and 228–229n38 compares the Greek, Arabic, and Latin versions.

225. aš-Šahrastānī 240.17–241.1.

of these conjunctions are not proved definitely to be valid, he makes it clear that the prophet Hermes must be accepted as the founder of astrology and astronomy. He then introduces quite a few *ḥikam* of Hermes, most of which also occur in the *Muntaḫab Ṣiwān al-ḥikma* and al-Mubaššir's *Muḫtār al-ḥikam*.[226] At their head he gives one *ḥikma* that lists the priorities of a virtuous and pious man: Hermes says that one must first glorify and thank God according to his understanding (*ʿalā maʿrifatihī*); then one must support the ruler (*as-sulṭān*) and obey the law (*an-nāmūs*); then one must look after his friends; finally, one must make sure to treat the common people (*al-ʿāmma*) well. Aš-Šahrastānī has a special purpose in relating this saying. He wants to point out what he sees as an inconsistency between the words of Hermes and what he understands to be the Ṣābian position. After this first *ḥikma*, he immediately interjects, "See, all you Ṣābians, how he [Hermes] emphasizes the topic of apostleship to the point that he ties obedience to the Apostle, whom he expresses by "the Law" (*an-nāmūs*), with understanding of God (*bi-maʿrifati llāh*), without mentioning here the glorification of the spiritual beings (*ar-rūḥānīyāt*) and making offerings to them, even if it is part of the religious obligations."[227]

Thus aš-Šahrastānī wants to say that the talismanic doctrines attributed to the Ṣābians, including their reverence of the planetary spirits, are not as important as prophethood, through which alone comes understanding of God. The implication is that the Ṣābians should heed later prophets, such as Muḥammad, and give up their irrelevant veneration of the spirits. As M. Levy-Rubin points out, this is part of aš-Šahrastānī's effort to shake the word *ḥanīf* free from its associations with the Ṣābians.[228] To do so, aš-Šahrastānī interprets the word *nāmūs* in the passage in its religious sense of prophetic law, not in its obviously intended meaning of mundane law. He has employed a maxim of Hermes from the gnomological genre in the different context of theological argument. He has taken the maxims of Hermes seriously as prophetic teachings.

Aš-Šahrastānī was dealing with traditions about the Ṣābians that had been developing for hundreds of years by his time and were more historically complicated than his understanding allowed. To achieve his goal of a comprehensive presentation of all the world's doctrines, he necessarily had to deal with the Ṣābians. Here the myth of Hermes' heavenly ascent, his wise maxims, and his character as an ancient prophet are all combined to further his theological argument. By the time that aš-Šahrastānī composed this anti-Ṣābian chapter, in the early twelfth century, there were no more Ḥarrānian Ṣābians in existence to be refuted.[229] Whoever had taken such pains to turn Hermes into a prophet, complete with a *šarīʿa* and ordinances for punishment, was gone. Muslims like aš-Šahrastānī, himself an Ismāʿīlī or at least a

226. *Muntaḫab* ed. Dunlop 1318–1323, ed. Badawī 188.14–189.2.

227. aš-Šahrastānī 241.12–15 (trans. Jolivet and Monnot 2.154–155).

228. Levy-Rubin 2003: 220–222. See further Lawrence 1976: 67–68.

229. See section 3.5 in this volume. The "Ṣābians of the Marshes," that is, the Mandaeans, were in this period a marginal people from the urban point of view and are seldom noticed in Arabic literature.

thinker with strong Ismā'īlī tendencies, could appropriate Hermes for his own sectarian purposes to the exclusion of the Ṣābians who had made him into a prophet, just as the Ismā'īlī Abū Ḥātim ar-Rāzī had used Hermes against the philosopher Abū Bakr ar-Rāzī in the early tenth century.

Aš-Šahrastānī is an early representative of, and a contributor to, the new, favorable attention toward Hermes among Arabic philosophers that becomes noticeable in Arabic scholarship of the twelfth century. In earlier centuries, Hermes was mostly irrelevant to Arabic philosophers, meaning those called *al-falāsifa*. Although astrologers, alchemists, and talisman-makers had long relied on Hermetica frequently in their works, most earlier Arabic philosophers, as such, scarcely mentioned Hermes, if at all. This is because Hermes and the Hermetica were extraneous to the philosophers' carefully constructed curriculum, which was based on models inherited from the late antique commentators on Aristotle and Plato. The exceptions have been encountered already; they are few. The Hermetic work that al-Kindī described seems to have left no trace. Hermes appears as an authority in *The Book of the Apple*, but in the background, and not as a participant in the dialogue. Al-'Āmirī merely mentions the heavenly ascent of Hermes, which he discussed in a chronographic context. The Iḫwān aṣ-Ṣafā' alone gave Hermes a greater role, not only putting him on a par with other prophets and philosophers, but also citing passages from some of the Pseudo-Aristotelian Talismanic Hermetica.[230] Their approach was to have influence in the long term. Nevertheless, important philosophers like al-Fārābī (d. 950), Yaḥyā ibn 'Adī (d. 974), Ibn Sīnā (d. 1037), and al-Ġazālī (d. 1111) had no use for Hermes in their systems. A major impetus for the increased attention to Hermes in later centuries was the gnomological histories of the ancient sages, the *Ṣiwān al-ḥikma* and al-Mubaššir's *Muḫtār al-ḥikam*, as for example in aš-Šahrastānī's treatment just examined. These two works put the prophetic Hermes in the same category as Plato and Aristotle, and integrated Hermes into the history of ancient philosophy, making Hermes one of the first, if not *the* first, and therefore a founding figure for philosophers to take into account.

Even if they were not especially interested in Hermetica, the earlier Arabic philosophers had inadvertently given Hermes a foothold in the history of philosophy with one of their most important doctrines. In their ongoing development of Aristotelian psychology and cosmology (particularly Aristotle's *De anima* 3.5), the Arabic Aristotelians from al-Fārābī onward upheld the existence of a special immortal and incorporeal entity called the Active Intellect or Agent Intellect (*al-'aql al-fa''āl*). This Active Intellect, conceived to be a result of the emanations from the One that gave rise to the celestial spheres, was regarded as the cause of all human understanding, frequently described by the visual metaphor of light. The way in which the philosophers arrived at such a view is a complicated story told in detail elsewhere.[231] Suffice it to say

230. *Rasā'il Iḫwān aṣ-ṣafā'* 4.428–445.
231. Davidson 1992. A concise account is provided by D. Black 2005: 317–322.

here that the doctrine of the Active Intellect was commonly accepted in various permutations after al-Fārābī and became especially widespread through the philosophical system of Ibn Sīnā. Some philosophers, such as al-Ġazālī, appear to have conceived of the Active Intellect as an angel;[232] some, including al-Fārābī and Ibn Sīnā, held that a prophet is in fact a person with the perfect disposition to receive the Active Intellect's effluescence, and thereby to have a supernormal, indeed prophetic, degree of knowledge.[233] It was easy for later philosophical authors to conclude that Hermes—preeminent as an ancient scholar and prophet, and reputed to have conversed with angels—was one of those who had obtained an especially close connection with this celestial source of knowledge. Thus the myth of Hermes could be fitted into and explained by the system of *falsafa* (philosophy), which was intended to take account of everything. For example, the Andalusian philosopher Ibn Bāǧǧa (d. 1139), much influenced by the writings of al-Fārābī, held that Hermes and Aristotle were somehow together through their contact with the Active Intellect.[234] Apparently on the basis of Ibn Bāǧǧa's work, the Andalusian Jewish author Yahūdā ha-Levi (d. 1141), in his Arabic dialogue *Kitāb ar-Radd wa-d-dalīl*, characterized his nameless philosopher as holding that those who obtain union with the angelic Active Intellect, after reaching the rank of "the perfect human" (*al-insān al-kāmil*), will become one with that angel and join the company of the other philosophers who previously obtained the same intellectual union, including Hermes, Asclepius, Socrates, Plato, and Aristotle.[235]

In this background, Hermes gained a special appeal for the twelfth- and thirteenth-century philosophers who blended spiritual exercises and elements of Ṣūfism—including an exaltation of transcendent, direct experience of the truth (*ḏawq*) over the truth obtained through the application of reason—with the Aristotelian philosophy systematized by Ibn Sīnā. The founder of the Išrāqī school of thought, Šihābaddīn Yaḥyā as-Suhrawardī (d. ca 1190–1192), writing in Arabic and Persian, appears to be the first Arabic philosopher as such to give Hermes a distinctive place as one of his most important intellectual forebears.[236] As-Suhrawardī insisted on Hermes as his spiritual ancestor, perhaps because he could criticize the Peripatetics (as he called the followers of Ibn Sīnā) and modify their system all the more persuasively if he claimed a more ancient, authoritative, and divine source than Aristotle. Like the Iḫwān aṣ-Ṣafā' two hundred years before, he held that the

232. Davidson 1992: 133–134.

233. Davidson 1992: 61–62, 116–123, and 139.

234. In *Fī ttiṣāl al-ʿaql bi-l-insān* (*On the Conjunction of the Intellect with Humans*): Asín Palacios 1942: 21.

235. ha-Levi 1977: 5.

236. For brief introductions to this philosopher, see Walbridge 2005 and Ziai 1992. Walbridge 2001: 42–50 discusses as-Suhrawardī as a "Hermetic thinker," albeit without giving evidence that this philosopher ever read any work of Hermes in Arabic or Persian. As Walbridge recognizes, "Hermetic" is an almost uselessly ill-defined category.

most accomplished ancient sages ascended psychically to the higher world, and that the first among them was Hermes: "All of [the ancient philosophers] have agreed that whoever has the power to doff his body and to reject his senses ascends to the higher world. They agreed that Hermes ascended in his soul to the upper world—and so did other sky-journeyers. None of the sages lacked the disposition to doff the body and to rise up. Pay no attention to those confused materalists who imitate the philosophers. For the matter is greater than they declare it to be."[237]

In his many writings, as-Suhrawardī frequently invokes Hermes, along with Empedocles, Agathodaemon, and Plato,[238] as his antecedents, whose doctrines he claims to uphold, but without making many specific references to the known works or doctrines of these teachers. For the most part, he demonstrates no real familiarity with actual Hermetica. He merely held the conviction, apparently based on his feelings rather than on readings, that his teaching was indeed a continuation or revival of the wisdom of these ancient philosophers. For him, Hermes was "the father of the sages."[239] But he had heard about Hermes *somewhere*. In two instances, a careful reading of the passages in which as-Suhrawardī makes reference to Hermes reveals some of his sources. Once, he indirectly states that some of his information about Hermes derives from a chronicle of ancient history.[240] As already shown, Arabic chronicles were one of the main sources of information about Hermes in Arabic. In the second instance, as-Suhrawardī cites a sentence of Hermes deriving from the opening pages of the Talismanic Pseudo-Aristotelian Hermetic text *Kitāb al-Isṭamāḫīs*, in which Hermes encounters and converses with his "perfect nature." It is unclear whether he had read the passage in the original talismanic work,[241] or through an intermediary that cited the passage, such as the magical compendium *Ġāyat al-ḥakīm* (*Picatrix*).[242] Interestingly, the biographical sources on as-Suhrawardī's life allege that he was known as a practitioner of magic.[243] When

237. as-Suhrawardī, *Maǧmūʿa* 1.113.1–6 (*Kitāb at-Talwīḥāt* §76): *wa-qadi ttafaqa kulluhum ʿalā anna man qadara ʿalā ḫalʿi ǧasadihī wa-rafḍi ḥawāssihī ṣaʿida ilā l-ʿālami l-aʿlā wa-ttafaqū ʿalā anna Hirmisa ṣaʿida bi-nafsihī ilā l-ʿālami l-aʿlā wa-ġayrahū min aṣḥābi l-maʿāriǧi wa-lā yakūnu l-insānu mina l-ḥukamāʾi mā lam yaḥṣul lahū malakatu ḫalʿi l-badani wa-t-taraqqī fa-lā taltafitū ilā hāʾulāʾi l-mutašabbihati bi-l-falāsifati l-muḫabbaṭīna l-māddīyīna fa-inna l-amra aʿẓama mimmā qālū.*

238. As-Suhrawardī understands the *Theology of Aristotle*, the Arabic adaptation of Plotinus' *Enneads IV-VI*, to have been written by Plato. This includes the passage in which the philosopher doffs his body and undergoes a spiritual ascent (cited in as-Suhrawardī, *Maǧmūʿa* 1.112.6–10 = *Kitāb at-Talwīḥāt* §86).

239. as-Suhrawardī, *Maǧmūʿa*, 111.4–5 (*Kitāb at-Talwīḥāt* §86): *wālid al-ḥukamāʾ ab al-ābāʾ*.

240. as-Suhrawardī, *Maǧmūʿa*, 1.502.15 (*Kitāb al-Mašāriʿ wa-l-muṭāraḥāt*: *mašraʿ* 7, *faṣl* 6, *fī sulūk al-ḥukamāʾ al-mutaʾallihīn*): *wa-baqiya smuhū fī t-tawārīḫi*. In the same passage, as-Suhrawardī refers to Plato as part of the *ṭabaqa* of ancient Greece, and to Kayūmarṯ (Gayōmard) as Gil-šāh (see chapter 4, note 146, in this volume); these terms also suggest a chronographic source relating the serial history of nations. On this passage see Ziai 1992: 328–329.

241. As in British Library Delhi Arabic MS 1946, 4b–5a, esp. 5a2–3.

242. Ed. Ritter 1933: 187–188 (*maqāla* 3, *faṣl* 6). Wherever he had read it, he was clearly paraphrasing the passage from memory, because he does not reproduce it in its exact words.

243. Ziai 1992: 338–339.

he cites this Hermetic passage, however, it is not for magical purposes, but rather to give a warning against a naïve interpretation of his expression "lord of the species" (*rabb an-nawʿ*), by which he means something like the immaterial and luminous ideal form of a species:

> When you hear Empedocles, Agathodaemon, and others alluding to the lords of the species, understand their meaning! Do not suppose that they say that the lord of the species is a body or a bodily thing or that it has a head and two legs. When you find that Hermes says, "A spiritual entity delivered understanding to me, so I said to it, 'Who are you?' and it said, 'I am your perfect nature,'" do not think that it [the spiritual entity] is like us. Everything that is attributed to [the ancient philosophers] on this subject is not correct. The subtleties of their words give evidence of it. But the blunder is due to the translators, to the natures of the languages, and to the attribution to them of [the statements of] those who do not understand their words.[244]

As-Suhrawardī recognizes that some of the works ascribed to the ancients are, to some degree, spurious.[245] He states the evidence for his doubt somewhat vaguely, but it seems that he thought the wording in these texts was not quite right. Nevertheless, he felt himself to be independently capable of understanding them correctly. At the same time, he appears to be unwilling to declare outright that they are forgeries. They are merely "incorrect" due to problems arising from translation and interference with the text by the ignorant. Importantly, in this sole instance in which as-Suhrawardī refers directly to a work of Hermes, he comes close to rejecting it. This is his normal procedure: he claims to represent the most ancient and pure philosophy, but disregards the sources available in his day, evidently because he felt that he knew better. Such a method of appropriation of the ancient teachers is typical of the inspired as-Suhrawardī. He expresses a similar doubt elsewhere about the authenticity of the reported views of Pythagoras.[246] The myth of Hermes' spiritual ascent, which attributed to Hermes a divine philosophy revealed to him from on

244. as-Suhrawardī, *Maǧmūʿa* 1.463.17–464.4 (*Kitāb al-Mašāriʿ wa-l-muṭāraḥāt. mašraʿ* 6, *faṣl* 9, *fī iṯbāt al-ʿuqūl allatī hiya arbāb al-anwāʿ*): *fa-iḏā samiʿta Anbāḏuqlīsa wa-Aġāṯāḏīmūna wa-ġayrahumā yušīrūna ilā aṣḥābi l-anwāʿi fa-fhum ġaraḍahum wa-lā taẓunnanna annahum yaqūlūna inna ṣāḥiba n-nawʿi ǧismun aw ǧusmānīyun aw lahū raʾsun wa-riǧlāni wa-iḏā waǧadta Hirmisa yaqūlu inna ḏātan rūḥānīyatan alqat ilayya l-maʿārifa fa-qultu lahā man anti fa-qāla* (sic ed.) *anā ṭibāʿuka t-tāmmatu* (sic ed.) *fa-lā taḥmilhu ʿalā annahā miṯlunā wa-kullu mā nusiba ilayhim fī hāḏā l-bābi laysa bi-ṣaḥīḥin wa-yadullu ʿalayhi laṭāʾifu kalimātihim wa-lākinna s-sahwa waqaʿa li-n-naqalati wa-li-ṭabāʾiʿi l-luġāti wa-li-ntisābi man lā yafhumu kalāmahum ilayhim.* On the "lords of the species," see Rahman 1975: 48. Walbridge 2001: 49 seems to cite the first half of the passage, ending with the words "perfect nature," as evidence of as-Suhrawardī's positive views about Hermes. The very next sentence, however, reveals as-Suhrawardī's reservations about its apparent meaning.

245. Although the context makes it appear that he is talking specifically about the "lord of the species," it is possible that he is rather referring to talismanic works when he mentions the problematic character of the writings of the ancients on "this subject" (*hāḏā l-bāb*).

246. Walbridge 2000: 62–63; Gutas 2003: 305–306.

high, appears to have been much more important for as-Suhrawardī than any specific Hermetic work.[247] In subsequent centuries, readers of as-Suhrawardī's books were exposed to this basic concept of Hermes, and Hermes' reputation thus increased often without any connection to the Hermetica themselves. Of course, the invocation of Hermes does not make as-Suhrawardī or his successors "Hermetic" in any useful sense. What as-Suhrawardī did accomplish, however, was to make Hermes into an ideal to which *philosophers* could call upon as their antecedent.

It was around this time that someone composed a new Arabic Hermetic work more suitable for philosophers and ascetics than for astrologers and alchemists, the only substantial Arabic Hermetic text that might be called philosophical as opposed to technical. Called the *Kitāb fī Zağr an-nafs* (also known as *Muʿāḏalat an-nafs* and *Muʿātabat an-nafs*), or *The Rebuke of the Soul*, it is today the best known Arabic work attributed to Hermes. In fact, there are more printed editions of the book today than independent studies of it.[248] Although several of the manuscripts attribute the work to Plato rather than to Hermes, and one even ascribes it to Agathodaemon, it seems that the author was originally named as Hermes. It first appears in the works of the Persian philosopher Bābā Afḍaladdīn Kāšānī (d. ca 1213), also known as Bābā Afḍal, who declares it a book of Hermes-Idrīs; he also translated the whole text into Persian.[249] Because Kāšānī gives the first evidence for the existence of the *Zağr an-nafs*, and his personal correspondence shows that he was sharing the work with others, there is the great possibility that he is in fact its original author and propagator.[250]

247. Ziai 1992: 344 puts it aptly when he refers to as-Suhrawardī's "*symbolic use* made of a host of ancient figures" (emphasis added), and his "metahistorical reformulation of commonly known myths and legends of ancient nations."

248. Fleischer 1840 (partial text, German), Fleischer 1870 (partial text, Arabic ed. and German trans.), Bardenhewer 1873 (Arabic ed. and Latin trans.), Scott 1936 (= Scott 4.277–352, English trans. of Bardenhewer's Latin), Badawī 1955: 51–116 (Arabic ed.). Another edition of 1903 in Beirut is reported, but I have not seen it (Cheikho 1925: 108: "Un religieux Salvatorien, Philémon Kâteb, l'a édité à Beyrouth, en 1903"). Other studies to mention this work include Affifi 1951: 853–854 and Genequand 1987–1988 (comparing elements of it with al-Kindī's *Fī n-nafs*). The work is registered as KZN in the catalog of Arabic Hermetica under preparation.

249. For his dates, which are uncertain, see Chittick 2001: 3 and J. Rypka, "Bābā Afḍal," *EI*², 1.838b–839b. The earliest known direct references to the text occur in two Persian epistles of Kāšānī; in one he cites a few sentences from it, as words of Idrīs, in a translation not matching his later, polished rendering (Kāšānī 1952: 2.806–809; trans. Chittick 2001: 143–145); in the other he says that he has sent a copy, perhaps of the Arabic text, to his correspondent, and he encourages him to study it (Kāšānī 1952: 2.810–812; trans. Chittick 2001: 157–159).

250. Several observations provide circumstantial evidence for this hypothesis. The contents of the *Zağr an-nafs* are especially suitable to Kāšānī's philosophy, while it resembles no other earlier Arabic Hermetic text, besides the Hermetic *ḥikam* of the gnomologia. The renown of Hermes had recently been increased by as-Suhrawardī, in Kāšānī's own lifetime; this situation was perhaps conducive to new Hermetic pseudepigraphy. If the work had been available earlier, one might expect that as-Suhrawardī (d. ca 1191) would have used it or referred to it, given its doctrine. Kāšānī wrote at least one of his philosophical treatises in Arabic and then subsequently translated it into Persian (his *Madāriğ al-kamāl*; see Chittick 2001: 20); he could have done the same with the *Zağr an-nafs*. Finally, one of the earliest manuscripts of *Zağr an-nafs*, Nuru Osmaniye 4931, in which are copies of the work in *both* the

In this *Rebuke of the Soul*, Hermes addresses the soul with a series of admonitions,[251] advocating a pure, otherworldly outlook whereby the philosophical contemplation of the world of intelligibles holds the promise of salvation from this baser, material world of bodies. As noted by Bardenhewer, a number of this book's passages closely resemble teachings of the Iḫwān aṣ-Ṣafā' and of the *Theology of Aristotle* (the Arabic adaptation of Plotinus' *Enneads*), suggesting that the work was composed later, in or after the eleventh century.[252] This surmise is supported by the observation that a few passages seem also to be liberal adaptations of wise sayings of Hermes collected by the *Ṣiwān al-ḥikma* and al-Mubaššir ibn Fātik. Compare the following saying from the *Muntaḫab* of the *Ṣiwān* with the subsequent passage from *The Rebuke of the Soul.*

> [Hermes] said: He who does not know and does not act dies a single death. He who knows but does not act has double the punishment! He who knows and acts is the one who enters the kingdom of heaven.[253]

> O Soul! These are the three grades; be in the highest and most beautiful of them: The one at the lowest rank is the man who knows without acting. The metaphor for this is the man who has weapons without courage in him. What can a coward do with weapons? The second rank is the man who acts without knowing. He is like the courageous man without weapons. For how can the one who has no weapons meet his enemy? But the courageous man is more a master of weapons than the coward is of courage. Thus the one who acts without knowing is nobler than the one who knows but does not act. The third rank is the man who knows and acts. He is like the man who has courage and weapons. This should be the noble rank.[254]

The similarity between the two passages is plain to see, but there are also significant differences. The saying in the *Muntaḫab* simply frames the lesson in the context of the reward and punishment due to the person at death. The passage in *Zağr an-nafs* emphasizes the three ranks, exhorts the soul to seek the highest rank, and adds a

---

Arabic and the Persian, contains also many other writings of Kāšānī; thus the Arabic and Persian versions of the *Zağr an-nafs* were circulating inserted within a collection of Kāšānī's works just half a century after his death. A future study should test this hypothesis by comparing the Hermetic text in Arabic and Persian with the doctrines and language of Kāšānī. In any case, the work cannot have been composed much earlier than this philosopher's lifetime.

251. The genre of addressing the soul directly in this way (*yā nafs!*) is an ancient and pre-Islamic one; an early example is found in Burzōy's autobiography in *Kalīla wa-Dimna* (ed. Cheikho 1905: 30.12–31.13), which is clearly another model of the work of Hermes under discussion. Goldziher 1913 collects some further references to works of this type.

252. Bardenhewer 1873: xv–xvi and 121–140. Cf. Scott 4.278–279.

253. *Muntaḫab* ed. Dunlop 1285–1287, ed. Badawī 186.2–4.

254. My translation from Badawī 1955: 72.14–73.3. Cf. the English translation of Bardenhewer's Latin translation of the latter's 1873 edition in Scott 4.304–305 (Cap. 4, § 13). Scott's note (4.305n1) adds, "This is a piece of proverbial wisdom that has no connexion with any particular school of philosophy."

martial metaphor to incite moral courage. The most important difference is in the second part of the message: where the *Muntaḫab* discusses one who knows but does not act, *Zağr an-nafs* describes one who acts without knowing. There is at least one other clear parallel between the maxims of Hermes (this time from al-Mubaššir) and the statements of Hermes in *The Rebuke of the Soul*:

> [Hermes] was asked about an old man who had a wife. He said: How can one who cannot swim in the sea carry another on his shoulders?[255]

> O Soul! How distracted from fishing is the man drowning in water! Likewise the dweller in this world: how much he is he distracted from worldly acquisitions and pleasures with the salvation of his soul, if he becomes aware of how bad his situation in this world is! O Soul! While you are in the world, it is sufficient for you to perceive the suffering you get from your instruments,[256] their contraries and their filth.[257] Do not add another person to your instrument so that you are like the drowning man, consigned to the sea, who has borne a stone on his shoulder. I do not think that a drowning man will be saved from the sea by himself alone, so how [will he be saved] when he has taken up on his shoulder another one?[258]

The maxim attributed to Hermes in al-Mubaššir's collection delivers approximately the same point as the much more elaborate passage in *The Rebuke of the Soul*. The former emphasizes that an old man does not need a wife, with the implication, in the context of the other sayings of Hermes, that an old man should devote himself to study. The passage in the *Zağr an-nafs* is rather a stern exhortation to avoid getting wrapped up in human relationships. But both use the metaphor of the man on the verge of drowning in the sea. Moreover, both end with the same rhetorical question: if one cannot swim, how can one carry another on his shoulders? Such parallels between the *Zağr an-nafs* and the *ḥikam* of Hermes, in which the former greatly elaborates on the latter, suggest that the *Zağr an-nafs* was inspired partly by the philosophical gnomologia, and therefore was composed no earlier than the eleventh century. This agrees with Bardenhewer's hypothesis. The Persian translation by Kāšānī (d. ca 1213), as well as two references to the text that he made in his correspondence,[259] along with the two earliest dated manuscripts of the text, copied in

255. al-Mubaššir 20.21–22 = aš-Šahrazūrī 154.4–5.

256. The notion that the body is the instrument of the soul is a commonplace of ancient Greek philosophy and medicine, well known in Arabic. For several late antique examples, see Blumenthal 1990: 321–323. For Kāšānī, *instruments* referred also to the sense organs, appropriate to the passage; on the importance of these *instruments*, see Chittick 2001: 84–86 (who translates the term as "tools").

257. *wa-aḍdādihā wa-awsāḫihā*. The Persian translation has *ālāt-i ḫwad-u aḍdād-i ān* (Kāšānī 1952: 353.6).

258. Badawī 1955: 79.3–8 (in *faṣl* 6).

259. Kāšānī 1952: 1.331–385. It is translated under the title *Yanbū ʿ al-ḥayāt dar muʿātabat-u naṣīḥat-i nafs*. The translation antedates all known manuscripts and is thus a crucial witness to the text; see note 250 above.

1267 and 1276, assure us that it must have been composed, at the latest, not long after 1200; the absence of any earlier references suggests a late twelfth-century composition at the earliest.[260]

In subsequent centuries, this *Rebuke of the Soul* found an especially wide readership. There are at least twenty-four surviving manuscripts of the Arabic Hermetic text, more than for any other with one exception (a work of Hermes on astral omens at the annual heliacal rising of Sirius). Some of the manuscripts were transmitted by Christian monks and clerics as well as Muslims; at least one manuscript of the text is preserved in Karšūnī (Arabic written in the Syriac script of Christians), and others bear the distinctive signs of Christian copyists. This demonstrates yet again that the appeal of the ancient sage Hermes in Arabic was not restricted to Muslims. Any Arabic reader seeking affirmation of and a protreptic toward a philosophically informed otherworldly attitude could find it in the *Rebuke of the Soul*.

The thirteenth-century successors of as-Suhrawardī maintained the same high opinion of Hermes. Late in the thirteenth century in Iran, his follower Šamsaddīn Muḥammad aš-Šahrazūrī copied the entirety of al-Mubaššir's sayings of Hermes, including *The Counsel of Hermes to Ammon*, into his own history of philosophy.[261] Hermes was for the Išrāqīs, as for the early Ismāʿīlīs, a founding figure as well as a prophet, their ancient predecessor and model in the search for truth. Similarly the Ṣūfī philosopher Ibn Sabʿīn (d. ca 1269–1271) considered Hermes to be his forebear. The first words after the opening *ḫuṭba* of his philosophical summa *Kitāb Budd al-ʿārif* are, "I prayed to great God to disclose the wisdom expressed in symbols (*ramazahā*) by the Hermeses of the first aeons (*Harāmisat ad-duhūr al-awwalīya*)."[262] In one of his poems, aš-Šuštarī (d. 1269 in Damietta), an Andalusian disciple of Ibn Sabʿīn, whom he met in Biğāya, follows him in putting "all the Hermeses" (*al-Harāmis kulluhum*)[263] at the head of his intellectual forebears in a list including not only Plato and Aristotle, but also as-Suhrawardī, Ibn Masarra, Ibn Sīnā, and Ibn Rušd.[264] None of these authors appears to have made much use of the actual works of Hermes in Arabic, the great majority of which are technical treatises on astrology, alchemy, and talismans. Instead, it was the myth of Hermes that appealed to them.

The thirteenth century also saw the composition of new bibliographical histories of science, based on earlier sources, in which Hermes figured prominently, such as those of Ibn al-Qifṭī and Ibn Abī Uṣaybiʿa, mentioned in chapter 4. These histories

260. See notes 249 and 250 above. The two manuscripts are Gotha 82 and Nuru Osmaniye 4931. The work is also attested fairly early by Ibn Abī Uṣaybiʿa (wr. 1268), who adds a *Kitāb Muʿātabat an-nafs* near to the end of his list of Plato's (not Hermes') works (1.54.3), and by the Copt Abū l-Barakāt Ibn Kabar (d. ca 1323–1335), who mentions it as a work of Hermes in his *Miṣbāḥ aẓ-ẓulma fī īḍāḥ al-ḫidma* (Bardenhewer 1873: xiv–xv; Badawī 1955: intro. 40).

261. aš-Šahrazūrī 139.3–160.3.

262. Translation adapted from Sirriyeh 2005, 11.

263. aš-Šuštarī 1960: 84.16.

264. Massignon 1949: 47–48.

of science made the integration of Hermes into the ranks of the philosophers, physicians, and other scholars, begun in the gnomological compendia, irreversible. They blended al-Mubaššir's Hermes with Abū Maʿšar's account of the three Hermeses so that it would be difficult to distinguish the original parts. For example, Ibn al-Qifṭī's *History of the Sages* puts this under the entry for Ammon: "*Ammon, King and Sage*: This is an epithet of his; his real name was Basīlūḫus. He is one of the four kings who took wisdom from the first Hermes. Hermes had made him governor over one quarter of the earth. This Ammon was counted among the philosophers, although nothing of his words have been translated into Arabic. When King Hermes made him governor, he bequeathed counsels to him, some of which have been excerpted and translated, including that he said..."[265] The *Testament of Hermes to King Ammon* immediately follows (as abbreviated by az-Zawzanī, who is responsible for shortening Ibn al-Qifṭī's work into the form extant today).

The treatment of Hermes here is a reworded version of the account by al-Mubaššir with a few slight but important modifications. First, we are told that the Hermes who counseled Ammon was specifically the *first* Hermes. Indeed, Ibn al-Qifṭī gives the stories of all three Hermeses, derived from Abū Maʿšar, elsewhere in his book. Second, Hermes here relates his counsel to Ammon at a particular time: when he made him governor. Ibn al-Qifṭī could have inferred this last piece of information from the contents of Hermes' counsel. Third, Hermes is himself a king, perhaps following the tradition of Ibn Nawbaḫt, according to whom the ancient Babylonian Hermes became king of Egypt. In this short text we can see the integration of numerous strands, each of separate origin, out of which the medieval Arabic Hermes was fashioned by well-meaning scholars who collected all the data at their disposal about the ancient Hermes and uncritically accepted them in synthesis.

Other offshoots of the gnomological tradition easily made their way into more popular literature, for one does not require technical training to understand the moral teachings of the *Ṣiwān al-ḥikma* or al-Mubaššir's *Muḫtār al-ḥikam*. This was the case with the European translations of al-Mubaššir, a work known in many languages and excerpted in later works. If a large portion of the *Counsel of Hermes to King Ammon*, along with other Hermetic sayings derived from al-Mubaššir, could survive in such florilegia as Palfreyman's *Treatise of Morall Philosophie*, printed in London in the early seventeenth century,[266] it should not be surprising to find Hermes passing from the Arabic wisdom literature and histories of science also into Persian song. This corresponds chronologically with the increasing efflorescence of Persian as a vehicle of literature.

265. Ibn al-Qifṭī 7.5–10. *Ammūnu l-maliku l-ḥakīmu. hāḏā laqabun lahū wa-smuhū l-ḥaqīqīyu bsylwḫs wa-huwa aḥadu l-mulūki l-arbaʿati llaḏīna aḫaḏū l-ḥikmata ʿan Hirmisa l-awwali wa-kāna Hirmisu qad wallāhu rubʿa l-arḍi wa-kāna Ammūnu hāḏā maʿdūdan fī l-ḥukamāʾi illā annahū lam yuḫraǧ min kalāmihī šayʾun ilā l-ʿarabīyati wa-lammā wallāhu Hirmisu l-maliku awṣāhu bi-waṣāyā ḫurriǧa baʿḍuhā wa-turǧima minhu annahū qāla...*

266. Palfreyman ca 1635: 60a–b.

The renowned poet Niẓāmī (d. 1209) of Ganǧa, Āḏarbayǧān, adopted Hermes, along with other ancient Greek sages—whom he learned about through al-Mubaššir's book, among other sources—into his epic romance of Alexander in verse, the *Šaraf-nāma* and its sequel *Iqbāl-nāma*.[267] In the *Iqbāl-nāma*, Niẓāmī relates tales of a timeless past in which King Alexander of the Greeks promotes the sciences and summons the philosophers together for disputation and instruction. Hermes appears before Alexander in a learned session as one of seven sages:

> Aristotle, who was the kingdom's vizier, young Apollonius, and old Socrates,
> Plato, Valens, and Porphyry, whose hands the Holy Spirit had kissed,
> And seventh of all, Hermes of good counsel, who had resided in the seventh heaven.[268]

Although it may appear strange to have Hermes side-by-side with philosophers of different periods such as Vettius Valens, Aristotle, and Porphyry, they are present not as figures in a historical narrative, but as representatives of ancient Greek wisdom in an idealized golden age of learning. Apollonius is young not because of any chronographic concern, but only because he appears as such in the autobiographical introduction of the pseudo-Apollonian *Sirr al-ḫalīqa*. Socrates is old because that is how he was idealized in the wisdom literature. Moreover, Niẓāmī had a literary model for the meetings of the ancient sages in works such as the alchemical *Muṣḥaf al-Ǧamāʿa* (*Turba Philosophorum*), written ca 900, in which the ancient philosophers gather for the exchange of views and discuss the nature of the material world.[269]

In another episode in Niẓāmī's *Iqbāl-nāma*, an assembly of Greek philosophers becomes jealous of Hermes' unparalleled knowledge. "Because of [his] many statements on the secrets of the Spirits, the Greeks envied him" (*zi bas-i guftan-i rāz-i rūḥānīyān bar ū rašk burdand Yūnāniyān*). The seventy other sages agree to spurn him in their learned sessions and ignore him, for which Hermes curses them: they are frozen in place until they die. Alexander heartily approves of this act of Hermes, who had remained intact in the midst of the "flood" (*ṭūfān*) of the seventy others. The tale is a lesson to those who do not heed the wise.[270]

The learned allusions that Niẓāmī makes—to Hermes' ascent to the celestial sphere of Saturn (encountered already in the *Epistles* of the Iḫwān aṣ-Ṣafāʾ: section 5.1 in this volume), to Hermes' teachings surviving the Flood (known from Abū Maʿšar: section 4.1), and to his expertise in the Spirits (*rūḥānīyāt*, as in the Talismanic

267. Southgate 1978: 173–179 and 185. P. Chelkowski, "Niẓāmī Gandjawī," *EI*², 8.76b–81a.
268. Niẓāmī 2001: section 80(19), p. 132 ll. 11–13.
269. On which, see Plessner 1954b.
270. Niẓāmī 2001: section 75(14), pp. 102–105.

Pseudo-Aristotelian Hermetica)—tell the educated listener that the poet does indeed know what the scholars have to say about Hermes. Such nods to erudition are typical of his poetry. But Niẓāmī's purpose was primarily to entertain and to edify, not to teach an obscure history. The figures of Alexander, Hermes, and the rest are for him merely vehicles for conveying allegorical higher truths.

The Persian poet Maḥmūd-i Šabistarī of Āḏarbayǧān used the heavenly ascent of Idrīs, without naming him as Hermes, in his famous *Gulšān-i rāz* (*Rosegarden of the Secrets*), composed in 1317: "He becomes pure of reproachable acts, like the prophet Idrīs in the fourth celestial sphere."[271] For Šabistarī, the prophet's journey to the heavenly sphere of the sun is a metaphor for the purification required for the successful spiritual journey to perfection that his work describes.

Jāmī (d. 1492), the celebrated and learned poet of Harāt, took Niẓāmī's *Iqbāl-nāma* for his model in composing his own book of wisdom in Persian verse, the *Ḫirad-nāma-yi Sikandarī* or "Alexandrine Gnomologium."[272] Jāmī adopts the same fictive setting, in which all the ancient Greek philosophers relate their maxims and teachings in Alexander's court. Here again Hermes appears with his own short section of two pages, the *Ḫirad-nāma-yi Hirmis*, out of the one hundred pages of the entire edited poem. Yet Hermes' wisdom here, like that of the other sages in the work, is more suitable for the Ṣūfism of Jāmī, who studied under a teacher in the Naqšbandī order. It bears no real resemblance to the Greek *Corpus Hermeticum*, published in Latin translation not long before, in 1471, in Italy. At just that time Jāmī was setting off on his pilgrimage to Mecca. Although reportedly invited to the court of Mehmet II the Conqueror at Istanbul, he returned to his home in what is today Afǧānistān, where he would write this *Ḫirad-nāma-yi Sikandarī* in about 1485.[273]

Hermes' fame thus reached far beyond the special disciplines of astrology and alchemy. Through the compilation of information over successive generations, Hermes gained a reputation in a variety of other genres of writing, including heresiography, wisdom literature, philosophy, works on ṣūfism, history, biography, philosophy, and even poetry. Knowledge about the ancient Hermes, as with any number of other things, became an important token of learning. Informed by all these traditions, in which a concept of Hermes was blended from separate strands into one fabric, later authors simply accepted that Hermes was an ancient prophet as well as the first philosopher. It was not always necessary even to mention that Hermes was the same as the Qur'ānic prophet Idrīs. That was common knowledge. For example, one short Arabic Hermetic manuscript begins "A treatise composed by Hermes—peace be upon him—to his son Ṭāṭā."[274] Muslims

271. Whinfield 1880: 30.333: *zi af'āl-i nakūhīda šud pāk ču Idrīs-i nabī dar čārom-i aflāk.*

272. Southgate 1978: 179–181.

273. Cl. Huart and Ḥ Massé, "Djāmī," *EI*², 2.421b–422b; Heer 1979: 1–4.

274. Istanbul University MS A 6156, 88a12: *Risālatun ṣannafahā Hirmisu 'alayhi s-salāmu li-waladihī Ṭāṭā*. This alchemical work claims to be a translation from Greek by the Šī'ite imām Ǧa'far aṣ-Ṣādiq.

customarily reserve the benediction "peace be upon him" for prophets. Here the name Hermes, without any other qualifier, warranted that recognition. The ancient Egyptian god of scribes, Thoth, identified in antiquity with the Olympian Hermes, had been transformed after long centuries, and through many omissions and accretions of tradition, into a philosopher-prophet widely recognized by Muslims.

Conclusion

The Making of the Arabic Hermes

Much of this book has been concerned with clearing the overgrown fields of speculation about the Arabic Hermetica. This has required lengthy analyses and close readings, along with a fair amount of detailed, source-based debunking of historical errors, so much so that the sum of positive gains made may not be readily apparent. In fact, much has been established throughout, and it is now time to summarize these advances in a synthesis: what is clearly and confidently known about the figure of Hermes and his myth in Arabic.

Arabic scholars' understanding of Hermes had its earliest roots in the previous Sasanian reception of Hermetica in Middle Persian translation. Although the precise character and complete contents of these Middle Persian Hermetica are not entirely known, there is no doubt that they existed and that they were, to a large extent, if not exclusively, astrological. When Iranian astrologers of the caliphal court began to write Arabic works in the second half of the eighth century, they already knew Middle Persian translations of Greek Hermetica concerned with their discipline. Ibn Nawbaḫt attests to a pre-Islamic Persian tradition alleging that the ancient Egyptian sage Hermes was originally from Babylon, and that he had taught the sciences of Iran to the Egyptians. The adoption of Hermes as Babylonian made him part of a complex of narratives well known among scholars in Sasanian Iran designed to promote the acceptance of "foreign" sciences as part of an ancient Iranian heritage resuscitated by the early Sasanian emperors themselves.

A second tradition about Hermes as a teacher of astrology comes from Ḥarrān. The polytheists in Byzantine Ḥarrān around 600, like other Hellenized polytheists up until that time, had revered Hermes Trismegistus as one of many ancient philosophers, their pagan holy men. The development of their beliefs in the seventh and eighth centuries is unknown, but by the beginning of the ninth century, the status of Hermes among the Ḥarrānians had greatly increased. Theodore Abū Qurra is the first to attest to Hermes as the *prophet* (*nabīy*) of the planet-worshipping Ḥarrānian pagans. From this testimony it is clear that Hermes had developed this status at Ḥarrān already before the arrival in Baġdād of Ṯābit ibn Qurra (ca 826–901), and indeed before Ṯābit was born. Nevertheless, no Arabic Hermetic text has been shown yet definitely to have been handed down from antiquity by the Ḥarrānians. For all we know, Hermes may have been acknowledged as their prophet without recourse to any lively tradition of Hermetic *texts* at Ḥarrān.

Al-Kindī (d. ca 870) testifies that some Hermetic dialogues, either those known today as the *Corpus Hermeticum* or related works now lost, did exist in an Arabic translation by the mid-ninth century. He says that he examined *maqālāt*, chapters or sections of a book, that Hermes wrote to his son on the subject of theology. He explicitly states that the work is regarded as authoritative among the Ḥarrānians, but he does *not* say that he received the text *from* a Ḥarrānian. Four hundred years later, the historian Ibn al-Qifṭī had read about Hermes' "chapters (*maqālāt*) to his disciple Ṭāṭī by way of question and answer between the two." He said that "they lack organization and continuity because the original was decayed and divided into separate parts." In Chapter 3 it was assumed that these are the same *maqālāt* of Hermes that al-Kindī saw. Ibn al-Qifṭī was probably just repeating a much earlier report; these Hermetic dialogues were in all likelihood lost in Arabic by his time. A few decades later, however, Ibn al-ʿIbrī claimed that it was extant in Syriac. Perhaps one day the Arabic or Syriac text will come to light in an overlooked manuscript. Only then would it be possible to assess fully any debt owed by Arabic philosophers to the Greek Hermetic dialogues. Given what is now known about Arabic philosophy, overwhelmingly based as it is on Aristotle, Ptolemy, and Galen, with an infusion of Plotinus, these Hermetic chapters appear to have been almost unknown, and we should expect such a debt, if it existed, to be minor. Otherwise they would not have totally evaded the detection of modern scholars.

Around the same time, the middle of the ninth century, the astrologer Abū Maʿšar of Balḫ designed his chronological framework for historical astrology. In this work he drew on both the Iranian and the Ḥarrānian accounts of Hermes. He also had a Christian chronographic source containing information about Hermes. This was derived from the antediluvian history of the Alexandrian monk Annianus. With these collected materials, and one more report about Hermes taken from al-Kindī, his contemporary at Baġdād, Abū Maʿšar composed a legend of three different figures named Hermes: the first, antediluvian Hermes, prophet of the sciences; the second, postdiluvian Hermes who recovered the inscribed teachings of the first; and a third Hermes of later times, an Egyptian sage. Although the book in which Abū Maʿšar created this synthesis is lost, its treatment of Hermes remained decisively influential, effectively the standard account of Hermes, cited with considerable variations for centuries thereafter. Most later historians in Arabic derived some of their information about Hermes from Abū Maʿšar.

In the tenth century, an unidentified scholar, supposed here to be a Ḥarrānian Ṣābian employed at the court at Baġdād, composed a different and implicitly apologetic account of Hermes, elaborating on his role as the prophet of the pagan religion. This author, who knew from some ancient source that Tat, Asclepius, and King

Ammon were disciples of Hermes, nevertheless did not borrow from any known Hermetic texts in composing this work, but had free recourse to existing Arabic gnomologia translated from Greek, Middle Persian, and even Sanskrit. He borrowed and adapted sentences from these diverse ancient sources to produce instructions of Hermes suitable for the learned courtly audience of his own day. Hermes is portrayed here as the founder of *al-milla al-ḥanīfīya*, roughly "the pagan religious community." This religion, which, the text claims, was originally followed by all peoples of the world, is characterized as monotheistic, not polytheistic. Hermes has a *sunna* and a *šarīʿa* and uses the terms of Islamic law. King Ammon receives from Hermes instructions suitable for a caliph. The wise commands of Hermes teach a generic piety acceptable to all. This work is lost in its original form, but in Egypt, in 1048, al-Mubaššir ibn Fātik excerpted and adapted a large section of it for his gnomology of ancient sages. Although he or an intermediary abbreviated the account of Hermes' life and religious teaching, he appears to have taken over the maxims of Hermes entire. Al-Mubaššir's work was subsequently highly successful. Through it, this account of Hermes became a new source of information that later historians blended freely with Abū Maʿšar's accounts of three Hermeses. Together, the two accounts from Abū Maʿšar and al-Mubaššir remained the bases of information about the figure and history of Hermes within Arabic literary tradition and its offshoots until relatively recent times.

Al-Mubaššir's work was also translated into Spanish and Latin in the thirteenth century, into French around 1400, and into Provençal and English in the fifteenth century. Through these translations, al-Mubaššir's account of Hermes and his teachings, including Hermes' *Testimony to King Ammon*, became known popularly in Europe, albeit in still more abbreviated form. It remains to be shown what role these versions of the text had, if any, in the later Western European reception of the Hermetic dialogues of the Greek *Corpus Hermeticum*. There is at least one example of its reception in the florilegium of William Baldwin (1547), which combined Ficino's and al-Mubaššir's accounts of Hermes together.

Along with al-Mubaššir's *Muḫtār al-ḥikam*, the *Ṣiwān al-ḥikma*, originally composed around 1000 but known now only from several abridgments and derivatives, also helped to popularize Hermes as a source of wise sayings. The story of Hermes found here is based on Abū Maʿšar's. Between these two important gnomological histories of philosophy, Hermes was assured an important place in many future Arabic gnomologia of the ancient philosophers and in the Arabic histories of philosophy in general, where he is often mentioned as the first scholar, or one of the first, to exist. Later philosophers, and would-be philosophers, could claim Hermes as their pristine forebear and the founder of their way.

In the 840s, al-Ǧāḥiẓ had already heard a claim that Hermes was Idrīs-Enoch, the prophet who had ascended to the heavens to behold visions and to learn secrets from the angels. Not long afterward, Abū Maʿšar's account in the *Thousands* stated definitely that Hermes was indeed Idrīs-Enoch, and the identification of the two was generally accepted from then on. By the tenth century, the identification of Hermes with Enoch had given rise to a myth about the heavenly ascent of Hermes, through which Hermes learned from the angels the secrets of his natural science and astrology. In the first decades of the tenth century, authors working in the Ismāʿīlī mission in Qayrawān and in northern Iran both already knew and accepted a form of this story, and used it to further their argument that all human knowledge was built from scientific principles obtained originally through revelation, not reason. The appearance of this idea in lands far apart make it appear likely that it was propagated through the Ismāʿīlī *daʿwa* in the late ninth century, while its headquarters were still in Syria. Several decades later, the myth of Hermes' ascent became better known through its treatment in the *Book of the Apple*, a widely read adaptation of Plato's *Phaedo*.

From these and other sources, the Arabic myth of Hermes Triplicate-in-Wisdom reached scholars of various inclinations from Central Asia and India to Andalusia. An overview of those who used and referred to works of Hermes from late antiquity onward shows that they include pagans, Christians, Jews, Magians, Manichaeans, and Muslims. Members of all of these groups had recourse to Hermes as an authority. What made this possible was his identification as a sage of primordially ancient times, before the founders of the other religions. The antediluvian authority of the Arabic Hermes was a direct inheritance from the late antique Greek chronographic tradition. Thanks to this estimation of his date, Hermes' name was not exclusive to any sect. He was a universally respected teacher. Yet Hermes would not have acquired the specific character of a prophet if it had not been for the Ḥarrānian Ṣābians, descendants of the ancient pagans, living in a milieu in which every sect required a prophetic founder. Even after the Ṣābians had entirely disappeared, Hermes was remembered as the prophet of paganism and the founder of the sciences. This was accepted by Abū Maʿšar, and immediately after him by the Ismāʿīlī *daʿwa*, and eventually by scholars everywhere.

The philosophical dialogues of Hermes Trismegistus, although so influential in Europe, were practically unknown in premodern Arabic. Instead, Hermes was famous more strictly as a teacher of astrology, alchemy, and talisman making. This is reflected in the hundreds of Arabic manuscripts of works on these subjects attributed to Hermes extant today. So far, surviving Greek antecedents have been discovered for only two of these Arabic Hermetica: the *Kyranis* (Arabic *Ǧiranīs*)[1] and the *Kitāb Asrār an-nuǧūm*, translated from a Middle Persian translation of a Greek original source shared with Rhetorius.[2] Future research may well discover extant Greek antecedents for more of the Arabic Hermetica. By contrast, some of these texts are rather original compositions in Arabic, written with a knowledge of the legends of the Hermes described above, as with the alchemical *ar-Risāla al-falakīya*

1. Toral-Niehoff 2004.
2. Kunitzsch 2001.

al-kubrā (RFK) and *The Rebuke of the Soul* (KZN). In the richly developed philosophical and magical Arabic pseudepigrapha of the ninth and tenth centuries, Hermes might be described as the prince of the "pseudos." His works are cited by pseudo-Aristotle, pseudo-Apollonius of Tyana,[3] and by authors writing under the alchemical pseudonym Ǧābir ibn Ḥayyān.[4] Characteristic of these Hermetica first composed in Arabic are some of the motifs common also to ancient Hermetic texts, but with their own distinctive character: they claim to have been discovered on books or stelae in underground tunnels or in Egyptian ruins, or to be translations of ancient "hidden books"; they ask their readers to guard the texts from the unworthy; they teach techniques for deriving magical powers from the occult properties of terrestrial things and their celestial sympathies. Several of them are decorated with cryptographic scripts, at times in imitation of Egyptian hieroglyphs. Much work is required to explain the interrelationships of all these Hermetic and pseudepigraphical texts. Editions of the texts will be an inevitable prerequisite for any definite conclusions. As editions become available, it should be possible to trace roughly the origins of the texts through citations of them by known authors, as well as their citations of each other, and to establish their chronology more exactly. It may well be that a large part of this Arabic pseudepigraphical literature, and not only the works Hermes, is due to a small number of individuals. If historians are lucky, it will be possible eventually to discover clusters of pseudepigraphical texts of common authorship, and even to identify individual authors. The search for authorship is not as unimportant as many interpreters of literature today think: it matters whether a book was written in the eighth or the tenth or the twelfth century, or in Toledo, Cairo, or Baġdād. Finding the source often means finding the context, and context provides a basis for interpretation. Without a context, these texts are more likely to appear as floating representatives of a hidden "school of thought," that vague "Islamic Hermeticism" that we should avoid using as the basis of discussion from now on.

When more of the Arabic texts are available, it will also be possible to tell even more clearly the history of their use and reception in later times. For now it is certain that Arabic Hermetica were used and cited in Arabic tradition continuously at least to the eighteenth century, and manuscripts of Arabic Hermetica continued to be copied down to the end of the nineteenth century. Translations of Arabic Hermetica are found in Latin, Persian, Hebrew, and other languages, in each case illustrating the appeal of the ancient Egyptian sage, and the promise of the teachings he offered, across confessional, ethnic, and regional boundaries. The development of the Arabic Hermetica, along with the Arabic myth of Hermes, came to an end only relatively recently, with European colonization, the adoption of the printing press, and modern systems of education, all of which together have effectively

3. Weisser 1980 (index pp. 245–246).

4. Kraus 1942–1943 (index: 1.212).

ended the Arabic manuscript tradition as a whole—though not yet absolutely everywhere.

This study has touched on a large number of Arabic authors who had something to say about Hermes. Well-known examples include al-Ǧāḥiẓ, Abū Maʿšar, the two Rāzīs Abū Ḥātim and Abū Bakr, Agapius of Manbiǧ, Sinān ibn Ṯābit, al-ʿĀmirī, Ibn Ǧulǧul, the Iḫwān aṣ-Ṣafāʾ, Miskawayh, al-Mubaššir ibn Fātik, Ibn Bāǧǧa, Yahūdā ha-Levi, as-Suhrawardī, aš-Šahrastānī, Mūsā ibn Maymūn, Ibn al-Qifṭī, Ibn Abī Uṣaybiʿa, Ibn al-ʿIbrī, Ibn Sabʿīn, aš-Šuštarī, and aš-Šahrazūrī, not to mention poets writing in Persian such as Niẓāmī and Jāmī. Even while excluding the very numerous astrologers and alchemists who continued to cite Hermes as an authority in works within their special disciplines (as practically all of them did), and the countless anonymous manuscript copyists who preserved the Arabic Hermetica, often in countries where Arabic was not the prevailing spoken language, this roster suffices to prove how widespread and influential the myth of Hermes became from the ninth to the fifteenth century over thousands of miles of territory. Many of these authors—who made their living as physicians, secretaries, missionaries, merchants, and still other ways—would have disagreed, if they could have met (as some actually did), on fundamental philosophical and religious issues. What they all have in common is some kind of interest in ancient Greek learning. That interest manifested in different ways in each of them: some of them are historians of philosophy, some of them philosophers, some of them Ṣūfīs distinguished by a wish to participate in the transmission of an idealized ancient wisdom. In a few of them, all of these interests overlapped. What these authors were *not* is "Hermetists." Their references to Hermes demonstrate not membership in a Hermetic school or sect, but merely their shared participation in and knowledge of Arabic scholarship concerned with the recovery and preservation of ancient science and wisdom. In this Eastern classical tradition, rooted in late antiquity, Hermes Triplicate-in-Wisdom came to play an important role as a figure of legend, an ideal of wisdom to be emulated, the founder of learning, a master of the spirits, the prophet of science who had touched the stars and heard the words of angels, the first sage.

Bibliography

Note: Bipartite surnames are alphabetized by their first part, regardless of traditional practice. Hence "van Bladel" is under V, "de Blois" under D, and so forth. The Arabic definite article is disregarded; for example, find al-Mubaššir under M.

ʿAbdalǧabbār ibn Aḥmad al-Asadābādī. *Kitāb al-Muġnī fī abwāb at-tawḥīd wa-l-ʿadl*, 20 vols., Cairo, Wizārat aṯ-Ṯaqāfa wa-l-Iršād al-Qawmī, 196–.

Abū l-Fidāʾ, 1831 = Abulfedae Historia Anteislamica, Arabice, ed. H. Fleischer, Leipzig, Vogel.

Abū Maʿšar. *The Great Introduction to the Science of Astrology / Al-Mudkhal al-kabīr ilā ʿilm aḥkām al-nujūm*, facsimile of MS 1508 Carullah collection, Süleymaniye Library, Istanbul, published by the Institute for the History of Arabic-Islamic Science, Frankfurt am Main, 1985.

Acampora-Michel, Elsbeth, 2001. *Liber de Pomo / Buch vom Apfel: eingeleitet, übersetzt und kommentiert*, Frankfurt am Main, Vittorio Klostermann.

Adamson, Peter, 2000. "Two Early Arabic Doxographies on the Soul: al-Kindī and the 'Theology of Aristotle,'" *Modern Schoolman* 77, 105–125.

——, 2002. *The Arabic Plotinus*, London, Duckworth.

Adang, Camilla, 1996. *Muslim Writers on Judaism and the Hebrew Bible*, Leiden, E. J. Brill.

Adler, William, 1989. *Time Immemorial: Archaic History and Its Sources in Christian Chronography from Julius Africanus to George Syncellus*, Washington, D.C., Dumbarton Oaks.

——, 1992. "Eusebius' *Chronicle* and Its Legacy," in Harold W. Attridge and Goei Hata, eds., *Eusebius, Christianity, and Judaism*, Detroit, Wayne State University Press, 467–491.

Affifi, A. E., 1951. "The Influence of Hermetic Literature on Moslem Thought," *Bulletin of the School of Oriental and African Studies*, 13, 840–855.

Afnan, Soheil M., 1964. *Philosophical Terminology in Arabic and Persian*, Leiden, E. J. Brill.

Agapius (al-Maḥbūb) of Manbiǧ = *Kitab al-ʿUnvan: Histoire Universelle écrite par Agapius (Mahboub) de Menbidj*, in Alexandre Vasiliev, ed. and trans., *Patrologia Orientalis* 5.4, 7.4, 8.3, and 11.1, 1910–1915.

Aḫbār az-zamān, 1983. [author named as al-Masʿūdī] ed. aṣ-Ṣāwī, fifth printing, Beirut, Dār al-Andalus.

Alexander of Lycopolis = Alexandri Lycopolitani *Contra Manichaei opiniones disputatio*, ed. Augustus Brinkman, Teubner, Stutgard, 1989 (reprinted from 1895). French translation and commentary by André Villey, *Contre la doctrine de Mani*, Éditions du Cerf, Paris,

1985. English translation and commentary by P. W. van der Horst and J. Mansfield, *An Alexandrian Platonist against Dualism*, Leiden, Brill, 1974.

Alexander, Philip S., 2000. "Jewish Tradition in Early Islam: the Case of Enoch/Idrīs," in G. R. Hawting, J. A. Mojaddedi, and A. Samely, eds., *Studies in Islamic and Middle Eastern Texts and Traditions in Memory of Norman Calder* (*Journal of Semitic Studies* Supplement 12), Oxford, Oxford University Press, 11–29.

Altmann, Alexander, 1963. "The Delphic Maxim in Medieval Islam and Judaism," in Alexander Altmann, ed., *Biblical and Other Studies*, Cambridge (Mass.), Harvard University Press, 196–232. Reprinted in *Studies in Religious Philosophy and Mysticism*, London, Routledge and K. Paul, 1969, 1–40.

Ambrose, Elizabeth Ann, 1992. *The* Hermetica*: An Annotated Bibliography*, Sixteenth Century Bibliography 30, St. Louis, Center for Reformation Research.

Andersen, F. I., 1983. "2 (Slavonic Apocalypse of) Enoch," in James H. Charlesworth, *The Old Testament Pseudepigrapha, Volume 1: Apocalyptic Literature and Testaments*, New York, Doubleday.

Andreas, F. C., and Henning, W., 1933. "Mitteliranische Manichaica aus Chinesisch-Turkestan II," *Sitzungsberichte der Preussischen Akademie der Wissenschaften. Philosophisch-historische Klasse*, 292–363.

Anklesaria, B. T., 1956. *Zand-Ākāsīh: Iranian or Greater Bundahišn*, Bombay, published for the Rahnumae Mazdayasnan Sabha by Its Honorary Secretary Dastur Framroze A. Bode.

al-Anṣārī = Ḥunayn ibn Isḥāq [*sic*], *Ādāb al-falāsifa, iḫtaṣarahū Muḥammad ibn ʿAlī ibn Ibrāhīm ibn Aḥmad ibn Muḥammad al-Anṣārī*, ed. ʿAbd-ar-Raḥmān Badawī, Kuwait, Maʿhad al-Maḫṭūṭāt al-ʿArabīya, 1406/1985.

Aouad, Maroun, 1989. "Aṯāwālīs," in Pierre Goulet, ed., *Dictionnaire des Philosophes Antiques*, vol. 1, 637–639.

Asín Palacios, Miguel, 1942. "Tratado de Avempace sobre la Unión del Intelecto con el Hombre," *al-Andalus* 7, 1–47.

Athanassiadi, Polymnia, 1993. "Persecution and Response in Late Paganism: The Evidence of Damascius," *Journal of Hellenic Studies* 113, 1–29.

Back, Michael, 1978. *Die sassanidischen Staatsinschriften*, Acta Iranica 18, Textes et Mémoires VIII, Leiden, E. J. Brill.

Badawī, ʿAbdarraḥmān, 1948. *Aflūṭīn ʿinda l-ʿArab*, Cairo, L'Institut Français.

——, 1954. *al-Uṣūl al-yūnānīya li-n-naẓarīyāt as-siyāsīya fī l-Islām*, Part 1, Cairo, Maktabat an-Nahḍa al-Miṣrīya.

——, 1955. *al-Aflāṭūnīya al-muḥdaṯa ʿind al-ʿArab*, vol. 1, Cairo, Maktabat an-Nahḍa al-Miṣrīya.

Bagnall, Roger, 1993. *Egypt in Late Antiquity*, Princeton, Princeton University Press.

Bailey, H. W., 1943. *Zoroastrian Problems in the Ninth-Century Books*, Oxford, Clarendon Press.

Bar Bahlūl, *Lexicon Syriacum auctore Hassano Bar-Bahlule*, 3 vols., ed. Rubens Duval, Paris, 1888–1901 (reprinted Amsterdam, Philo Press, in 2 vols., 1970).

——, [Ibn Bahlūl, al-Ḥasan]. *Kitāb ad-Dalāʾil*, Publications of the Institute for the History of Arabic-Islamic Sciences, ed. Fuat Sezgin, Series C: Facsimile Editions, vol. 10, reproduced from MS Hekimoğlu 572, Süleymaniye Library, Istanbul.

Bardenhewer, Otto, 1873. *Hermetis Trismegisti qui apud Arabes fertur de castigatione Animae libellum*, Bonn, A. Marcus.

Bar ʿEbrāyā, see Ibn al-ʿIbrī.

Barry, Catherine, Wolf-Peter Funk, Paul-Hubert Poirier, and John D. Turner, 2000. *Zostrien (NH VIII, 1)*, Bibliothèque copte de Nag Hammadi, Section "Textes" 24, Louvain, Peeters.

Bartholomae, Christian, 1920. "Zur Kenntnis der mitteliranischen Mundarten III," *Sitzungsberichte der Heidelberger Akademie der Wissenschaften, phil.-hist. Klasse.*

Baumstark, Anton, 1922. *Geschichte der syrischen Literatur*, Bonn, A. Marcus und E. Weber.

Beck, Roger, 1991. "Thus spake not Zarathuštra: Zoroastrian pseduepigrapha of the Greco-Roman world," in Mary Boyce and Franz Grenet, *History of Zoroastrianism, Volume III: Zoroastrianism under Macedonian and Roman Rule*, Leiden, Brill, 491–565.

Beeston, A. F. L., 1962–1967. "An Arabic Hermetic Manuscript," *The Bodleian Library Record* 7.1, 11–23.

Bell, Richard, 1926. *The Origin of Islam in Its Christian Environment*, London, Macmillan.

Berthelot = Berthelot, Marcellin, 1893. *La Chimie au moyen âge*, 3 vols., Paris, Imprimerie nationale.

Berthelot, Marcellin, and Ch.-Em. Ruelle, eds., 1888. *Collection des anciens alchemistes grecs*, Paris, G. Steinheil.

Bidez, Joseph, and Franz Cumont, 1938. *Les Mages hellénisés: Zoroaster, Ostanès et Hystaspe d'après la tradition grecque*, 2 vols., Paris, Societé d'éditions.

Bielawski, Jozef, 1974. "*Phédon* en version arabe et le *Risālat at-Tuffāḥa*," in J. M. Barral, ed., *Orientalia Hispanica sive Studia F.M. Pareja Octogenario Dicata. Studia Islamica, pars prior*, vol. 1, Leiden, E. J. Brill, 120–134.

al-Bīrūnī, Abū r-Rayḥān, *Āṯār = al-Āṯār al-bāqīya ʿan al-qurūn al-ḫāliya = Chronologie orientalischer Völker von Albêrûnî*, ed. C. Eduard Sachau, Leipzig, A. F. Brockhaus, 1878; reprinted in *Islamic Mathematics and Astronomy*, vol. 30, Publications for the Institute for the History of Arabic-Islamic Science, ed. Fuat Sezgin; translated into English and annotated by C. E. Sachau as *The Chronology of Ancient Nations*, London, W. H. Allen, 1879; translation reprinted in *Islamic Mathematics and Astronomy*, vol. 31.

——, *Fihrist-i Kitābhā-yi Rāzī va-nāmhā-yi kitābhā-yi Bīrūnī*, ed. Mahdī Mohaqqeq, Tehran University Publications, 1992.

——, *al-Qānūn al-masʿūdī*, 3 vols., Hyderabad, Osmania Oriental Publications Bureau, 1954–1956.

——, *Taḥdīd nihāyāt al-amākin*, ed. P. Bolgakoff, Cairo, Maṭbaʿat Lağnat at-Ta'līf wa-t-Tarǧama wa-n-Našr, 1964. Translation by J. Ali, Beirut, 1967.

——, 1948. *Rasā'il al-Bīrūnī*, Hyderabad, Osmania Oriental Publications Bureau.

Black, Deborah L., 2005. "Psychology," in Peter Adamson and Richard C. Taylor eds., *The Cambridge Companion to Arabic Philosophy*, Cambridge, Cambridge University Press, 308–326.

Black, Matthew, ed., 1970. *Apocalypsis Henochi Graece*, Leiden, Brill.

——, 1985. *The Book of Enoch, or I Enoch: A New English Edition with Commentary and Textual Notes*, Leiden, E. J. Brill.

Blochet, E., 1909–1915. "Études sur le Gnosticisme musulmane," parts 1–4, *Rivista degli Studi Orientali*, (part 1) 2, 1909, 717–756, (part 2) 3, 1910, 177–203, (part 3) 4, 1911–1912, 47–79 and 267–300, (part 4) 6, 1914–1915, 5–67.

Blumenthal, Henry J. 1990. "Neoplatonic elements in the *de Anima* commentaries," in Richard Sorabji, ed., *Aristotle Transformed*, London, Duckworth, 305–324.

Bosworth, Clifford Edmund, trans., 1968. *The Book of Curious and Entertaining Information: The Laṭā'if al-maʿārif of Thaʿālibī*, Edinburgh, Edinburgh University Press.

——, 1996. *The* New *Islamic Dynasties. A Chronological and Genealogical Manual*, New York, Columbia University Press.

Boyce, Mary, 1968a. *The Letter of Tansar*, Rome Oriental Series XXXVIII, Rome, Istituto Italiano per il Medio ed Estremo Oriente.

——, 1968b. "Middle Persian Literature," in *Handbuch der Orientalistik* I *der nahe und der mittlere Osten* IV *Iranistik* 2 *Literatur* 1, Leiden, E. J. Brill, 31–66.

Boyce, Mary, 1975. *A Reader in Manichaean Middle Persian and Parthian*, Acta Iranica 9.

——, 1977. *A Word-List of Manichaean Middle Persian and Parthian*, Acta Iranica 9a.

Boylan, P., 1922. *Thoth the Hermes of Egypt: A Study of Some Aspects of Theological Thought in Ancient Egypt*, London, Oxford University Press.

Briant, Pierre, 2002. *From Cyrus to Alexander: A History of the Persian Empire*, trans. Peter T. Daniels, Winona Lake, Eisenbrauns.

Brittain, Charles, and Tad Brennan, 2002. *Simplicius* On Epictetus' "Handbook 1–26," Ithaca, Cornell University Press.

Brock, Sebastian, 1983. "A Syriac Collection of Prophecies of the Pagan Philosophers," *Orientalia Lovaniensia Periodica* XIV, 203–246. Reprinted in S. Brock, *Studies in Syriac Christianity*, Hampshire, Variorum, 1992.

Buck, Christopher, 1984. "The Identity of the Ṣābi'ūn: An Historical Quest," *Muslim World* 74, 172–186.

Buckley, Jorunn Jacobsen, 1992. "The Colophons in *The Canonical Prayerbook of the Mandaeans*," *Journal of Near Eastern Studies* 51.1, 33–50.

Burnett, Charles S. F., 1976. "The Legend of the Three Hermes and Abū Maʿshar's *Kitāb al-ulūf* in the Latin Middle Ages," *Journal of the Warburg and Courtauld Institutes* 39, 231–234.

——, 1981. "Hermann of Carinthia and the *Kitāb al-isṭamāṭīs*: Further Evidence for the Transmission of Hermetic Magic," *Journal of the Warburgand Courtauld Institutes*, 44, 167–169.

——, 1986. "Arabic, Greek, and Latin Works on Astrological Magic Attributed to Aristotle," in Jill Kraye, W. F. Ryan, and C.B. Schmitt, eds., *Pseudo-Aristotle in the Middle Ages: The* Theology *and Other Texts*, Warburg Institute Surveys and Texts XI, London, The Warburg Institute, University of London, 84–96.

——, 1996. "The *kitāb al-Isṭamāṭīs* and a Manuscript of Astrological and Astronomical Works from Barcelona (Biblioteca de Catalunya, 634)," in *Magic and Divination in the Middle Ages*, Ashgate, Variorum (article VII, 19 pages).

——, 2001a. "Aristoteles/Hermes: Liber Antimaquis," in Paolo Lucentini et al., eds., *Hermetis Trismegisti Astrologica et Divinatoria*, Turnhout, Brepols, 179–221.

——, 2001b. "The Establishment of Medieval Hermeticism," in Peter Linehan and Janet L. Nelson, eds., *The Medieval World*, London, Routledge, 111–129.

——, 2002. "Remarques paléographiques et philologiques sur les noms d'anges et d'esprits dans les traités de magie traduits de l'arabe en latin," *Mélanges de l'École Française de Rome. Moyen Âge* 114, 657–668.

——, 2004. "Lunar Astrology: The Varieties of Texts Using Lunar Mansions, with Emphasis on *Jafar Indus*," *Micrologus* 12, 43–133.

——, 2007. "Ṭābit ibn Qurra the Ḥarrānian on Talismans and the Spirits of the Planets," *La Corónica* 36, 13–40.

Burnett, Charles, and Ahmed al-Hamdi, 1991–1992. "Zādānfarrūkh al-Andarzaghar on Anniversay Horoscopes," in *Zeitschrift für Geschichte der arabisch-islamischen Wissenschaften* 7, 294–399.

Burnett, Charles, and David Pingree, 1997. *The* Liber Aristotelis *of Hugo of Santalla*, ed. Charles Burnett, Jill Kraye, and W. F. Ryan, Warburg Institute Surveys and Texts 26, London, Warburg Institute.

Burnett, Charles, and Keiji Yamamoto, 2000. *Abū Maʿšar on Historical Astrology: The Book of Religions and Dynasties (On the Great Conjunctions)*, 2 vols., Leiden, Brill.

Carra de Vaux, Bernard, 1900. *Avicenne*, Paris, Félix Alcan.

——, 1902. *Gazali*, Paris, Félix Alcan.

Cassius Dio, 1895–1901. *Historiarum Romanarum quae supersunt*, ed. U. P. Boissevain, 3 vols., Berlin, Weidmann.

CCAG = *Catalogus Codicum Astrologorum Graecorum*, 12 vols., ed. F. Cumont et al., Brussels, 1898–1953.

Cereti, Carlo G., 2001. *La Letteratura Pahlavi*, Milan, Mimesis.

CH = *Corpus Hermeticum*.

Chabot, J. B., ed., 1927–1933. *Incerti Auctoris Chronicon Pseudo-Dionysianum Vulgo Dictum* [= *Zuqnin Chronicle*] 2 vols., Corpus Scriptorum Christianorum Orientalium 91 and 103 (Scriptores Syri 43 and 53), Paris.

Chassinat, Emile, 1934–. Le Temple de Dendara, 11 vols., Cairo, Institut français d'archéologie orientale.

Cheikho, Louis, 1905. *La version arabe de* Kalîlah et Dimnah, *d'après le plus ancien manuscrit arabe daté*, Beirut, Imprimerie Catholique.

——, 1925. "Catalogue raisonné des manuscrits de la bibliothèque orientale IV: Philosophie et écriture sainte," *Mélanges de l'Université Saint-Joseph* 10, 107–179.

CHI = *Cambridge History of Iran*, 6 vols., Cambridge, Cambridge University Press, 1968–1986.

Chittick, William C. 2001. *The Heart of Islamic Philosophy: The Quest for Self-Knowledge in the Teachings of Afḍal al-Dīn Kāshānī*. Oxford, Oxford University Press.

Chuvin, Pierre, 1990. *A Chronicle of the Last Pagans*, trans. B. A. Archer, Cambridge (Mass.), Harvard University Press.

Chwolsohn, Daniel, 1856. *Die Ssabier und die Ssabismus*, 2 vols., St. Petersburg, Kaiserlichen Akademie der Wissenschaften.

Ciancaglini, Claudia A., 1998. "Gli Antecedenti del *Romanzo* Siriaco di Alessandro," in G. Bolognesi and U. Cozzoli, eds., *L'eredità classica nel mondo orientali nell'età tardoantica e medievale. Il "Romanzo di Alessandro" e altri scritti. Atti del Seminario internazionale di studio (Roma-Napoli, 25–27 settembre 1997)*, Alessandria, Orso, 55–93.

——, 2001. "The Syriac Version of the Alexander Romance," *Le Muséon* 114, 121–140.

Conrad, Lawrence I., 1990. "Theophanes and the Arabic Historical Tradition: Some Indications of Intercultural Transmission," *Byzantinische Forschungen* 15, 1–44.

——, 1992. "The Conquest of Arwād: A Source-Critical Study in the Historiography of the Early Medieval Near East," in Averil Cameron and Lawrence I. Conrad eds., *The Byzantine and Early Islamic Near East I: Problems in the Literary Source Material*, Studies in Late Antiquity and Early Islam 1, Princeton, Darwin, 317–401.

Copenhaver, Brian P., 1990. "Natural Magic, Hermetism, and Occultism in Early Modern Science," in David C. Lindberg and Robert S. Westman, eds., *Reappraisals of the Scientific Revolution*, Cambridge, Cambridge University Press, 261–301.

——, 1992. *Hermetica. The Greek* Corpus Hermeticum *and the Latin* Asclepius *in a New English Translation with Notes and Introduction*, Cambridge, Cambridge University Press.

Corpus Hermeticum, 4 vols., eds. A. D. Nock and A. J. Festugière, Paris, 1946–1954.

Cowley, A. E., 1923. *Aramaic Papyri of the Fifth Century B.C.*, Oxford, Clarendon.

Croke, Brian, 1982. "The Originality of Eusebius' *Chronicle*," *American Journal of Philology* 103.2, 195–200.

Daftary, Farhad, 1990. *The Ismāʿīlīs: their history and doctrines*, Cambridge, Cambridge University Press.

Daiber, Hans, 1984. "Der Ṣiwān al-ḥikma und Abū Sulaimān al-Manṭiqī as-Siǧistānī in der Forschung," *Arabica* 31.1, 36–68.

Dandamaev, Muhammad A., 1992. *Iranians in Achaemenid Babylonia*, Columbia Lectures on Iranian Studies 6, ed. Ehsan Yarshater, Costa Mesa and New York, Mazda.

Daniels, Peter T., 1996. "The Study of Writing Systems," in Peter T. Daniels and William Bright, eds., *The World's Writing Systems*, Oxford, Oxford University Press, 3–17.

Dannenfeldt, K. H., 1960. "Hermetica Philosophica," with supplements by Marie-Thérèse d'Alverny and Theodore Silverstein, in *Catalogus Translationum et Commentariorum: Mediaeval and Renaissance Latin Translations and Commentaries. Annotated Lists and Guides*, vol. 1, ed. Paul Oskar Kristeller, Washington, D.C., Catholic University of America Press.

Darling, Linda. 2002. "'Do Justice, Do Justice, for That Is Paradise': Middle Eastern Advice for Indian Muslim Rulers," *Comparative Studies on South Asia, Africa, and the Middle East* 22, 3–19.

Daryaee, Touraj, 2002. *Šahrestānīhā ī Ērānšahr: A Middle Persian Text on Late Antique Geography, Epic, and History*, Costa Mesa, Mazda.

Davidson, Herbert A. 1992. *Alfarabi, Avicenna, and Averroes, on Intellect: Their Cosmologies, Theories of the Active Intellect, and Theories of Human Intellect*, Oxford, Oxford University Press.

de Blois, François, 1995. "The Sabians (Ṣābi'ūn) in Pre-Islamic Arabia," *Acta Orientalia* 56, 39–61.

——, 2002. "*Naṣrānī* (Ναζαραῖος) and *ḥanīf* (ἐθνικός): Studies on the Religious Vocabulary of Christianity and of Islam," *Bulletin of the School of Oriental and African Studies* 65.1, 1–30.

de Boer, T. J., 1901. *Geschichte der Philosophie im Islam*, Stuttgart, F. Frommans. Translated into English by E. R. Jones as *The History of Philosophy in Islam*, London, Luzac, 1903.

——, 1912. "اوراني" (*Ūrānī*), *Zeitschrift für Assyriologie* 17, 8–15.

de Goeje, M. J., and R. Dozy, 1885. "Nouveaux documents pour l'étude de la religion des Ḥarraniens," *Actes du Sixième Congrès international des orientalistes tenu en 1883 à Leide*, part 2, section 1, Leiden, E. J. Brill.

de Goeje, M. J., and M. Th. Houtsma, 1888–1907. *Catalogus Codicum Arabicorum Bibliothecae Academiae Lugduno-Batavae*, 2 vols., Leiden, Brill.

Delp, Mark M., 2006. "A Twelfth Century Cosmology: The *Liber de sex rerum principiis*," in *Corpus Christianorum Continuatio Mediaevalis* 142: *Hermetis Trismegisti* De sex rerum principiis (Hermes Latinus 2), 1–124.

Description de l'Égypte, 1809. Paris, Imperial Press.

Dieleman, Jacco, 2005. *Priests, Tongues, and Rites: The London-Leiden Magical Manuscripts and Translation in Egyptian Ritual (100–300 CE)*, Leiden, Brill.

Dignas, Beate, and Engelbert Winter, 2007. *Rome and Persia in Late Antiquity: Neighbors and Rivals*, Cambridge, Cambridge University Press.

Dodge, Bayard, 1970. *The Fihrist of al-Nadīm: A Tenth-Century Survey of Muslim Culture*, 2 vols., New York, Columbia University Press.

Donner, Fred M., 1998. *Narratives of Islamic Origins: The Beginnings of Islamic Historical Writing*, Princeton, Darwin.

——, 1999. "Muhammad and the Caliphate: Political History of the Islamic Empire up to the Mongol Conquest," *The Oxford History of Islam*, Oxford, Oxford University Press, 1–61.

——, 2002–2003. "From Believers to Muslims: Confessional Self-Identity in the Early Islamic Community," *al-Abḥāṯ* 50–51, 9–53.

Dorotheus, *Carmen Astrologicum*, ed. David Pingree, Leipzig, Teubner, 1976.

Dostourian, Ara Edmond, 1993. *Armenia and the Crusades, Tenth to Twelfth Centuries: The Chronicle of Matthew of Edessa, Translated from the Original Armenian with a Commentary and Introduction*, Lanham/New York/London, University Press of America.

Doufikar-Aerts, F., 2003. *Alexander Magnus Arabicus. Zeven eeuwen Arabische Alexandertraditie: van Pseudo-Callisthenes tot Ṣūrī*, doctoral diss., Universiteit Leiden.

Drijvers, H. J. W., 1965. *The Book of the Laws of the Countries. Dialogue on Fate of Bardaiṣan of Edessa*, Assen, van Gorcum.

——, 1966. *Bardaiṣan of Edessa*, Assen, van Gorcum.

Drower, E. S., and R. Macuch, 1963. *A Mandaic Dictionary*, Oxford, Clarendon.

Dunlop, D. M., 1971. "The *Mudhākarāt fī ʿIlm an-Nujūm* (Dialogues on Astrology) attributed to Abū Maʿshar al-Balkhī (Albumasar)," in *Iran and Islam*, ed. C. E. Bosworth, Edinburgh, Edinburgh University Press, 229–246.

Durkin-Meisterernst, Desmond, 2004. *Dictionary of Manichaean Middle Persian and Parthian* (*Dictionary of Manichaean Texts, vol. 3: Texts from Central Asian and China, Part 1*), series ed. Nicholas Sims-Williams, Turnhout, Brepols.

Ebeling, Florian, 2007. *The Secret History of Hermes Trismegistus: Hermeticism from Ancient to Modern Times*, trans. from German by David Lorton, Ithaca, Cornell University Press.

Effendi, Mahmoud, 1858. "Mémoire sur la calandrier arabe avant l'Islamisme," *Journal Asiatique* sér. 5, 11, 109–192.

*EI*² = *Encyclopaedia of Islam*, 2nd ed.

Eilers, Wilhelm, 1940. *Iranische Beamtennamen in der Keilsinschriftlichen Überlieferung*, Abhandlungen für die Kunde des Morgenlandes 25.5.

EIr = *Encyclopaedia Iranica.*

El Daly, Okasha, 2005. *Egyptology: The Missing Millennium. Ancient Egypt in Medieval Arabic Writings*, London, UCL.

Elias of Nisibis, *Opus Chronologicum*, Corpus Scriptorum Christianorum Orientalium 62–63, Scriptores Syri Series 3, 21–24, vols. 7–8 (both volumes in two parts: part 1 text and part 2 trans.) ed. and trans. E. Brooks (vol. 7) and I.-B. Chabot (vol. 8), Leipzig, Harrassowitz, 1909–1910.

Endress, Gerhard, 1987–1992. "Die wissenschaftliche Literatur," in *Grundriss der arabischen Philologie*, 3 vols., Wiesbaden, Ludwig Reichert, 2.400–506 and 3.3–152.

——, 1990. "The Defense of Reason: The Plea for Philosophy in the Religious Community," *Zeitschrift für Geschichte der arabisch-islamischen Wissenschaften* 6, 1–49.

——, 2007. "Building the Library of Arabic Philosophy: Platonism and Aritotelianism in the Sources of al-Kindī," in Cristina D'Ancona, ed., *The Libraries of the Neoplatonists*, Leiden, Brill, 319–350.

Evans, James, 1998. *The History and Practice of Ancient Astronomy*, Oxford, Oxford University Press.

Fahd, Toufic, 1966. *La Divination arabe. Études religieuses, sociologiques et folkloriques sure le milieu natif de l'Islam*, Leiden, Brill.

Faivre, Antoine, 1995. *The Eternal Hermes: From Greek God to Alchemical Magus*, trans. Joscelyn Godwin, Grand Rapids, Phanes.

al-Farġānī, Aḥmad ibn Muḥammad ibn Kaṯīr, *Ǧawāmiʿ ʿilm an-nuǧūm wa-uṣūl al-ḥarakāt al-samāwīya*, ed. and trans. Jacob Golius, Amsterdam 1669.

Festugière = Festugière, A.-J., 1944–1954. *La Révèlation d'Hermès Trismégiste*, 4 vols., Paris, Lecoffre; vol. 1, 2nd ed., 1950.

Fihrist = Ibn an-Nadīm, Muḥammad ibn Isḥāq, *Kitāb al-Fihrist*, ed. Riẓā Taǧaddud, Tehran, 1981.

Fihrist (Flügel) = Ibn an-Nadīm, Muḥammad ibn Isḥāq, *Kitāb al-Fihrist*, 2 vols., ed. G. Flügel, Leipzig, 1871–1872.

Fleischer, Heinrich L., 1840. "Hermes Trismegistus an die menschliche Seele," *Zeitschrift für historische Theologie*, 10.1, 87–117.

——, 1870. *Hermes Trismegistus an die menschlichen Seele, arabisch und deutsch*, Leipzig, F. A. Brockhaus.

Fodor, A., 1970. "The Origins of the Arabic Legends of the Pyramids," *Acta Orientalia Academiae Scientarum Hungaricae*, 23, 335–363.

Fowden, Garth, 1993a. *The Egyptian Hermes: A Historical Approach to the Late Pagan Mind*, Princeton, Princeton University Press, 2nd ed. (first ed. 1986).

——, 1993b. *Empire to Commonwealth: Consequences of Monotheism in Late Antiquity*, Princeton, Princeton University Press.

Franceschini, E., 1931–1932. "Il 'Liber philosophorum moralium antiquorum,'" *Atti del Reale Istituto Veneto di scienze, lettere ed arti* 91.2, 393–597.

Frankfurter, David, 1998. *Religion in Roman Egypt: Assimilation and Resistance*, Princeton, Princeton University Press.

Frendo, David, 2006. "Cassius Dio and Herodian on the First Sasanian Offensive against the Eastern Provinces of the Roman Empire," *Bulletin of the Asia Institute* 16, 25–36.

Frye, Richard, 1985. "Two Iranian Notes," in *Papers in Honour of Professor Mary Boyce*, vol. 2 (Hommages et Opera Minora Volume X), Leiden, E. J. Brill, 185–190.

Fück, J. W., 1951. "The Arabic Literature on Alchemy according to an-Nadīm (A.D. 987)," *Ambix* 4, 81–144.

al-Ǧāḥiẓ, *Kitāb at-Tarbīʿ wa-t-tadwīr*, ed. Charles Pellat, Damascus, Institut Français de Damas, 1955.

GAL = Brockelmann, Carl, 1937–1949. *Geschichte der Arabischen Literatur*, 2 vols., 2nd ed., Leiden, E. J. Brill, 1943–1949; *Supplement*, 3 vols., Leiden, E. J. Brill, 1937–1942.

Gardner, Iain, 1995. *The Kephalaia of the Teacher. Edited Coptic Manichaean Texts in Translation with Commentary*, Nag Hammadi and Manichaean Studies XXXVII, Leiden, E. J. Brill.

Gardner, Iain, and Samuel Lieu, 2004. *Manichaean Texts from the Roman Empire*, Cambridge, Cambridge University Press.

GAS = Sezgin, Fuat, *Geschichte des arabischen Schrifttums*, 13 vols. to date plus index volume, Leiden, E. J. Brill, 1967–2007.

Ġāyat al-ḥakīm wa-aḥaqq an-natīǧatayn bi-t-taqdīm = Das Ziel des Weisen von Pseudo-Maǧrīṭī, ed. H. Ritter, Studien der Bibliothek Warburg 12, Berlin, Teubner, 1933.

Gelzer, H., 1885–1888. *Sextus Julius Africanus und die byzantinische Chronographie*, 2 vols., Leipzig, Gerstenberg, reprinted in 1 vol., New York, B. Franklin, 1967.

Genequand, Charles, 1987–1988. "Platonism and Hermetism in al-Kindī's *Fī al-nafs*," *Zeitschrift für Geschichte der arabisch-islamischen Wissenschaften* 4, 1–18.

——, 1999. "Idolâtrie, astrolâtrie, et sabéisme," *Studia Islamica* 47, 109–128.

Giannakis, Elias, 2002–2003. "The Quotations from John Philoponus' *De aeternitate mundi contra Proclum* in al-Biruni's *India*," *Zeitschrift für Geschichte der Arabisch-Islamischen Wissenschaften* 15, 185–195.

Gignoux, Philippe, 1972. *Glossaire des Inscriptions Pehlevies et Parthes*, Corpus Inscriptionum Iranicarum Supplementary Series Vol. I, London, Lund Humphries.

——, 1987. *Incantations magiques syriaques*, Collection de la revue des Études Juives, Louvain–Paris, E. Peters.

——, 2001. "L'Apport scientifique des chrétiens syriaques à l'Iran sassanide," *Journal Asiatique* 289, 217–236.

Gill, J. S., 1984. "How Hermes Trismegistus Was Introduced to Renaissance England: The Influences of Caxton and Ficino's 'Argumentum' on Baldwin and Palfreyman," *Journal of the Warburg and Courtauld Institutes* 47, 222–225.

Gnomologium Vaticanum e codice Vaticano Graeco 743, ed. Leo Sternbach, Berlin, Walter de Gruyter, 1963.

Goldziher, Ignaz, 1913. "Die Zurechtweisung der Seele," in *Studies in Jewish Literature Issued in Honor of Professor Kaufmann Kohler, Ph.D.*, Berlin, Georg Reimer, 128–133.

Gonzáles Blanco, A., 1984. "Hermetism: A Bibliographical Approach," *Aufstieg und Niedergang der römischen Welt*, II.17.4, 2240–2281.

Graf, Georg, 1944–1953. *Geschichte der christlichen arabischen Literatur*, 5 vols., Studi e Testi 118, 133, 146, 147, 172, Vatican City, Biblioteca Apostolica Vaticana.

Green, Tamara, 1992. *The City of the Moon God: Religious Traditions of Harran*, Leiden, E. J. Brill.

Grignaschi, Mario, 1967. "Le Roman épistolaire classique conservé dans la version arabe de Sâlim Abû-l-'alâ'," *Le Muséon* 80, 211–253.

——, 1975. "Le 'Siyāsatu-l-'âmmiyya' et l'influence iranienne sure la pensée politique islamique," in *Monumentum H. S. Nyberg*, vol. III (Acta Iranica 6), Leiden, Brill, 33–287.

Grundriss der Arabischen Philologie, Wiesbaden, L. Reichert; vol. 1: *Sprachwissenschaft*, ed. Wolfdietrich Fischer, 1982; vol. 2: *Literaturwissenschaft*, ed. Helmut Gätje; vol. 3: *Supplement*, ed. W. Fischer, 1992.

Gundel, Wilhelm, 1978. *Neue astrologische Texte des Hermes Trismegistos. Funde und Forschungen auf dem Gebiet der antiken Astronomie und Astrologie*, Abhandlungen der Bayerischen Akademie der Wissenschaften, Philosophicsch-historische Abteilung, New Series, 12, Munich, 1936, reprinted with corrections and additions Hildesheim, 1978.

Gündüz, Şinasi, 1994. *The Knowledge of Life: The Origins and Early History of the Mandaeans and Their Relation to the Sabians of the Qur'ān and to the Harranians*, Journal of Semitic Studies Supplement 3, Oxford, Oxford University Press.

Gutas, Dimitri, 1975. *Greek Wisdom Literature in Arabic Translation: A Study of the Graeco-Arabic Gnomologia*, New Haven, American Oriental Society.

——, 1981. "Classical Arabic Wisdom Literature: Nature and Scope," *Journal of the American Oriental Society* 101, 49–86.

——, 1982. "The *Ṣiwān al-Ḥikma* Cycle of Texts," *Journal of the American Oriental Society* 102, 645–650.

——, 1983. "Paul the Persian on the Classification of the Parts of Aristotle's Philosophy: A Milestone between Alexandria and Baġdād," *Der Islam* 60, 231–267.

——, 1986. "The Spurious and the Authentic in the Arabic Lives of Aristotle," in J. Kraye, W. F. Ryan, and C. B. Schmitt, eds., *Pseudo-Aristotle in the Middle Ages: The Theology and Other Texts*, Warburg Institute Surveys and Texts 11, London, 15–36. Reprinted in D. Gutas, *Greek Philosophers in the Arabic Tradition*, Aldershot, Variorum, 2000.

——, 1988. "Plato's *Symposium* in the Arabic Tradition," *Oriens* 31, 36–60.

——, 1994. "Pre-Plotinian Philosophy in Arabic (Other than Platonism and Aristotelianism): A Review of the Sources," in *Aufstieg und Niedergang der Römischen Welt* II.36.7, 4939–4973.

——, 1998. *Greek Thought, Arabic Culture. The Graeco-Arabic Translation Movement in Baghdad and Early 'Abbāsid Society (2nd–4th/8th–10th centuries)*, London/New York, Routledge.

——, 1999. "The 'Alexandria to Baghdad' Complex of Narratives. A Contribution to the Study of Philosophical and Medical Historiography among the Arabs," *Documenti e Studi sulla Tradizione Filosofica Medievale* 10, 155–193.

——, 2002. "The Study of Arabic Philosophy in the Twentieth Century," in *British Journal of Middle Eastern Studies* 29.1, 5–25.

——, 2003. "Essay-Review: Suhrawardī and Greek Philosophy," *Arabic Sciences and Philosophy*, 13, 303–309.

——, forthcoming. "The Works of Ṯābit b. Qurra."

al-Ǧuzǧānī, *Ṭabaqāt-i Nāṣirī*, ed. 'Abdalḥayy Ḥabībī, Tehran, Dunyā-yi Kitāb, 1363 [1984].

Haas, E., 1876. "Über die Ursprünge der Indischen Medizin, mit besonderem Bezug auf Suṣruta" (*sic*), *Zeitschrift der deutschen morgenländischen Gesellschaft* 30, 617–670.

Habby, Joseph, 1998. "Alessandro incendiario di libri?," in G. Bolognesi and U. Cozzoli, eds., *L'eredità classica nel mondo orientali nell'età tardoantica e medievale. Il "Romanzo di Alessandro" e altri scritti. Atti del Seminario internazionale di studio (Roma-Napoli, 25–27 settembre 1997)*, Alessandria, Orso, 135–141.

Hachmeier, Klaus, 2002. *Die Briefe Abū Isḥāq Ibrāhīm al-Ṣābi's (st. 384/994 A.H./A.D.) : Untersuchungen zur Briefsammlung eines berühmten arabischen Kanzleischreibers mit Erstedition einiger seiner Briefe*, Hildesheim/New York: Georg Olms.

Hadot, Ilsetraut, 1987. "La vie et l'œvre de Simplicius d'après des sources grecques et arabes," in Ilsetraut Hadot, ed., *Simplicius: sa vie, son œvre, sa survie. Actes du colloque international de Paris (28 Sept.–1er Oct. 1985)*, Berlin, de Gruyter, 1987, 3–39. Published in English translation with minor revisions in Richard Sorabji, ed., *Aristotle Transformed: The Ancient Commentators and Their Influence*, London, Duckworth, 1990, 275–303.

——, 1990. "The Life and Works of Simplicius in Greek and Arabic Sources," in Richard Sorabji, ed., *Aristotle Transformed: The Ancient Commentators and Their Influence*, London, Duckworth, 275–303.

——, 1996. *Simplicius. Commentaire sur le* Manuel *d'Épictète. Introduction et édition critique du texte grec*, Leiden, Brill.

——, ed., 2001. *Simplicius: Commentaire sur le* Manuel *d'Épictète, t. 1, chapitres I-XXIX*, Paris, Les Belles Lettres.

——, 2007. "Dans quel lieu le néoplatonicien Simplicius a-t-il fondé son école de mathématiques, et où a pu avoir lieu son entretien avec un manichéen?" *The International Journal of the Platonic Tradition* 1, 42–107.

ha-Levi, Yahūdā, 1977. *Kitāb ar-Radd wa-d-dalīl fī d-dīn aḏ-ḏalīl*, ed. David Baneth, Jerusalem, Magnes Press.

Halleux, Robert, 1981. *Les alchimistes grecs I: Papyrus de Leyde, Papyrus de Stockholm, Recettes*, Paris, Les Belles Lettres.

Halm, Heinz, 1996a. *The Empire of the Mahdi: The Rise of the Fatimids*, trans. Michael Bonner, Leiden, Brill.

Halm, Heinz, 1996b. "The Isma'ili Oath of Allegiance (*'ahd*) and the 'sessions of wisdom' (*majālis al-ḥikma*) in Fatimid times," in *Medieval Isma'ili History and Thought*, ed. Farhad Daftary, Cambridge, Cambridge University Press, 91–115.

al-Ḥamd, Muḥammad 'Abdalḥamīd, 1998. *Ṣābi'at Ḥarrān wa-Iḫwān aṣ-Ṣafā'*, Damascus, al-Ahālī.

Hämeen-Anttila, Jaakko, 2006. *The Last Pagans of Iraq: Ibn Waḥshiyya and His* Nabataean Agriculture, Leiden, Brill.

Ḥamza al-Iṣfahānī, 1921/1922. *Kitāb Sinī mulūk al-arḍ wa-anbiyā'*, ed. as-Sayyid Ǧawād al-Īrānī at-Tibrīzī, Berlin, Kaviani.

Harrak, Amir, trans., 1999. *Chronicle of Zuqnin, Parts III and IV, A.D. 488–775*, Mediaeval Sources in Translation 36, Toronto, Pontifical Institute of Mediaeval Studies.

——, 2004. "Anti-Manichaean Propaganda in Syriac Literature," *Journal of Eastern Christian Studies* 56, 49–67.

Hartman, Sven S., 1953. *Gayōmart: Étude sur le syncretisme dans l'ancien Iran*, Uppsala, Almqvist & Wiksells.

al-Hāšimī, *The Book of the Reasons Behind the Astronomical Tables (Kitāb 'ilal al-zījāt)*, facsimile, trans. Fuad I. Haddad and E. S. Kennedy, commentary by David Pingree and E. S. Kennedy, Delmar, Scholar's Facsimiles and Reprints, 1981.

Ḫayrallāh, Amīn Ẓāhir 1919–1920. "Kitāb at-Tuffāḥa," *al-Muqtaṭaf* 55, 1919, 475–484 and 56, 1920, 18–22, 105–110, and 217–221.

Heer, Nicholas, trans., 1979. *The Precious Pearl: Al-Jāmī's al-Durrah al-Fākhirah*, Albany, State University of New York.

Henning, Walter, 1934. "Ein manichäisches Henochbuch," *Sitzungsberichte der preussischen Akademie der Wissenschaften*, 27–35.

——, 1942a. "An Astronomical Chapter of the Bundahishn," *Journal of the Royal Asiatic Society*, 229–248.

——, 1942b. "The Disintegration of the Avestic Studies," *Transactions of the Philological Society*, 40–56.

Hermetis Trismegisti Astrologica et Divinatoria, ed. Paolo Lucentini et al., Turnhout, Brepols, 2001.

Herodian, *Ab excessu divi Marci libri octo*, ed. K. Stavenhagen, Leipzig, Teubner, 1922.

Hinz, Walther, 1975. *Altiranisches Sprachgut der Nebenüberlieferungen*, Göttinger Orientforschungen, III. Reihe: Iranica, Band 3, Wiesbaden, Harrassowitz.

Historia Augusta, ed. D. Magie, Loeb Classical Texts, 1922–1932.

Hjärpe, Jan, 1972. *Analyse critique des traditions arabes sur les Sabéens Ḥarraniens*, doctoral thesis, University of Uppsala.

Holden, James Herschel, 1996. *A History of Horoscopic Astrology*, Tempe, American Federation of Astrologers.

Honigmann, Ernst, 1929. *Die sieben Klimata und die* Poleis Episêmoi. *Eine Untersuchung zur Geschichte der Geographie und Astrologie im Altertum und Mittelalter*, Heidelberg, Carl Winter's Universitätsbuchhandlung.

——, 1952. "Notes sur trois passages d'al-Mas'oudi," *Annuaire de l'Institut de philologie et d'histoire orientales et slaves* 12, 177–184.

Honigmann, Ernst, and André Maricq, 1953. *Recherches sur les* Res Gestae Divi Saporis, Académie royale de Belgique, Mémoires, 2. Série, 47.

Howard-Johnston, James, 1995. "The Two Great Powers in Late Antiquity: A Comparison," in Averil Cameron and Lawrence I. Conrad, eds., *The Byzantine and Early Islamic Near East III: States, Resources and Armies*, Princeton, Darwin, 157–222.

Hoyland, Robert, 1997. *Seeing Islam as Others Saw It: A Survey and Evaluation of Christian, Jewish and Zoroastrian Writings on Early Islam*, Princeton, Darwin.

Hunter, Erica C. D., 2002. "Beautiful black bronzes: Zosimos' treatises in Cam. Mm.6.29," in Alessandra Giumlia-Mair, ed., *I bronzi antichi: Produzione e tecnologia. Atti del XV Congresso Internazionale sui Bronzi Antichi*, Montagnac, Monique Mergoil, 655–660.

Huyse, Philip, 1999. *Die Dreisprachige Inschrift Šābuhrs I. an der Kaʿba-i Zardušt (ŠKZ)*, Corpus Inscriptionus Iranicarum III: Pahlavi Inscriptions, vol. 1 (in 2 vols.), London.

——, 2002, "La Revendication de territoires achéménides par les Sassanides: une réalité historique?," in Philip Huyse, ed., *Iran. Questions et Connaissances. Actes du IVᵉ congrès européen des études iraniennes, organisé par la Societas Iranologica Europaea, Paris, 6–10 septembre 1999*, vol. 1: *La période ancienne*. Paris, Peeters/AAEI (Cahiers de Studia Iranica 25), 297–311.

——, 2008. "Late Sasanian Society between Orality and Literacy," in Vesta Sarkhosh Curtis and Sarah Stewart, eds., *The Sasanian Era: The Idea of Iran, Volume III*, London, I.B. Tauris, 140–155.

al-Ḫwārizmī, Muḥammad ibn Aḥmad, *Mafātīḥ al-ʿulūm*, ed. G. van Vloten, Leiden, Brill, 1895.

Iamblichus, *De mysteriis Aegyptiorum*, ed. E. des Places, Paris, Les Belles Lettres, 1966.

Ibn ʿAbd Rabbih, 1940–1968. *Kitāb al-ʿIqd al-farīd*, 7 vols., ed. Aḥmad Amīn, Aḥmad az-Zayn, and Ibrāhīm al-Abyārī, Cairo, Maṭbaʿat Lağnat at-Ta'līf wa-t-Tarğama wa-n-Našr.

Ibn Abī Uṣaybiʿa, Aḥmad ibn al-Qāsim, *ʿUyūn al-anbā' fī ṭabaqāt al-aṭibbā'*, ed. A. Müller, 2 vols., Cairo-Königsberg i. Pr., al-Maṭbaʿ al-Wahbīya, 1299–1301/1882–1884.

Ibn al-Ğawzī, *Kitāb al Muntaẓam fī ta'rīḫ al-mulūk wa-l-umam*, 10 vols., Baġdād, ad-Dār al-Waṭanīya, 1990.

Ibn Ǧubayr = *The Travels of Ibn Jubayr*, ed. W. Wright, 2nd ed. rev. M.J. de Goeje, Leiden, Brill and London, Luzac, 1907.

Ibn Ǧulǧul, Sulaymān ibn Ḥassān, *Ṭabaqāt al-aṭibbā' wa-l-ḥukamā'*, ed. Fu'ād Sayyid, Cairo, Institut Français d'Archéologie Orientale, 1955.

Ibn Ḥabīb, 'Abdalmalik, *Kitāb at-Ta'rīḫ*, ed. Jorge Aguadé, Madrid, Consejo Superior de Investigaciones Científicas, 1991.

Ibn Ḫaldūn, *The Muqaddimah: An Introduction to History*, trans. F. Rosenthal, 3 vols., Bollingen Series XLIII, Kingsport, Pantheon, 1958.

Ibn Ḫallikān, *Kitāb Wafayāt al-a'yān wa-anbā' abnā' az-zamān*, 8 vols., ed. Iḥsān 'Abbās, Beirut, Dār aṯ-Ṯaqāfa, 1968.

Ibn Ḥazm, 1996. *Kitāb al-Faṣl fī l-milal wa-l-ahwā' wa-n-niḥal*, 3 vols., Beirut, Dār al-al-Kutub al-'Ilmīya.

Ibn Ḫurradāḏbih, *Muḫtār min Kitāb al-lahw wa-l-mahālī*, Nuṣūṣ wa-durūs 17, ed. Iġnāṭīyūs 'Abdū Ḫalīfa, Beirut, al-Maṭba'a al-Kāṯūlīkīya [The Catholic Press], 1961.

Ibn al-'Ibrī, *Ta'rīḫ* = Ibn al-'Ibrī, *Ta'rīḫ muḫtaṣar ad-duwal*, ed. A. Ṣāliḥānī, Beirut, 1890 (reprinted Beirut, Dār ar-Rā'id al-Lubnānī, 1983).

——, Syriac Chronicle trans. = Barhebraeus, *Secular History*, vol. 1, trans. into English by E. A. W. Budge, London, Oxford University Press, 1932.

——, Bodleian MS Hunt. 52 = Barhebraeus, *Secular History*, vol. 2, ed. E. A. W. Budge, London, Oxford University Press, 1932.

Ibn al-Mu'tazz, 1982. *Kitāb al-Ādāb*, ed. Ṣabīḥ Radīf, Baġdād, al-Hawadi.

Ibn an-Nadīm, Muḥammad ibn Isḥāq, *Kitāb al-Fihrist*, ed. G. Flügel, 2 vols., Leipzig, F.C.W. Vogel, 1871–1872.

Ibn Nubāta, Ǧamāladdīn, *Šarḥ al-'uyūn fī šarḥ risālat Ibn Zaydūn*, ed. M. Abū l-Faḍl Ibrāhīm, Cairo, Dār al-Fikr al-'Arabī, 1964.

Ibn al-Qifṭī, 'Alī ibn Yūsuf, *Ta'rīḫ al-ḥukamā'*, ed. J. Lippert, Leipzig, Dieterich, 1903.

Ibn Qutayba, *Kitāb al-Ma'ārif*, ed. Saroite Okacha [Ṯarwat 'Ukāša], Cairo, 1960.

Ibn Rustah, *Kitāb al-a'lāq an-nafīsa*, Bibliotheca Geographorum Arabicorum VII, Leiden, Brill, 1892, 1–229.

Ibn Waḥšīya, *Kitāb al-Filāḥa an-Nabaṭīya*, ed. Toufic Fahd, 3 vols., Damascus, Institut Français de Damas, 1993–1998.

Ideler, Iulius Ludovicus, 1841. *Physici et Medici Graeci Minores*, vol. 1, Berlin, G. Reimer.

Ihm, Sibylle, 2001. *Ps.-Maximus Confessor: Erste kritische Edition einer Redaktion des sacro-profanen* Florilegiums Loci communes *nebst einer vollständigen Kollation einer zweiten Redaktion und weiterem Material*, Palingenesia 73, Stuttgart, Franz Steiner.

Iḫwān aṣ-Ṣafā', *Rasā'il Iḫwān aṣ-Ṣafā'*, 4 vols., ed. Ḫayraddīn az-Zarkalī, Egypt, al-Maṭba'a al-'Arabīya, 1928.

Iversen, E., 1984. *Egyptian and Hermetic Doctrine*, Copenhagen, Museum Tusculanum.

Jacob of Edessa, *Chronicon Iacobi Eddeseni*, ed. and trans. E. W. Brooks in *Chronica Minora*, Corpus Scriptorum Christianorum Orientalium: Scriptores Syri 3.4, 1903, Syriac text pp. 261–330, Latin trans. pp. 197–258.

——, *Hexaemeron*, ed. I.-B. Chabot and A. Vaschalde, Corpus Scriptorum Christianorum Orientalium, Series Secunda, 56.1 Louvain, 1928; trans. A. Vaschalde, Corpus Scriptorum Christianorum Orientalium, Series Secunda, 56.2 Louvain, 1932.

Jasnow, Richard, and Karl-Theodor Zauzich, 1998. "A Book of Thoth?," *Orientalia Lovaniensia Analecta* 82 (Proceedings of the Seventh International Congress of Egyptologists, Cambridge, September 3–9, 1995), 607–618.

——, 2005. *The Ancient Egyptian Book of Thoth: A Demotic Discourse on Knowledge and Pendant to the Classical Hermetica. Volume 1: Text. Volume 2: Plates*, Wiesbaden, Harrassowitz.

Johnson, Aaron P., 2006. *Ethnicity and Argument in Eusebius'* Praeparatio Evangelica, Oxford, Oxford University Press.
Jolivet, Jean, and Guy Monnot, trans. and comment., 1986 = Shahrastani, *Livre des religions et des sectes*, 2 vols., Peeters/UNESCO.
Joosse, N. Peter, 1993. "An Example of Medieval Arabic Pseudo-Hermetism: The Tale of Salāmān and Absāl," *Journal of Semitic Studies* 38/2, 279–293.
Justi, Ferdinand, 1895. *Iranisches Namenbuch*, Marburg, N.G. Elwert.
Kartanegara, R. Muhyiddin, 1996. *The Mukhtaṣar Ṣiwān al-Ḥikma of 'Umar b. Sahlān al-Sāwī: Arabic Text and Introduction*, doctoral diss., University of Chicago.
Kāšānī, Afḍaladdīn, 1952. *Muṣannafāt*, vol. 1, ed. M. Minovi and Y. Mahdavi, Tehran, Dānišgāh-i Tihrān.
Kennedy, E. S., and David Pingree, 1971. *The Astrological History of Māshā'allāh*, Cambridge (Mass.), Harvard University Press.
Kennedy, E. S., and B. L. van der Waerden, 1963. "The World-Year of the Persians," *Journal of the American Oriental Society* 83.3, 315–327.
Kennedy, Hugh, 1999. "Islam," in G. Bowersock, P. Brown, and O. Grabar, eds., *Late Antiquity: A Guide to the Post-Classical World*, Cambridge (Mass.), Harvard University Press, 219–237.
Khoury, Raif Georges, 1972. *Wahb b. Munabbih. Teil 1. Der Heidelberger Papyrus PSR Heid Arab 23. Leben und Werk des Dichters*, Wiesbaden, Harrasowitz.
al-Kindī, 1950. *Rasā'il al-Kindī al-falsafiya*, ed. M. 'Abdalhādī Abū Rīda, 2 vols., Cairo, Dār al-Fikr al-'Arabī.
al-Kindī, 1974 = *al-Kindi's Metaphysics*, introduction and trans. Alfred L. Ivry, Albany, State University of New York Press.
Kindstrand, Jan Fredrik, 1991. *Gnomica Basileensia*, Acta Universitatis Upsaliensis, Studia Byzantina Uppsalensia 2, Uppsala.
King, David, and Julio Samsó, 2001. "Astronomical Handbooks and Tables from the Islamic World (750–1900): An Interim Report," *Suhayl* 2, 9–105.
Kingsley, Peter, 1993. "Poimandres: The Etymology of the Name and the Origins of the Hermetica," *Journal of the Warburg and Courtauld Institutes* 56, 1–24.
——, 1994. "From Pythagoras to the *Turba Philosophorum*: Egypt and Pythagorean Tradition," *Journal of the Warburg and Courtauld Institutes* 57, 1–13.
——, 1995. "Meetings with Magi: Iranian Themes among the Greeks, from Xanthus of Lydia to Plato's Academy," *Journal of the Royal Asiatic Society*, 173–209.
Klein-Franke, Fritz, 1973. "The Geomancy of Aḥmad b. 'Ali Zunbul: A Study of the Arabic Corpus Hermeticum," *Ambix*, 20, 26–35.
Klijn, A. F. J., 2003. *The Acts of Thomas: Introduction, Text, and Commentary*, 2nd rev. ed., Leiden, Brill.
Knibb, Michael A., 1978. *The Ethiopic Book of Enoch: A New Edition in the Light of the Aramaic Dead Sea Fragments*, 2 vols., Oxford, Clarendon.
Kornfeld, Walter, 1978. *Onomastica Aramaica aus Ägypten*, Sitzungsberichte der Österreichische Akademie der Wissenshaften, philosophisch-historische Klasse, 333, Vienna.
Kraemer, Jörg, 1956. "Das arabische Original des pseudo-aristotelischen *Liber de Pomo*," *Studi Orientalistici in onore di Giorgio Levi della Vida*, vol. 1, Rome, Istituto per l'Oriente, 484–506.
Kraemer, Joel, 1986. *Humanism in the Renaissance of Islam: The Cultural Revival during the Buyid Age*, Leiden, Brill.
Kratchkowsky, Ignace, ed., 1997. *Histoire de Yaḥyā ibn Sa'īd d'Antioche* (with French translation by Françoise Micheau and Gérard Troupeau), Patrologia Orientalis 47.4 (n. 212), Turnhout, Brepols.

Kraus, Paul, 1936. "Raziana II," *Orientalia* 5, 38–56 and 358–378.

——, 1942–1943. *Jābir ibn Ḥayyān. Contribution à l'Histoire des Idées scientifiques dans l'Islam. 1. Le Corpus des écrits jābiriens. 2. Jābir et la science grecque*, Cairo (Mémoires présentés à l'Institut d'Égypte, 44, 45).

Krause, Max, 1936. "Stambuler Handschriften islamischer Mathematiker," in *Quellen und Studien zur Geschichte der Mathematik, Astronomie und Physik*, Abteilung B, Studien 3, 437–531.

Kühlmann, Wilhelm, 1999. "Der 'Hermetismus' als literarische Formation: Grundzüge seiner Rezeption in Deutschland," *Scientia et Poetica* 3, 145–157.

Kunitzsch, Paul, 1968. "Zum 'liber hermetis de stellis beibeniis,'" *Zeitschrift der deutschen morgenländischen Gesellschaft* 118, 62–74.

——, 1970. "Neues zum 'liber hermetis de stellis beibeniis,'" *Zeitschrift der Deutschen Morgenländischen Gesellschaft* 120, 126–130.

——, 1974. *Der Almagest. Die Syntaxis Mathematica des Claudis Ptolemäus in arabisch-lateinischer Überlieferung*, Wiesbaden, Harrassowitz.

——, 1975. "Über das Frühstadium der arabischen Aneignung antiken Gutes," *Saeculum* 26, 268–282.

——, 1981. "Stelle beibenie—*al-kawākib al-biyābānīya*—Ein Nachtrag," *Zeitschrift der Deutschen Morgenländischen Gesellschaft* 131, 263–267.

——, 1993. "The Chapter on the Fixed Stars in Zarādusht's *Kitāb al-Mawālīd*," *Zeitschrift für Geschichte der Arabisch-Islamischen Wissenschaften* 8, 241–249.

——, 2001. "Liber de stellis beibeniis," in Paolo Lucentini et al., eds., *Hermetis Trismegisti Astrologica et Divinatoria*, Turnhout, Brepols, 11–81.

Lameer, Joep, 1997. "From Alexandria to Baghdad: Reflections on the Genesis of a Problematical Tradition," in G. Endress and Remke Kruk, eds., *The Ancient Tradition in Christian and Islamic Hellenism*, Leiden, CNWS, 181–191.

Lamoreaux, John C., 2002. "The Biography of Theodore Abū Qurrah Revisited," *Dumbarton Oaks Papers* 56, 2002, 25–40.

——. 2005. *Theodore Abū Qurrah*, Library of the Christian East 1, Provo, Brigham Young University Press.

Lane, E. W., 1863–1893. *Arabic-English Lexicon*, 8 vols., London, Williams and Norgate.

Lane Fox, Robin, 1986. *Pagans and Christians*, New York, Alfred A. Knopf.

——, 2005. "Appendix: Harran, the Sabians and the Late Platonist 'Movers,'" in Andrew Smith, ed., *The Philosopher and Society in Late Antiquity: Essays in Honour of Peter Brown*, Swansea, Classical Press of Wales, 231–244.

Lawrence, Bruce B., 1976. *Shahrastānī on the Indian Religions*, The Hague/Paris, Mouton.

Lechner, Kilian, 1955. *Hellenen und Barbaren im Weltbild der Byzantiner: Die alten Bezeichnungen als Ausdruck eines neues Kulturbewusstseins*, Munich.

Letrouit, Jean, 1995. "Chronologie des alchimistes grecs," in Didier Kahn and Sylvain Matton, eds., *Alchimie. Art, Histoire et Mythes*, Textes et Travaux de Chrysopoiea 1, 11–93.

Levy-Rubin, Milka, 2003. "Praise or Defamation? On the Polemical Usage of the Term Ḥanīf among Christians and Muslims in the Middle Ages," *Jerusalem Studies in Arabic and Islam* 28, 202–224.

Lexicon der Ägyptologie, 7 vols., ed. Wolfgang Helck and Eberhard Otto, Wiesbaden, Harrassowitz, 1972–1992.

Liber Hermetis Trismegisti = Gundel 1936.

Lindsay, Jack, 1970. *The Origins of Alchemy in Graeco-Roman Egypt*, London, Muller.

Lippmann, Edmund O. von, 1918–1954. *Entstehung und Ausbreitung der Alchemie*, vols. 1–2, Berlin, Julius Sprenger, 1918–1931; vol. 3, Meinheim, Chemie, 1954.

Lloyd, Seton, and William Brice, 1955. "Harran," *Anatolian Studies* 1, 77–111.

Lopilato, Robert, 1998. *The Apotelesmatika of Manetho*, doctoral diss., Brown University, Providence.

Lory, Pierre, 1988. "Hermès/Idris: Prophète et Sage dans la tradition islamique," in Antoine Faivre, ed., Présence d'Hermès Trismégiste, Paris, Albin Michel, 100–109.

Lucentini, Paolo, and Vittoria Perrone Compagni, 2001. *I testi e i codici di Ermete nel Medioevo*, Hermetica Mediaevalia I, Florence, Polistampa.

Luna, C., 2001. Review of R. Thiel, *Simplikios und das Ende der neuplatonischen Schule in Athen*, in *Mnemosyne* 54, 482–504.

MacKenzie, D. N., 1971. *A Concise Pahlavi Dictionary*, London, Oxford University Press.

MacMullen, Ramsay, 1984. *Christianizing the Roman Empire (A.D. 100–400)*, New Haven/ London, Yale University Press.

——, 1997. *Christianity and Paganism in the Fourth to the Eighth Centuries*, New Haven/ London, Yale University Press.

Macuch, Rudolf, 1965. "Anfänge der Mandäer," in *Die Araber in der alten Welt*, vol. 2, eds. F. Altheim and R. Stiehl, Berlin, Walter de Gruyter, 76–191.

Madelung, Wilfred, and Paul E. Walker, 2000. *The Advent of the Fatimids: A Contemporary Shiʿi Witness*, London, I.B. Tauris.

Mahé, Jean-Pierre, 1978–1982. *Hermès en Haute-Egypt*, 2 vols., Quebec, Presses de l'Université Laval.

——, 1984. "Fragments Hermétiques dans les papyri vindobonenses graecae 29456 r° et 29828 r°," in E. Lucchesi and H. D. Saffrey, eds., *Mémorial André-Jean Festugière. Antiquité païenne et chrétienne*, Geneva, P. Kramer, 51–64.

——, 1996. "Preliminary Remarks on the *Demotic Book of Thoth* and the Greek *Hermetica*," *Vigiliae Christianae* 50, 353–363.

——, 2007. "Hermetic Religion," in *The Nag Hammadi Scriptures: The International Edition*, ed. Marvin Meyer, New York, HarperOne, 795–798.

Maimonides [Mūsā ibn Maymūn], *Kitāb Dalālat al-ḥā'irīn*, ed. Ḥusayn Ātāy, Maktabat aṯ-Ṯaqāfa ad-Dīnīya, Cairo, 198– (not dated).

Mango, Cyril, 1978. "Who Wrote the Chronicle of Theophanes?" *Zbornik Radova Vizantološkog Instituta* 18, 9–17. Reprinted in Cyril Mango, *Byzantium and Its Image*, Variorum, Ashgate, 1984, article XI.

Mango, Cyril, and Roger Scott, 1997. *The Chronicle of Theophanes the Confessor*, Oxford, Clarendon.

Manzaloui, Mahmoud, 1974, "The Pseudo-Aristotelian *Kitāb sirr al-asrār*. Facts and Problems," *Oriens* 23–24, 147–257.

al-Maqdisī, Muṭahhar ibn Ṭāhir, *Kitāb al-Bad' wa-t-ta'rīḫ*, 6 vols., ed. and trans. Cl. Huart, Paris, Leroux, 1899–1919.

Margoliouth, D. S., 1892. "The Book of the Apple, Ascribed to Aristotle," *Journal of the Royal Asiatic Society*, 187–252.

——, "Ḥarrānians," in James Hastings, ed., *Encyclopaedia of Religion and Ethics*, vol. 6, Edinburgh, T. & T. Clark, 519–520.

Markwart, J., 1931. *A Catalogue of the Provincial Captials of Eranshahr (Pahlavi Text, Version, and Commentary)*, ed. G. Messina, Rome, Pontificio Istituto Biblico.

Marquet, Yves, 1966. "Sabéens et Ikhwan al-Safa," *Studia Islamica* 24, 35–80, and 25, 77–109.

Massignon, Louis, 1949. "Investigaciones sobre Šuštarī, poeta andaluz, enterrada en Damieta," *al-Andalus* 14, 29–57.

——. 1950. "Inventaire de la littérature hermetique arabe," in A. M. J. Festugière, *La Révèlation d'Hermès Trismégiste*, vol. 1, 2nd ed., Paris, Lecoffre, 384–400; reprinted in Louis Massignon, *Opera Minora*, vol. 1, ed. Youakim Moubarac, Beirut, Dār al-Maʿārif, 1963, 650–666.

al-Mas'ūdī, *Kitāb at-Tanbīh wa-l-išrāf*, ed. M. J. de Goeje, Bibliotheca Geographorum Arabicorum VIII, Leiden, E. J. Brill, 1894.

——, *Murūǧ aḏ-ḏahab wa-ma'ādin al-ǧawhar*, ed. Charles Pellat, 7 vols., Beirut, al-Ǧāmi'a al-Lubnānīya, 1965–1979.

Menasce, Jean de, trans., 1973. *Le troisième livre du Denkart*. Paris, C. Klincksieck.

Michael the Elder = *Chronique de Michel le syrien*, ed. and trans. J.-B. Chabot, 3 vols., Paris 1899–1910. Reprinted in 4 volumes (cited here), Brussels, Culture et Civilization, 1963.

Miskawayh, Aḥmad ibn Muḥammad, *al-Ḥikma al-ḫālida (Ǧāwidān ḫirad)*, ed. 'Abdurraḥmān Badawī, Cairo, Maktabat an-Nahḍa al-Miṣrīya 1952.

Monnot, Guy, 1986. " Les Sabéens de Šahrastānī," in Šahrastānī, *Livre des religions et des sectes*, 2 vols., trans. and comment. Jean Jolivet and Guy Monnot, Peeters/UNESCO, vol. 2, 3–13.

Montgomery, James A., 1913. *Aramaic Incantation Texts from Nippur*, The Museum Publications of the Babylonian Section, vol. 3, Philadelphia, University of Pennsylvania.

Montgomery, James E., 2007. "Al-Ǧāḥiẓ and Hellenizing Philosophy," in Cristina D'Ancona, ed., *The Libraries of the Neoplatonists*, Leiden, Brill, 443–456.

Morony, Michael, 1984. *Iraq after the Muslim Conquest*, Princeton, Princeton University Press.

Moses Khorenats'i, 1978. *History of the Armenians*, translated and annotated with introduction by Robert Thomson, Harvard Armenian Texts and Studies 4, Cambridge (Mass.), Harvard University Press.

Mosshammer, Alden A., 1979. *The* Chronicle *of Eusebius and Greek Chronographic Tradition*, London, Associated University Presses.

al-Mubaššir ibn Fātik, Abū l-Wafā', *Muḫtār al-ḥikam wa-maḥāsin al-kalim* = *Los Bocados de oro (Mujtār al-ḥikam). Edición crítica del texto árabe con prólogo y notas*, ed. 'Abdarraḥmān Badawī, Madrid, Publicaciones del Instituto Egipcio de Estudios Islámicos, 1958.

Müller, August, 1880. "Arabische Quellen zur Geschichte der indischen Medizin," *Zeitschrift der deutschen morgenländischen Gesellschaft* 34, 465–556.

Muntaḫab Ṣiwān al-ḥikma (a). Ed. 'Abd-ar-Raḥmān Badawī, *Ṣiwān al-ḥikma wa-ṯalāṯ rasā'il ta'līf Abū* (sic) *Sulaymān al-Manṭiqī as-Siǧistānī*, Tehran, 1974.

Muntaḫab Ṣiwān al-ḥikma (b). Ed. D. M. Dunlop, *The Muntakhab Ṣiwān al-Ḥikma of Abū Sulaimān as-Sijistānī*, Near East and Middle East Monographs 4, The Hague, 1979.

Nag Hammadi Codices V,2–5 and VI with Papyrus Berolinensis 8502,1 and 4, Douglas M. Parrott et alii, Leiden, E. J. Brill, 1979.

Nallino, Carlo A., 1922. "Tracce di opera greche giunte agli arabi per trafila pehlevica," in T. W. Arnold and R. A. Nicholson, eds., *A Volume of Oriental Studies Presented to E.G. Browne*, Cambridge, Cambridge University Press, 345–363.

Nasr, Seyyed Hossein, 1967. "Hermes and Hermetic Writings in the Islamic World," chapter 6 in *Islamic Studies: Essays on Law and Society, the Sciences and Philosophy and Sufism*, Beirut, Librairie du Liban, 63–89.

an-Naššār, 'Alī Sāmī, and 'Abbās aš-Širbīnī, 1965. *Fīdūn wa-Kitāb at-Tuffāḥa al-mansūb li-Suqrāṭ*, Cairo, Dār al-Ma'ārif.

Nau, F., 1900. "Lettre de Jacques d'Édesse sur la chronologie biblique et la date de naissance du Messie," *Revue de l'orient chrétien* 5, 581–596.

Naveh, Joseph, and Shaul Shaked, 1985. *Amulets and Incantation Bowls*, Jerusalem, Magnes Press.

an-Nawawī, Muḥyīddīn Abū Zakarīyā' Yaḥyā ibn Šaraf, 1977. *Tahḏīb al-asmā'*, Beirut, Dār al-Kutub al-'Ilmīya.

Nemoy, Leon, 1939. "The Treatise on the Egyptian Pyramids (*Tuḥfat al-kirām fi khabar al-ahrām*) by Jalāl al-Dīn al-Suyūṭī," *Isis* 30, 17–37.

Neugebauer, Otto, 1975. *A History of Ancient Mathematical Astronomy*, 3 vols., New York/Heidelberg/Berlin, Springer.

——, 1979. *Ethiopic Astronomy and Computus*, Sitzungsberichte der österreichischen Akademie der Wissenschaften, philosophisch-historische Klasse 347, Vienna.

——, 1989. *Chronography in Ethiopic Sources*, Sitzungsberichte der österreichischen Akademie der Wissenschaften, philosophisch-historische Klasse 512, Vienna.

Niẓāmī, Ǧamāladdīn Abū Muḥammad Ilyās ibn Yūsuf, 2001 [1379]. *Iqbāl-nāma*, ed. Bihruz Sarvatiyān, Tehran, Tus.

Nöldeke, Theodor, 1890. *Beiträge zur Geschichte des Alexanderromans*, Denkschriften der kaiserlichen Akademie der Wissenschaften in Wien, Philosophisch-historische Classe, 38.V, Vienna.

——, 1904. *Compendious Syriac Grammar*, trans. James A Crichton, with an appendix of the handwritten additions in Theodor Nöldeke's personal copy, ed. Anton Schall, trans. Peter T. Daniels, Winona Lake, Eisenbrauns, 2001 (first German edition 1898, first English edition, 1904).

Palfreyman, Thomas, ca. 1635. *A treatise of morall philosophie: wherein is contained the worthy sayings of philosophers emperours, kings, and orators: their liues and answers: of what linage they came: and of what countrey they were: whose worthy sentences. notable precepts, counsels, and parables, doe hereafter follow. First gathered & set forth by William Bauldwin, and now the ninthtime since inlarged by Thomas Palfreyman, Gentleman*, London, William Stansby.

Palmer, Andrew, with Sebastian Brock and Robert Hoyland, 1993. *The Seventh Century in the West-Syrian Chronicles*, Translated Texts for Historians 15, Liverpool, Liverpool University Press.

Panaino, Antonio, 2001. "L'influsso greco nella letteratura e nella cultura medio-persiana," in *Autori classici in lingue del Medio e Vicino Oriente*, Rome, Istituto poligrafico e Zecca dello Stato, 29–45.

Pappacena, Massimo, 2004. "La figura di Ermete Trismegisto nella tradizione araba," in Paolo Lucentini, Iliaria Parri, and Vittoria Perrone Compagni, eds., *Atti del Convegno Internazionale sulla Tradizione Ermetica dal Mondo Tardo-Antico all'Umanesimo*, Turnhout, Brepols, 263–283.

Papyri Graecae Magicae, 2 vols. Ed. Karl Preisendanz, Leipzig, Teubner, 1928–1931.

Paulus = *Kitāb 'Unṣur al-mūsīqā wa-mā ftaraqat 'alayhi l-falāsifa min tarkībihī wa-mā'īyatihī*, MS Manisa Genel Kütüphane 1705, 123 verso–130 verso.

Pedersen, J., 1922. "The Ṣābians," in T. W. Arnold and R. A. Nicholson, eds., *'Aǧab-nāma: A Volume of Oriental Studies Presented to Edward G. Browne on His 60th Birthday*, Cambridge, Cambridge University Press, 383–391.

Perlmann, Moshe, trans., 1971. *Ibn Kammūna's Examination of the Three Faiths*, Berkeley/Los Angeles/London, University of California Press.

Perry, B. E., 1952. *Aesopica, Vol. 1: Greek and Latin Texts*, Urbana/Chicago, University of Illinois.

Peters, Francis E., 1990. "Hermes and Harran: The Roots of Arabic-Islamic Occultism," in Michel M. Mazzaoui and Vera B. Moreen, eds., *Intellectual Studies on Islam: Essays Written in Honor of Martin B. Dickson*, Salt Lake City, University of Utah Press, 185–215.

Pines, Shlomo, 1970. "Philosophy," in *The Cambridge History of Islam*, vol. 2, Cambridge, Cambridge University Press, 780–823.

——, 1990. "A Parallel between Two Iranian and Jewish Themes," in *Irano-Judaica II*, Jerusalem, Ben-Zvi Institute, 41–51.

Pingree, David, 1962. "Historical Horoscopes," *Journal of the American Oriental Society* 82, 487–502.

Pingree, David, 1963a. "Astronomy and Astrology in India and Iran," *Isis* 54, 229–246.

——, 1963b. "The Indian Iconography of the Decans and Horâs," *Journal of the Warburg and Courtauld Institutes* 26, 223–254.

——, 1968. *The Thousands of Abū Ma'shar*, London, Warburg Institute.

——, 1970. "The Fragments of the Works of al-Fazārī," *Journal of Near Eastern Studies* 29, 103–123.

——, 1971. "The Astrological School of John Abramius," *Dumbarton Oaks Papers* 25, 189–215.

——, 1973. "The Greek Influence on Early Islamic Mathematical Astronomy," *Journal of the American Oriental Society* 93, 32–43.

——, 1975. "Māshā'allāh: Some Sasanian and Syriac Sources," in G. F. Hourani, ed., *Essays on Islamic Philosophy and Science*, Albany, Statue University of New York Press, 5–14.

——, 1976. "The Indian and Pseudo-Indian Passages in Greek and Latin Astronomical and Astrological Texts," *Viator* 7, 141–195.

——, 1978. *The Yavanajātaka of Sphujidhvaja*, 2 vols., Cambridge (Mass.), Harvard University Press.

——, 1980. "Some of the Sources of the *Ghāyat al-ḥakīm*," *Journal of the Warburg and Courtauld Institutes* 43, 1–15.

——, 1989a. "Classical and Byzantine Astrology in Sasanian Persia," *Dumbarton Oaks Papers* 43, 227–239.

——, 1989b. "Indian Planetary Images and the Tradition of Astral Magic," *Journal of the Warburg and Courtauld Institutes* 52, 1–13 + 5 plates.

——, 1992. "Al-Ṭabarī on the Prayers to the Planets," in *Bulletin d'études orientales* 44, 105–117.

——, 1997. *From Astral Omens to Astrology: From Babylon to Bīkāner*, Serie Orientale Romae LXXVIII, Rome, Istituto Italiano per l'Africa e l'Oriente.

——, 1999. "Māshā'allāh's (?) Arabic Translation of Dorotheus," in Rika Gyselen, ed., *La Science des Cieux. Sages, mages, astrologues = Res Orientales* 12, Bures-sur-Yvette, Group pour l'Etude de la Civilisation du Moyen-Orient, 191–209.

——, 2001. "From Alexandria to Baghdād to Byzantium: The Transmission of Astrology," *International Journal of the Classical Tradition* 8.1, 3–37.

——, 2002. "The Ṣābians of Ḥarrān and the Classical Tradition," in *International Journal of the Classical Tradition* 9.1, 8–35.

——, 2006. "From Hermes to Jābir and the *Book of the Cow*," in Ch. Burnett and W. F. Ryan, eds., *Magic and the Classical Tradition*, Warburg Institute Colloquia 7, London, 19–28.

Pitra, Joannes Baptista Cardinalis, 1888. *Analecta Sacra et Classica V: Spicilegio Solesmensi*, Paris, Roger and Chernowitz.

Plessner, Martin, 1927. "Neue Materialien zur Geschichte der Tabula Smaragdina," *Der Islam* 16, 77–113.

——, 1954a. "Hermes Trismegistus and Arab Science," *Studia Islamica* 2, 45–59.

——, 1954b. "The Place of the Turba Philosophorum in the Development of Alchemy," *Isis* 45, 331–338.

——, "Hirmis," *EI*², vol. 3, 463–465.

——, 1972. Essay review of F. Sezgin, GAS 3–4, *Ambix* 19, 209–215.

Pliny the Elder. *Natural History*, vol. 7 (*Libri* xx14-xxvii), trans. W. H. S. Jones, Loeb Classical Library 393, Cambridge (Mass.), Harvard University Press, 1956.

Polotsky, H.-J., and A. Böhlig, 1940. *Kephalaia*, Manichäische Handschriften der Staatlichen Museen Berlin 1, Stuttgart, W. Kohlhammer.

Porphyry. *Epistola ad Anebonem=Lettera ad Anebo*, ed. A. R. Sodano, Naples, L'Arte Tipografica, 1958.

——. *In Platonis Timaeum Commentaria*, ed. A. R. Sodano, Naples, n.p., 1964.
Porreca, David, 2001. *The Influence of Hermetic Texts on Western European Philosophers and Theologians (1160–1300)*, Ph.D. Diss., Warburg Institute, University of London.
Posener, G., 1936. *La Première Domination Perse en Égypte. Recueil d'inscriptions hiéroglyphiques*, Cairo, Bibliotheque d'Étude, L'Institut Français d'archéologie orientale.
Pourshariati, Parvaneh, 2008. *Decline and Fall of the Sasanian Empire: The Sasanian-Parthian Confederacy and the Arab Conquest of Iran*, London, I. B. Tauris.
Pseudo-Apollonius, *Sirr al-ḫalīqa* = Weisser 1979.
PW = Pauly, A., G. Wissowa, and W. Kroll, *Paulys Real-Encyclopädie der klassischen Altertumswissenschaft*, Stuttgart, J. B. Metzler, 1894–1963.
Q = Ibn al-Qifṭī, ʿAlī ibn Yūsuf, *Taʾrīḫ al-ḥukamāʾ*, ed. J. Lippert, Leipzig, Dieterich'sche Verlagsbuchhandlung, 1903.
al-Qāḍī, Wadād, 1981. "*Kitāb Ṣiwān al-Ḥikma*: Structure, Composition, Authorship, and Sources," *Der Islam* 58, 87–124.
Rahman, Fazlur, 1975. *The Philosophy of Mullā Ṣadrā (Ṣadr al-Dīn al-Shīrāzī)*, Albany, State University of New York Press.
Reed, Annette Yoshiko, 2005. *Fallen Angels and the History of Judaism and Christianity: The Reception of Enochic Literature*, Cambridge, Cambridge University Press.
Reeves, John C., 2004. "A Manichaean 'Blood-Libel'?," *Aram* 16, 217–232.
Reisman, David, 2000. *A Catalogue of the Arabic and Syriac Manuscripts on Microfilm in the Library of Franz Rosenthal*, privately published in limited number.
Reitzenstein, R., 1904. *Poimandres. Studien zur griechisch-ägyptischen und frühchristlichen Literatur*, Leipzig, Teubner.
Reitzenstein, R., and H. H. Schaeder, 1926. *Studien zum antiken Synkretismus aus Iran und Griechenland*, Studien der Bibliothek Warburg VII, Leipzig, Teubner.
Rhetorius Aegyptius, 2009. *Compendium Astrologicum*, ed. David Pingree and Stephan Heilen, Leipzig, Teubner.
Rice, D. S., 1952. "Medieval Ḥarrān: Its Topography and Monuments, I," *Anatolian Studies* 2, 36–84.
——, 1955. "A Muslim Shrine at Ḥarrān," *Bulletin of the School of Oriental and African Studies*, 17.3, 436–448.
Rigo, Antonio, 2002. "Da Costantinopoli alla Biblioteca di Venezia: I Libri Ermetici di Medici, Astrologi e Maghi dell'Ultima Bisanzio" ["From Constantinople to the Library of Venice: The Hermetic Books of Late Byzantine Doctors, Astrologers and Magicians"], in Carlos Gilly and Cis van Heertum, eds., *Magia, Alchimia, Scienze dal '400 al '700: L'influsso di Ermete Trismegisto* [Magic, Alchemy and Science 15th–18th Centuries: The Influence of Hermes Trismegistus], Venice/Amsterdam, Centro Di, 69–84.
Ritter, Hellmut, 1921–1922. "Picatrix. Ein arabisches Handbuch hellenistischer Magie," in *Vorträge der Bibliothek Warburg*, 94–124.
——, 1950. "Philologika. XIII Arabische Handschriften in Anatolien und İstanbul," *Oriens* 3.1, 31–107.
Ritter, Hellmut, and Martin Plessner, 1962. "*Picatrix.*" *Das Ziel des Weisen von Pseudo-Mağrīṭī. Translated into German from the Arabic*, Studies of the Warburg Institute 27, London, Warburg Institute.
Rosenthal, Franz, 1940. "On the Knowledge of Plato's Philosophy in the Islamic World," *Islamic Culture* 14, 387–422.
——, 1942. "A Short Treatise on the Meaning of the Names of Some Greek Scholars Attributed to al-Fārābī," *Journal of the American Oriental Society* 62, 73–74.
——, 1943 *Aḥmad b. aṭ Ṭayyib as-Saraḫsī*, American Oriental Series 26, New Haven, American Oriental Society.

Rosenthal, Franz, 1958. "The Sayings of the Ancients from Ibn Durayd's *Kitâb al-Mujtanâ*," *Orientalia* 27, 29–54 and 150–83.

——, 1960–1961. "Al-Mubashshir ibn Fâtik. Prolegomena to an Abortive Edition," *Oriens* 13–14, 132–158.

——, 1962. "The Prophecies of Bâbâ the Ḥarrânian," in *A Locust's Leg: Studies in honour of S.H. Taqizadeh*, London, Percy, Lund, Humphries, & Co.

——, 1963. "From Arabic Books and Manuscripts IX–XI," *Journal of the American Oriental Society* 83.

——, 1966. "Two Graeco-Arabic Works on Music," *Proceedings of the American Philosophical Society* 110.4, 261–268.

——, 1975a. *The Classical Heritage in Islam*, London, Routledge and Kegan Paul.

——, 1975b. "From Arabic Books and Manuscripts XIII–XIV," *Journal of the American Oriental Society* 95, 209–213.

Rosenthal's notes = Franz Rosenthal's unpublished, handwritten edition of al-Mubaššir ibn Fātik's *Kitāb Muḫtār al-ḥikam*.

Rowson, Everett, 1988. *A Muslim Philosopher on the Soul and Its Fate: Al-ʿĀmirī's* Kitāb al-Amad ʿalā l-abad, New Haven, American Oriental Society.

——, forthcoming. "The Futility of Empricism: Muslim Arguments for the Prophetic Origin of the Sciences."

Rubin, Zeev, 2002. "*Res Gestae Divi Saporis*: Greek and Middle Iranian in a Document of Sasanian Anti-Roman Propaganda," in J. N. Adams, Mark Janse, and Simon Swain, eds., *Bilingualism in Ancient Society: Language Contact and the Written Text*, Oxford, Oxford University Press, 267–297.

Rudolph, Ulrich. 1995. "*Kalām* in antiken Gewand. Das theologische Konzept des *Kitāb Sirr al-Ḫalīqa*," *Proceedings of the 14th Congress of the Union Européenne des Arabisants et Islamisants (Budapest, 29th August–3rd September 1988) = The Arabist: Budapest Studies in Arabic* 13–14, ed. Alexander Fodor, 123–136.

Ruelle, C.-E., 1908. "Hermès Trismégiste. Le Livre sacré sur les décans," *Revue de Philologie* 32, 247–277.

Ruska, Julius, 1911. *Untersuchungen über das Steinbuch des Aristoteles*, Heidelberg, Carl Winter.

——, 1926. *Tabula Smaragdina. Ein Beitrag zur Geschichte der hermetischen Literatur*, Heidelberg, Carl Winter.

Sabra, A. I., 1968. "Thābit b. Qurra on Euclid's Parallels Postulate," *Journal of the Warburg and Courtauld Institutes* 31, 12–32.

——, 1969. "Simplicius' Proof of Euclid's Parallels Postulate," *Journal of the Warburg and Courtauld Institutes* 32, 1–24.

Saffrey, H. D., 1971. "Abamon, Pseudonyme de Jamblique," in Robert B. Palmer and Robert Hamerton-Kelly, eds., *Philomathes: Studies and Essays in the Humanities in Memory of Philip Merlan*, The Hague, Nijhoff, 227–239.

aš-Šahrastānī, *Kitāb al-Milal wa-niḥal = Book of Religious and Philosophical Sects* by Muhammad al-Shahrastáni, ed. William Cureton, London 1842–1846. Reprinted by Gorgias Press, 2002.

aš-Šahrazūrī, *Kitāb Nuzhat al-arwāḥ wa-rawḍat al-afrāḥ (Tawārīḫ al-ḥukamāʾ)*, ed. Muḥammad ʿAlī Abū Rayyān, Alexandria, Dār al-Maʿrifa al-Ǧāmiʿīya, 1414/1993.

Ṣāʿid al-Andalusī, al-Qāḍī Abū l-Qāsim, *Ṭabaqāt al-umam*, ed. L. Cheikho, Beirut, Imprimerie Catholique, 1912.

Salaman, Clement, Dorine van Oyen, William Wharton, and Jean-Pierre Mahé, 2000. *The Way of Hermes: New Translations of* The Corpus Hermeticum *and* The Definitions of Hermes Trismegistus to Asclepius, Rochester (Vt.), Inner Traditions.

Sargologos, Étienne 1990. *Un traité de vie spirituelle et morale du XIe siècle: le florilège sacro-profane du manuscrit 6 de Patmos*, Thessaloniki, Asprovalta.

Schoeler, Gregor, 1990. *Arabische Handschriften II*, Verzeichnis der orientalischen Handschriften in Deutschland, Band XVII, Reihe B, Stuttgart, Franz Steiner.

Schott, Jeremy M., 2008, "'Living Like a Christian, but Playing the Greek': Accounts of Apostasy and Conversion in Porphyry and Eusebius," *Journal of Late Antiquity* 1.2, 258–277.

Scott = Scott, W., *Hermetica: The Ancient Greek and Latin Writings Which Contain Religious or Philosophic Teachings Ascribed to Hermes Trismegistus*, 4 vols., Oxford, Clarendon, 1924–1936.

Searby, D., 1999. "The Sources of the So-called *Gnomologium Parisinum Ineditum*," *Revue d'Histoire des Textes* 29, 295–300.

Sezgin, Fuat, *Geschichte des Arabischen Schrifttums*, 13 vols. to date plus index volume, Leiden, E. J. Brill, 1967–2007.

SH = Stobaei Hermetica = *Corpus Hermeticum*, ed. Nock and Festugière, vols. 3–4.

Shahid, Irfan, 2004. "The Last Sasanid-Byzantine Conflict in the Seventh Century: The Causes and Its Outbreak," *La Persia e Bisanzio*, Accademia Nazionale dei Lincei, Atti dei Convegni Lincei 201, Rome, 223–244.

Shaked, Shaul, 1969. "Esoteric Trends in Zoroastrianism," *Proceedings of the Israel Academy of Sciences and Humanities* 3, 175–221.

——, ed. and trans., 1979. *Wisdom of the Sasanian Sages (Dēnkard VI)*, Persian Heritage Series 34, Boulder, Westview Press.

——, 1980. "Mihr the Judge," *Jerusalem Studies in Arabic and Islam* 2, 1–31.

——, 1984. "From Iran to Islam: Notes on Some Themes in Transmission," *Jerusalem Studies in Arabic and Islam* 4, 31–67.

——, 1986. "From Iran to Islam: On Some Symbols of Royalty," *Jerusalem Studies in Arabic and Islam* 7, 75–91.

——, 1987a. "*Paymān*: An Iranian Idea in Contact with Greek Thought and Islam," in *Transition Periods in Iranian History. Actes du Symposium de Fribourg-en-Brisgau (22–24 mai 1985), Studia Iranica* 5, Paris, Association pour l'Avancement des Études Irannienes, 217–240.

——, 1987b. "First Man, First King: Notes on Semitic-Iranian Themes in Muslim Garb," in Sh. Shaked, D. Shulman, and G. G. Stroumsa eds., *Gilgul: Essays on Transformation, Revolution, and Permanence in the History of Religions Dedicated to R. J. Zwi Werblowsky*, Leiden, E. J. Brill.

——, 1992. "The Myth of Zurvan: Cosmology and Eschatology," in I. Gruenwald, Sh. Shaked and G. G. Stroumsa, eds., *Messiah and Christos: Studies in Jewish Origins of Christianity presented to David Flusser* (Texte und Studien zum Antiken Judentum 32), Tübingen, Mohr, 219–240.

——, 1994. *Dualism in Transformation: Varieties of Religion in Sasanian Iran*, London, School of Oriental and African Studies.

Shaki, Mansour, 1981. "The Dēnkard Account of the History of the Zoroastrian Scriptures," *Archív Orientální* 49, 114–125.

Shayegan, Rahim M., 2000. *Aspects of Early Sasanian History and Historiography*, doctoral diss., Harvard University.

Sheppard, Anne, 2000. "Philosophy and Philosophical Schools," in Averil Cameron, Bryan Ward-Perkins, and Michael Whitby, eds., *The Cambridge Ancient History*, Vol 14: *Late Antiquity: Empire and Successors, A.D. 425–600*, Cambridge, Cambridge University Press, 835–854.

Sims-Williams, Ursula, 1982. "The Arabic and Persian Collections in the India Office Library," in Paul Auchterlonie, ed., *Collections in British Libraries on Middle Eastern and Islamic Studies* 12, 47–52.

Sirriyeh, Elizabeth, 2005. *Sufi Visionary of Ottoman Damascus: 'Abd al-Ghanī al-Nābulusī, 1641–1731*, London and New York, RoutledgeCurzon.

Sladek, Mirko, 1988. "Mercurius triplex, Mercurius termaximus et les 'trois Hermès,'" trans. from German by Paul Kessler, in *Présence d'Hermès Trismégiste*, Paris, Albin Michel, 88–99.

Socrates of Constantinople = *Socrates' Ecclesiastical History*, 2nd ed., ed. W. Bright, Oxford, Clarendon, 1893.

Southgate, Minoo S., 1978. *Iskandarnamah: A Persian Medieval Alexander-Romance*, New York, Columbia University Press.

Stapleton, H. E., G. L. Lewis, and F. Sherwood Taylor, 1949. "The Sayings of Hermes quoted in the *Mā' al-waraqī* of Ibn Umail," *Ambix* 3, 69–90.

Stern, Samuel M., 1946. "The Authorship of the Epistles of the Ikhwān-aṣ-Ṣafā," *Islamic Culture* 20, 367–372 and 403–404.

——, 1972. "Abū 'Īsā ibn al-Munajjim's Chronography," in S. Stern, Albert Hourani, and Vivian Brown, eds., *Islamic Philosophy and the Classical Tradition: Essays Presented by His Friends and Pupils to Richard Walzer on His Seventieth Birthday*, Columbia, University of South Carolina Press, 437–466.

——, 1983a. "The Earliest Cosmological Doctrines of Ismā'īlism," in *Studies in Early Ismā'īlism*, Jerusalem, Magnes Press, 3–29.

——, 1983b. "New Information about the Authors of the 'Epistles of the Sincere Brethren,'" in *Studies in Early Ismā'īlism*, Jerusalem, Magnes Press, 155–176 (originally in *Islamic Studies* 3, 1964, 405–428).

——, 1983c. "Cairo as the Centre of the Ismā'īlī Movement," in *Studies in Early Ismā'īlism*, Jerusalem, Magnes Press, 234–256.

——, 1983d. "The Early Ismā'īlī Missionaries in North-West Persia and in Khurāsān and Transoxania," in *Studies in Early Ismā'īlism*, Jerusalem, Magnes Press, 189–233.

Sternbach, Leo, 1894. "Gnomologium Parisinum ineditum," *Rozprawy Akademii Umiejętności, Wydział Filologiczny*, ser. 2, vol. 5, 135–171.

Sternbach, Ludwik, 1963. *Cāṇakya-rāja-nīti: Maxims on Rāja-nīti Compiled from Various Collections of Maxims Attributed to Cāṇakya*, Adyar Library Series vol. 92, Adyar, Vasanta Press.

——, 1981. "Indian Wisdom and Its Spread Beyond India," *Journal of the American Oriental Society* 101, 97–131.

Stobaeus = *Ioannis Stobaei Anthologium*, ed. C. Wachsmuth and O. Hense, 4 vols., Berlin, Weidmann, 1884–1912 (2nd ed. 1958).

Strohmaier, Gotthard, 1996. "Die ḥarrānischen Sabier bei Ibn an-Nadīm und al-Bīrūnī," in *Ibn an-Nadīm und die mittelalterliche arabische Literatur*, Wiesbaden, 51–56, reprinted in Gotthard Strohmaier, *Hellas im Islam: Interdisziplinäre Studien zur Ikonograhpie, Wissenschaft und Religionsgeschichte*, Diskurse der Arabistik 6, Wiesbaden, Harrassowitz, 2003, 167–169.

Stroumsa, Sarah, 2004. "Sabéens de Ḥarrān et Sabéens de Maïmonide," in Tony Lévy and Roshdi Rashed, eds., *Maïmonide: Philosophe et savant (1138–1204)*, Leuven, Peeters, 335–352.

Sundermann, Werner, 1989. "Mittelpersisch," in Rüdiger Schmitt, ed., *Compendium Linguarum Iranicarum*, Wiesbaden, Ludwig Reichert, 138–164.

aš-Šuštarī, 1960. Dīwān Abī l-Ḥasan aš-Šuštarī, ed. 'Alī Sāmī an-Naššār, Alexandria, Dār al-Ma'ārif.

Syncellus = Georgius Syncellus, *Ecloga Chronographica*, ed. A. A. Mosshammer, Leipzig, Teubner, 1984. English translation with introduction and notes: *The Chronography of*

George Synkellos by William Adler and Paul Tuffin, Oxford, Oxford University Press, 2002.

aṯ-Ṯaʿālibī, *Laṭā'if al-maʿārif*, ed. I. al-Abyārī and Ḥ. aṣ-Ṣayrafī, Cairo, Dār Iḥyā' al-Kutub al-ʿArabīya, 1960.

——, *Kitāb Yatīmat ad-dahr fī maḥāsin ahl al-ʿaṣr*, 4 vols., ed. Muḥammad Muḥyīddīn ʿAbdalḥamīd, Beirut, Dār al-Fikr, 1947.

aṭ-Ṭabarī, *Ta'rīḫ ar-rusul wa-l-mulūk*, ed. M. J. de Goeje et al., Leiden, E. J. Brill, 1879–1901.

aṯ-Ṯaʿlabī, *ʿArā'is al-majālis fī qiṣaṣ al-anbiyā'*, trans. William M. Brinner, Leiden, E. J. Brill, 2002.

Tardieu, Michel, 1986. "Ṣābiens coraniques et 'Ṣābiens' de Ḥarrān," *Journal Asiatique* 274, 1–44.

——, 1987. "Les calandriers en usage a Ḥarrān d'après les sources arabes et le commentaire de Simplicius a la physique d'Aristote," in Ilsetraut Hadot, ed., *Simplicius: Sa vie, son œvre, sa survie. Actes du colloque international de Paris (28 Sept.–1er Oct. 1985)*, Berlin, de Gruyter, 40–57.

Tavernier, J., 2007. *Iranica in the Achaemenid Period (ca. 550–330 B.C.): Lexicon of Old Iranian Proper Names and Loanwords, Attested in Non-Iranian Texts*, Leuven/Paris/Dudley (Mass.), Peeters.

Theodore Abū Qurra, *Maymar fī Wuǧūd al-ḫāliq wa-d-dīn al-qawīm / Traité de l'existence du Créateur et le la vraie religion*, ed. Ignace Dick, Jounieh, Librairie Saint-Paul/Rome, Pontificio Istituto Orientale, 1982.

Theodore Bar Koni = Théodore bar Koni, *Livre des Scolies (recension de Séert)*, Corpus Scriptorum Christianorum Orientalium, Scriptores Syri, vols. 187–188, trans. Robert Hespel and René Draguet, Louvain, Peeters, 1981–1982.

Thiel, Rainer, 1999. *Simplikios und das Ende der neuplatonischen Schule in Athen*, Akademie der Wissenschaften und der Literatur, Abhandlungen der Geistes- und Sozialwissenschaftlichen Klasse, Nr. 8.

Thomson, Robert W., 1992. "A Medieval Armenian View of the Physical World: The Cosmology of Vardan Arewelcʿi," *Revue des Études arméniennes* 23, 192–208.

Toomer, G. J., 1970. Review of *The Thousands of Abū Maʿshar* by David Pingree, *Speculum* 45, 156–157.

——, 1998. *Ptolemy's Almagest*, Princeton, Princeton University Press.

Toral-Niehoff, Isabel, 2004. *Kitāb Ǧiranīs. Die arabischen Übersetzung der ersten Kyranis des Hermes Trismegistos und die griechischen Parallelen herausgegeben, übersetzt und kommentiert*, München, Herbert Utz.

Trombley, Frank, 1993–1995. *Hellenic Religion and Christianization c. 370–529*, 2 vols. (2nd ed. vol. 2 in 1995), Leiden, Brill.

aṭ-Ṭurṭūšī, Abū Bakr, 1872/3. *Kitāb Sirāǧ al-mulūk*, Alexandria, al-Maṭbaʿa al-Waṭanīya.

Ullmann, Manfred, 1970. *Die Medizin im Islam*, Handbuch der Orientalistik, Erste Abteilung, Ergänzungsband VI,1, Leiden, E. J. Brill.

——, 1972. *Die Natur- und Geheimwissenschaften im Islam*, Handbuch der Orientalistik, Erste Abteilung, Ergänzungsband VI,2, Leiden, E. J. Brill.

——, 1975. "Die arabische Überlieferung der Kyranis des Hermes Trismegistos," *Proceedings of the VIth Congress of Arabic and Islamic Studies* [1972], Visby-Stockholm, 196–200.

——, 1980. Review of Weisser 1979, *Journal for the History of Arabic Science* 4, 91–94.

——, 1994. *Das Schlangenbuch des Hermes Trismegistos*, Wiesbaden, Harrassowitz.

——, 2002–2007. *Wörterbuch zu den griechisch-arabischen Übersetzungen des 9. Jahrhunderts*, 1 vol. plus 2 supplement volumes, Wiesbaden, Harrassowitz.

Vahman, Fereydun, 1986. *Ardā Wirāz Nāmag*, Scandinavian Institute of Asian Studies Monograph Series 53, London, Curzon Press.

Vallat, Philippe, 2004. *Farabi et l'École d'Alexandrie. Des prémisses de la connaissance à la philosophie politique*, Paris, J. Vrin.

van Bladel, Kevin, 2003. "Sources and History of the Hermes Legend in Arabic," in Paolo Lucentini, Iliaria Parri, and Vittoria Perrone Compagni, eds., *Atti del Convegno Internazionale sulla Tradizione Ermetica dal Mondo Tardo-Antico all'Umanesimo*, Turnhout, Brepols, 285–293.

——, 2004. "The Iranian Characteristics and Forged Greek Attributions in the Arabic *Sirr al-asrār* (*Secret of Secrets*)," *Mélanges de l'Université Saint-Joseph* 57, 151–172.

——, 2007a. "The Syriac Sources of the Early Arabic Narratives of Alexander," in Himanshu Prabha Ray and Daniel T. Potts, eds., *Memory as History: The Legacy of Alexander in South Asia*, New Delhi, Aryan International, 54–75.

——, 2007b. "Heavenly Cords and Prophetic Authority in the Quran and Its Late Antique Context," *Bulletin of the School of Oriental and African Studies* 70/2, 223–246.

——, forthcoming. "The Bactrian Background of the Barmakids."

van Koningsveld, P. S., 1998. "Greek Manuscripts in the Early Abbasid Empire: Fiction and Facts about Their Origin, Translation and Destruction," *Bibliotheca Orientalis* 55, 345–372.

Verbrugghe, Gerald P., and John Wickersham, 1996. *Berossos and Manetho, Introduced and Translated*, Ann Arbor, University of Michigan Press.

Vereno, Ingolf, 1992. *Studien zum ältesten alchemistischen Schrifttum. Auf der Grundlage zweier erstmals edierter arabischer Hermetica*, Islamkundliche Untersuchungen 155, Berlin, Klaus Schwarz.

Waddell, William G., trans., 1940. *Manetho*, Loeb Classical Library, London, William Heinemann.

Walbridge, John, 1998. "Explaining Away the Greek Gods in Islam," *Journal of the History of Ideas* 59.3, 389–403.

——, 2000. *The Leaven of the Ancients: Suhrawardī and the Heritage of the Greeks*, Albany, State University of New York Press.

——, 2001. *The Wisdom of the Mystic East: Suhrawardī and Platonic Orientalism*, Albany, State University of New York Press.

——, 2005. "al-Suhrawardī," in Adamson, Peter, and Richard C. Taylor, eds., *The Cambridge Companion to Arabic Philosophy*, Cambridge, Cambridge University Press, 201–223.

Walker, Paul E., 1993. *Early Philosophical Shiism: The Ismaili Neoplatonism of Abū Yaʿqūb al-Sijistānī*. Cambridge, Cambridge University Press.

——, 1997. "Fatimid Institutions of Learning," *Journal of the American Research Center in Egypt* 34, 179–200.

Walzer, Richard, 1950. "Two Fragments of Galen in Arabic Translation," *Journal of the American Oriental Society* 70, 182–187.

Watts, Edward, 2005. "Where to Live the Philosophical Life in the Sixth Century? Damascius, Simplicius, and the Return from Persia," in *Greek, Roman, and Byzantine Studies* 45, 285–315.

Weisser, Ursula, 1978. "Hellenistisch Offenbarungsmotive und das Buch 'Geheimnis der Schöpfung,'" *Journal for the History of Arabic Science* 2, 101–125.

——, ed., 1979. *Buch über das Geheimnis der Schöpfung und die Darstellung der Natur*, Aleppo, Institute for the History of Arabic Science, University of Aleppo.

——, 1980. *Das "Buch über das Geheimnis der Schöpfung" von Pseudo-Apollonios von Tyana*, Berlin, Walter de Gruyter.

Wellmann, Max, 1921. *Die Georgika des Demokritos*, Abhandlungen der Deutschen Akademie der Wissenschaften zu Berlin, philosophisch-historische Klasse.

West, E. W., 1892. *Pahlavi Texts, Part IV: Contents of the Nasks*, Sacred Books of the East 37, Oxford, Oxford University Press.

Whinfield, E. H., 1880. *Gulshan i Raz. The Mystic Rose Garden of Sa'd ud Din Muhammad Shabistari. The Persian Text, with an English Translation and Notes, Chiefly from the Commentary of Muhammad bin Yahya Lahiji*, London, Trubner.

Wiedemann, Eilhard, 1920–1921. "Beiträge zur Geschichte der Naturwissenschaften. LXIV. Über Ṭâbit ben Qurra, sein Leben und Wirken," *Sitzungsberichte der Physicalisch-Medizinisches Sozietät zu Erlangen* 52–53, 189–219. Reprinted in Eilhard Wiedemann, *Aufsätze zur Arabischen Wissenschaftsgeschichte*, 2 vols., Hildesheim/New York, Georg Olms, vol. 2, 548–578.

——, 1922. *Zur Alchemie bei den Arabern*, Abhandlungen zur Geschichte der Naturwissenschaften und der Medizin V, Erlangen, Max Mencke.

Wiesehöfer, Josef, 1986. "Iranische Ansprüche an Rom auf ehemals achaeimenidische Territorien," *Archaeologische Mitteilungen aus Iran* 19, 177–185.

Witakowski, Witold, 1987. *The Syriac Chronicle of Pseudo-Dionysius of Tel-Maḥrē: A Study in the History of Historiography*, Acta Universitatis Upsaliensis, *Studia Semitica Upsaliensia* 9, Uppsala.

——, 1999–2000. "The Chronicle of Eusebius: Its Type and Continuation in Syriac Historiography," *Aram* 11–12, 419–437.

——, 2008. "The *Chronicle* of Jacob of Edessa," in Bas Ter Haar Romeny, ed., *Jacob of Edessa and the Culture of His Day*, Leiden, Brill, 25–47.

Yar-Shater, Ehsan, 1971. "Were the Sasanians Heirs to the Achaemenids?," in *Atti del Convengno internazionale sul tema: La Persia nel Medioevo (Roma, 31 marzo-5 aprile 1970)*, Rome, Accademia nazionale dei Lincei, 517–531.

al-Yaʿqūbī, Aḥmad ibn Abī Yaʿqūb, *at-Taʾrīḫ = Ibn Wādhih qui dicitur al-Jaʿqūbī Historiae*, ed. M. Th. Houtsma, 2 vols., Leiden, E. J. Brill, 1883.

Yāqūt, *Iršād al-arīb ilā maʿrifat al-adīb*, London, Luzac, 1923–1931.

——, *Muʿǧam al-buldān*, ed. F. Wüstenfeld, 6 vols., Leipzig, 1866–1872. Partial translation by W. Jwaideh, Leiden, 1959.

Yates, Frances A., 1964. *Giordano Bruno and the Hermetic Tradition*, Chicago, University of Chicago Press.

——, 1967. "The Hermetic Tradition in Renaissance Science," in Charles S. Singleton, ed., *Art, Science, and History in the Renaissance*, Baltimore, Johns Hopkins, 255–274.

Zachariae, Theodor, 1914. "Die Weisheitssprüche des Šānāq bei aṭ-Ṭorṭūšī (Übersetzung und Kommentar)," *Wiener Zeitschrift für die Kunde des Morgenlandes* 28, 182–210.

Zaehner, R. C., 1955. *Zurvan: A Zoroastrian Dilemma*, Oxford, Clarendon.

——, 1961. *The Dawn and Twilight of Zoroastrianism*, New York, G. B. Putnam's Sons.

Zakeri, Mohsen, 2004. "Ādāb al-falāsifa: The Persian Content of an Arabic Collection of Aphorisms," *Mélanges de l'Université Saint-Joseph* 57, 173–190.

——, 2007. *Persian Wisdom in Arabic Garb. ʿAlī b. ʿUbayda al-Rayḥānī (d. 219/834) and His Jawāhir al-kilam wa-farāʾid al-ḥikam*, 2 vols., Leiden/Boston, Brill.

Ziai, Hossein, 1992. "The Source and Nature of Authority: A Study of al-Suhrawardī's Illuminationist Political Doctrine," in Charles E. Butterworth, ed., *The Political Aspects of Islamic Philosophy: Essays in Honor of Muhsin S. Mahdi*, Cambridge (Mass.), Harvard University Press, 304–344.

Zimmerman, F. W., 1981. Review of Ursula Weisser 1980, *Medical History* 25, 439–440.

——, 1986. "The Origins of the So-Called *Theology of Aristotle*," *Pseudo-Aristotle in the Middle Ages: The* Theology *and Other Texts*, Warburg Institute Surveys and Texts XI, London, The Warburg Institute, University of London, 110–240.

Zosimus = *Les Alchimistes Grecs IV.1. Zosime de Panopolis, Mémoires authentiques*, ed. and French trans. Michèle Mertens, Paris, Budé, 1995.

——, *Zosimus of Panopolis: On the Letter Omega*, ed. and English trans. Howard M. Jackson (Texts and Translations 14, Graeco-Roman Religion 5), Missoula, Scholars Press, 1978.

Index

Printed in the USA/Agawam, MA
November 20, 2019

743066.031